MECHANICAL BEHAVIOR OF MATERIALS
SECOND EDITION

This textbook fits courses on mechanical behavior of materials in mechanical engineering and materials science, and it includes numerous examples and problems. It emphasizes quantitative problem solving. This text differs from others because the treatment of plasticity emphasizes the interrelationship of the flow, effective strain, and effective stress, and their use in conjunction with yield criteria to solve problems. The treatment of defects is new, as is the analysis of particulate composites. Schmid's law is generalized for complex stress states. Its use with strains allows for prediction of R values for textures. Of note is the treatment of lattice rotations related to deformation textures. The chapter on fracture mechanics includes coverage of Gurney's approach. Among the highlights in this new edition are the treatment of the effects of texture on properties and microstructure in Chapter 7, a new chapter on discontinuous and inhomogeneous deformation (Chapter 12), and the treatment of foams in Chapter 21.

William F. Hosford is a Professor Emeritus of Materials Science at the University of Michigan. He is the author of numerous research publications, and textbooks including *Materials for Engineers; Metal Forming, Third Edition* (with Robert M. Caddell); *Materials Science: An Intermediate Text; Reporting Results* (with David C. Van Aken); *Mechanics of Crystals, and Textured Polycrystals*; and *Physical Metallurgy*.

Mechanical Behavior of Materials

SECOND EDITION

William F. Hosford
University of Michigan

CAMBRIDGE
UNIVERSITY PRESS

CAMBRIDGE UNIVERSITY PRESS
Cambridge, New York, Melbourne, Madrid, Cape Town, Singapore,
São Paulo, Delhi, Dubai, Tokyo

Cambridge University Press
32 Avenue of the Americas, New York, NY 10013-2473, USA

www.cambridge.org
Information on this title: www.cambridge.org/9780521195690

First published 2010

Printed in the United States of America

A catalog record for this publication is available from the British Library.

Library of Congress Cataloging in Publication data

Hosford, William F.
Mechanical behavior of materials / William F. Hosford. – 2nd ed.
    p.   cm.
Includes bibliographical references and index.
ISBN 978-0-521-19569-0 (hardback)
1. Materials – Mechanical properties.   I. Title.
TA405.H59   2010
620.1′1292–dc22   2009037740

ISBN   978-0-521-19569-0 Hardback

Contents

Preface

The term *mechanical behavior* encompasses the response of materials to external forces. This text considers a wide range of topics. These include mechanical testing to determine material properties; plasticity, which is needed for FEM analyses of automobile crashes; means of altering mechanical properties; and treatment of several modes of failure.

The two principal responses of materials to external forces are deformation and fracture. The deformation may be elastic, viscoelastic (time-dependent elastic deformation), or plastic and creep (time-dependent plastic deformation). Fracture may occur suddenly or after repeated applications of loads (fatigue). For some materials, failure is time dependent. Both deformation and fracture are sensitive to defects, temperature, and rate of loading.

Key to understanding these phenomena is a basic knowledge of the three-dimensional nature of stress and strain and common boundary conditions, which are covered in Chapter 1. Chapter 2 covers elasticity, including thermal expansion. Chapter 3 treats mechanical testing. Chapter 4 is focused on mathematical approximations to stress–strain behavior of metals, and how these approximations can be used to understand the effect of defects on strain distribution in the presence of defects. Yield criteria and flow rules are covered in Chapter 5. Their interplay is emphasized in problem solving. Chapter 6 treats temperature and strain rate effects and uses an Arrhenius approach to relate them. Defect analysis is used to understand both superplasticity and strain distribution.

Chapter 7 is devoted to the role of slip as a deformation mechanism. The tensor nature of stresses and strains are used to generalize Schmid's law. Lattice rotations caused by slip are covered. The effects of texture on properties and microstructure have been added. Chapters 8 and 9 treat dislocations: their geometry, their movement, and their interactions. There is a treatment of stacking faults in fcc metals and how they affect strain hardening. Hardening by intersections of dislocations is emphasized. Twinning and martensitic shears are treated in Chapter 10. Chapter 11 treats the various hardening mechanisms in metallic materials.

Chapter 12 is a new chapter that covers discontinous and inhomogeneous deformation. Chapter 13 presents phenomenological and qualitative treatment of ductility, whereas Chapter 14 focuses on quantitative coverage of fracture mechanics.

Viscoelasticity (time-dependent elasticity) is treated in Chapter 15. Mathematical models are presented and used to explain stress and strain relaxation as well as damping and rate dependence of the elastic modulus. Several mechanisms of damping are presented. Chapter 16 is devoted to creep (time-dependent plasticity) and stress rupture. The coverage includes creep mechanisms and extrapolation techniques for predicting life.

Failure by fatigue is the topic of Chapter 17. The chapter starts with a phenomenological treatment of the *S-N* curve and the effects of mean stress, variable stress amplitude, and surface condition. The important material aspects, Coffin's law and crack propagation rate, are treated. Chapter 18 covers residual stresses, their origins, their effects, their measurement, and their removal.

Chapters 19, 20, and 21 cover ceramics, polymers, and composites. Separate chapters are devoted to these materials because their mechanical behaviors are very different from that of metals, which were emphasized in the earlier chapters. Because ceramics and glass are brittle and their properties are variable, Weibull analysis is presented here. Chapter 19 also covers methods of improving toughness of ceramics and the role of thermally induced stresses. The most important aspect of the mechanical behavior of polymers is their great time dependence and the associated temperature dependence. The effects of pressure on yielding and the phenomenon of crazing are also unique. Rubber elasticity is very different from Hookean elasticity. Chapter 21 covers composites, including fiber, sheet, and particulate composites. Coverage of the structure and properties of foams has been added to this chapter. Chapter 22 on metal forming covers analyses of bulk-forming and sheet-forming operations.

This text differs from other books on mechanical behavior in several aspects. The treatment of plasticity has greater emphasis on the interrelationship of the flow, effective strain, and effective stress, and their use in conjunction with yield criteria to solve problems. The treatment of defects is new. Schmid's law is generalized for complex stress states. Its use with strains allows for prediction of *R* values for textures. Another feature is the treatment of lattice rotations and how they lead to deformation textures. Most texts treat only strain relaxation and neglect stress relaxation. The chapter on fracture mechanics includes coverage of Gurney's approach. Most texts omit any coverage of residual stresses. Much of the analysis of particulate composites is new. Few texts include anything on metal forming. Throughout the text, there is more emphasis on quantitative problem solving than in most other texts. The notes at the end of the chapters are included to increase reader interest in the subject.

As a consequence of the increased coverage in these areas, the treatment of some other topics is not as extensive as in competing texts. There is less coverage of fatigue failure and fracture mechanics.

This book may contain more material than can be covered in a single course. Depending on the focus of the course, various chapters, or portions of chapters, may be omitted. It is hoped that this book will be of value to mechanical engineers as well as materials engineers. If the book is used in a mechanical engineering course, the instructor may want to skip some chapters. In particular, Chapters 8 through 11 may be omitted. If the book is used in a materials science course, the instructor may want to omit Chapters 10, 18, and 22. Both may want to skip Chapter 11 on twinning and memory metals. Even though it was realized that most users may want to skip this chapter, it was included for completeness and in the hope that it may prove useful as a reference.

It is assumed that the students who use this book will have had both an introductory materials science course and a "strength of materials" course. From the strength of materials course, they can be expected to know basic concepts of stress and strain, how to resolve stresses from one axis system to another, and how to use Hooke's laws in three dimensions. They should be familiar with force and moment balances. From their materials science course, they should understand that most materials are crystalline and that crystalline materials deform by slip resulting from the movement of dislocations. They should also be familiar with such concepts as substitutional and interstitial solid solutions and diffusion. Appendices aI (Miller indices) and aII (Stereographic projection) are available for students not familiar with these topics.

The main difference between this and the first edition is the treatment of the effects of texture on properties and microstructure in Chapter 7, the addition of Chapter 12 on discontinuous and inhomogeneous deformation, and the treatment of foams in Chapter 21.

1 Stress and Strain

Introduction

This book is concerned with the mechanical behavior of materials. The term *mechanical behavior* refers to the response of materials to forces. Under load, a material may either deform or break. The factors that govern a material's resistance to deforming are quite different than those governing its resistance to fracture. The word *strength* may refer either to the stress required to deform a material or to the stress required to cause fracture; therefore, care must be used with the term *strength*.

When a material deforms under small stresses, the deformation may be *elastic*. In this case, when the stress is removed, the material will revert to its original shape. Most of the elastic deformation will recover immediately. There may be, however, some time-dependent shape recovery. This time-dependent elastic behavior is called *anelasticity* or *viscoelasticity*.

Larger stresses may cause *plastic* deformation. After a material undergoes plastic deformation, it will not revert to its original shape when the stress is removed. Usually, high resistance to deformation is desirable so that a part will maintain its shape in service when stressed. However, it is desirable to have materials deform easily when forming them by rolling, extrusion, and so on. Plastic deformation usually occurs as soon as the stress is applied. At high temperatures, however, time-dependent plastic deformation called *creep* may occur.

Fracture is the breaking of a material into two or more pieces. If fracture occurs before much plastic deformation occurs, we say the material is *brittle*. In contrast, if there has been extensive plastic deformation preceding fracture, the material is considered *ductile*. Fracture usually occurs as soon as a critical stress has been reached; however, repeated applications of a somewhat lower stress may cause fracture. This is called *fatigue*.

The amount of deformation that a material undergoes is described by *strain*. The forces acting on a body are described by *stress*. Although the reader should already be familiar with these terms, they will be reviewed in this chapter.

Stress

Stress, σ, is defined as the intensity of force at a point.

$$\sigma = \partial F/\partial A \quad \text{as } \partial A \to 0. \tag{1.1a}$$

If the state of stress is the same everywhere in a body,

$$\sigma = F/A. \tag{1.1b}$$

A *normal stress* (compressive or tensile) is one in which the force is normal to the area on which it acts. With a *shear stress,* the force is parallel to the area on which it acts. Two subscripts are required to define a stress. The first subscript denotes the normal to the plane on which the force acts, and the second subscript identifies the direction of the force.* For example, a tensile stress in the x-direction is denoted by σ_{xx} indicating that the force is in the x-direction and it acts on a plane normal to x. For a shear stress, σ_{xy}, a force in the y-direction acts on a plane normal to x.

Because stresses involve both forces and areas, they are not vector quantities. Nine components of stress are needed to fully describe a state of stress at a point, as shown in Figure 1.1. The stress component, $\sigma_{yy} = F_y/A_y$, describes the tensile stress in the y-direction. The stress component, $\sigma_{zy} = F_y/A_z$, is the shear stress caused by a shear force in the y-direction acting on a plane normal to z.

Repeated subscripts denote normal stresses (e.g., σ_{xx}, σ_{yy}), whereas mixed subscripts denote shear stresses (e.g., σ_{xy}, σ_{zx}). In *tensor* notation, the state of stress is expressed as

$$\sigma_{ij} = \begin{vmatrix} \sigma_{xx} & \sigma_{yx} & \sigma_{zx} \\ \sigma_{xy} & \sigma_{yy} & \sigma_{zy} \\ \sigma_{xz} & \sigma_{yz} & \sigma_{zz} \end{vmatrix} \tag{1.2}$$

where i and j are iterated over x, y, and z. Except where tensor notation is required, it is often simpler to use a single subscript for a normal stress and to denote a shear stress by τ:

$$\sigma_x = \sigma_{xx}, \quad \text{and} \quad \tau_{xy} = \sigma_{xy}. \tag{1.3}$$

A stress component expressed along one set of axes may be expressed along another set of axes. Consider the case in Figure 1.2. The body is subjected to a stress

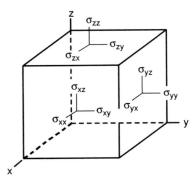

Figure 1.1. Nine components of stress acting on an infinitesimal element. Normal stress components are σ_{xx}, σ_{yy}, and σ_{zz}. Shear stress components are $\sigma_{yz}, \sigma_{zx}, \sigma_{xy}. \sigma_{zy}, \sigma_{xz}$, and σ_{yx}.

---

* Use of the opposite convention should not cause confusion because $\sigma_{ij} = \sigma_{ji}$.

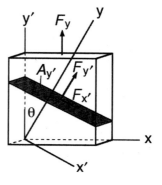

Figure 1.2. Stresses acting on an area, $A_{y'}$, under a normal force, F_y. The normal stress, $\sigma_{y'y'} = F_{y'}/A_{y'} = F_y \cos\theta/(A_y/\cos\theta) = \sigma_{yy} \cos^2\theta$. The shear stress, $\tau_{y'x'} = F_{x'}/A_{y'} = F_y \sin\theta/(A_{yx}/\cos\theta) = \sigma_{yy} \cos\theta \sin\theta$.

$\sigma_{yy} = F_y/A_y$. It is possible to calculate the stress acting on a plane whose normal, y', is at an angle θ to y. The normal force acting on the plane is $F_{y'} = F_y \cos\theta$, and the area normal to y' is $A_y/\cos\theta$, so

$$\sigma_{y'} = \sigma_{y'y'} = F_{y'}/A_{y'} = (F_y \cos\theta)/(A_y/\cos\theta) = \sigma_y \cos^2\theta. \tag{1.4a}$$

Similarly, the shear stress on this plane acting in the x'-direction, $\tau_{y'x'}(= \sigma_{y'x'})$, is given by

$$\tau_{y'x'} = \sigma_{y'x'} = F_{x'}/A_{y'} = (F_y \sin\theta)/(A_y/\cos\theta) = \sigma_y \cos\theta \sin\theta. \tag{1.4b}$$

Note: The transformation requires the product of two cosine and/or sine terms.

Sign Convention

When we write $\sigma_{ij} = F_i/A_j$, the term σ_{ij} is positive if i and j are either both positive or both negative. However, the stress component is negative for a combination of i and j in which one is positive and the other is negative. For example, in Figure 1.3, the terms σ_{xx} are positive on both sides of the element because both the force and normal to the area are negative on the left and positive on the right. The stress, τ_{yx}, is negative because on the top surface y is positive and x-direction force is negative, and on the bottom surface, x-direction force is positive and the normal to the area, y, is negative. Similarly, τ_{xy} is negative.

Pairs of shear stress terms with reversed subscripts are always equal. A moment balance requires that $\tau_{ij} = \tau_{ji}$. If they were not, the element would undergo an infinite rotational acceleration (Figure 1.4). For example, $\tau_{yx} = \tau_{xy}$. Therefore, we

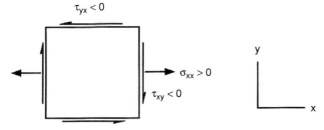

Figure 1.3. The normal stress, σ_{xx}, is positive because the direction of the force, F_x, and the normal to the plane are either both positive (*right*) or both negative (*left*). The shear stresses, τ_{xy} and τ_{yx}, are negative because the direction of the force and the normal to the plane have opposite signs.

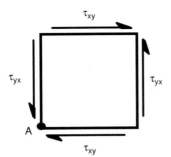

Figure 1.4. An infinitesimal element under shear stresses, τ_{xy} and τ_{yx}. A moment balance about A requires that $\tau_{xy} = \tau_{yx}$.

can write, in general, that $\Sigma M_A = \tau_{yx} = \tau_{xy} = 0$, so

$$\sigma_{ij} = \sigma_{ji,} \quad \text{or} \quad \tau_{ij} = \tau_{ji}. \tag{1.5}$$

This makes its stress tensor symmetric about the diagonal.

Transformation of Axes

Frequently, we must change the axis system on which a stress state is expressed. For example, we may want to find the shear stress on a slip system from the external stresses acting on a crystal. Another example is finding the normal stress across a glued joint in a tube subjected to tension and torsion. In general, a stress state expressed along one set of orthogonal axes (e.g., m, n, and p) may be expressed along a different set of orthogonal axes (e.g., i, j, and k). The general form of the transformation is

$$\sigma_{ij} = \sum_{n=1}^{3} \sum_{m=1}^{3} \ell_{im}\ell_{jn}\sigma_{mn}. \tag{1.6}$$

The term, ℓ_{im}, is the cosine of the angle between the i and m axes, and ℓ_{jn} is the cosine of the angle between the j and n axes. The summations are over the three possible values of m and n, namely, m, n, and p. This is often written as

$$\sigma_{ij} = \ell_{im}\ell_{jn}\sigma_{mn}, \tag{1.7}$$

with the summation implied. The stresses in the x, y, z coordinate system in Figure 1.5 may be transformed onto the x', y', z' coordinate system by

$$\begin{aligned}
\sigma_{x'x'} = {} & \ell_{x'x}\ell_{x'x}\sigma_{xx} + \ell_{x'y}\ell_{x'x}\sigma_{yx} + \ell_{x'z}\ell_{x'x}\sigma_{zx} \\
& + \ell_{x'x}\ell_{x'y}\sigma_{xy} + \ell_{x'y}\ell_{x'y}\sigma_{yy} + \ell_{x'z}\ell_{x'y}\sigma_{zy} \\
& + \ell_{x'x}\ell_{x'z}\sigma_{xz} + \ell_{x'y}\ell_{x'z}\sigma_{yz} + \ell_{x'z}\ell_{x'z}\sigma_{zz}
\end{aligned} \tag{1.8a}$$

and

$$\sigma_{x'y'} = \ell_{x'x}\ell_{y'x}\sigma_{xx} + \ell_{x'y}\ell_{y'x}\sigma_{yx} + \ell_{x'z}\ell_{y'x}\sigma_{zx}$$

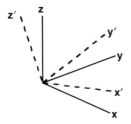

Figure 1.5. Two orthogonal coordinate systems, x, y, and z and x', y', and z'. The stress state may be expressed in terms of either.

$$+ \ell_{x'x}\ell_{y'y}\sigma_{xy} + \ell_{x'y}\ell_{y'y}\sigma_{yy} + \ell_{x'z}\ell_{y'y}\sigma_{zy}$$
$$+ \ell_{x'x}\ell_{y'z}\sigma_{xz} + \ell_{x'y}\ell_{y'z}\sigma_{yz} + \ell_{x'z}\ell_{y'z}\sigma_{zz}. \tag{1.8b}$$

These equations may be simplified with the notation in equation (1.3) using equation (1.5),

$$\sigma_{x'} = \ell_{x'x}2\sigma_x + \ell_{x'y}2\sigma_y + \ell_{x'z}2\sigma_z$$
$$+ 2\ell_{x'y}\ell_{x'z}\tau_{yz} + 2\ell_{x'z}\ell_{x'x}\tau_{zx} + 2\ell_{x'x}\ell_{x'y}\tau_{xy} \tag{1.9a}$$

and

$$\tau_{x'z'} = \ell_{x'x}\ell_{y'x}\sigma_{xx} + \ell_{x'y}\ell_{y'y}\sigma_{yy} + \ell_{x'z}\ell_{y'z}\sigma_{zz}$$
$$+ (\ell_{x'y}\ell_{1y'z} + \ell_{x'z}\ell_{y'y})\tau_{yz} + (\ell_{x'z}\ell_{y'x} + \ell_{x'x}\ell_{y'z})\tau_{zx}$$
$$+ (\ell_{x'x}\ell_{y'y} + \ell_{x'y}\ell_{y'x})\tau_{xy}. \tag{1.9b}$$

Now reconsider the transformation in Figure 1.2. Using equations (1.9a) and (1.9b), with σ_{yy} as the only finite term on the x, y, z axis system,

$$\sigma_{y'} = \ell_{y'y}^2\sigma_{yy} = \sigma_y \cos^2\theta \quad \text{and} \quad \tau_{x'y'} = \ell_{x'y}\ell_{y'y}\sigma_{yy} = \sigma_y \cos\theta \sin\theta \tag{1.10}$$

in agreement with equations 1.4a and 1.4b. These equations can be used together with Miller indices for planes and direction indices for crystals. The reader that is not familiar with these is referred to Appendix I.

EXAMPLE PROBLEM 1.1: A cubic crystal is loaded with a tensile stress of 2.8 MPa applied along the [210] direction, as shown in Figure 1.6. Find the shear stress on the (111) plane in the [10$\bar{1}$] direction.

Solution: In a cubic crystal, the normal to a plane has the same indices as the plane, so the normal to (111) is [111]. Also, in a cubic crystal, the cosine of the angle between two directions is given by the dot product of unit vectors in those

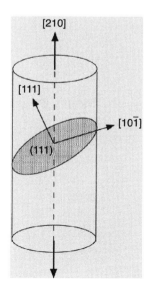

Figure 1.6. A crystal stressed in tension along [210] showing the (111) slip plane and the [10$\bar{1}$] slip direction.

directions. For example, the cosine of the angle between $[u_1 v_1 w_1]$ and $[u_2 v_2 w_2]$ is equal to $(u_1 u_2 + v_1 v_2 + w_1 w_2)/[(u_1^2 + v_1^2 + w_1^2)(u_2^2 + v_2^2 + w_2^2)]^{1/2}$. Designating [210] as x, [10$\bar{1}$] as d, and [111] as n, $\tau_{nd} = \ell_{nx}\ell_{dx}\sigma_{xx} = \{(2 \cdot 1 + 1 \cdot 1 + 0)/\sqrt{[(2^2 + 1^2 + 0)(1^2 + 1^2 + 1^2)]}\} \cdot \{(2 \cdot 1 + 1 \cdot 0 + 0 \cdot 0)/\sqrt{[(2^2 + 1^2 + 0)(1^2 + 0 + 1^2)]}\}2.8 \text{ MPa} = 2.8(6/5)\sqrt{6} = 1.372 \text{ MPa}$.

Principal Stresses

It is always possible to find a set of axes (1, 2, 3) along which the shear stress components vanish. In this case, the normal stresses, $\sigma_1, \sigma_2,$ and σ_3, are called *principal stresses*, and the 1, 2, and 3 axes are the *principal stress axes*. The magnitudes of the principal stresses, σ_p, are the three roots of

$$\sigma_p^3 - I_1\sigma_p^2 - I_2\sigma_p - I_3 = 0, \tag{1.11}$$

where

$$\begin{aligned} I_1 &= \sigma_{xx} + \sigma_{yy} + \sigma_{zz}, \\ I_2 &= \sigma_{yz}^2 + \sigma_{zx}^2 + \sigma_{xy}^2 - \sigma_{yy}\sigma_{zz} - \sigma_{zz}\sigma_{xx} - \sigma_{xx}\sigma_{yy}, \\ I_3 &= \sigma_{xx}\sigma_{yy}\sigma_{zz} + 2\sigma_{yz}\sigma_{zx}\sigma_{xy} - \sigma_{xx}\sigma_{yz}^2 - \sigma_{yy}\sigma_{zx}^2 - \sigma_{zz}\sigma_{xy}^2. \end{aligned} \tag{1.12}$$

The first invariant, $I_1 = -p/3$, where p is the pressure. $I_1, I_2,$ and I_3 are independent of the orientation of the axes and are therefore called *stress invariants*. In terms of the principal stresses, the invariants are

$$\begin{aligned} I_1 &= \sigma_1 + \sigma_2 + \sigma_3, \\ I_2 &= -\sigma_{22}\sigma_{33} - \sigma_{33}\sigma_{11} - \sigma_{11}\sigma_{22}, \\ I_3 &= \sigma_{11}\sigma_{22}\sigma_{33}. \end{aligned} \tag{1.13}$$

EXAMPLE PROBLEM 1.2: Find the principal stresses in a body under the stress state, $\sigma_x = 10, \sigma_y = 8, \sigma_z = -5, \tau_{yz} = \tau_{zy} = 5, \tau_{zx} = \tau_{xz} = -4,$ and $\tau_{xy} = \tau_{yx} = -8$, where all stresses are in MPa.

Solution: Using equation (1.13), $I_1 = 10 + 8 - 5 = 13$, $I_2 = 5^2 + (-4)^2 + (-8)^2 - 8(-5) - (-5)10 - 10 \cdot 8 = 115$, $I_3 = 10 \times 8(-5) + 2 \times 5(-4)(-8) - 10 \times 5^2 - 8(-4)^2 - (-5)(-8)^2 = -138$.

Solving equation (1.11), $\sigma_p^3 - 13\sigma_p^2 - 115\sigma_p + 138 = 0, \sigma_p = 1.079, 18.72, -6.82$.

Mohr's Stress Circles

In the special case where there are no shear stresses acting on one of the reference planes (e.g., $\tau_{zy} = \tau_{zx} = 0$), the normal to that plane, z, is a direction of principal stress, and the other two principal stress directions lie in the plane. This is illustrated in Figure 1.7. For these conditions, $\ell_{x'z} = \ell_{y'z} = 0, \tau_{zy} = \tau_{zx} = 0, \ell_{x'x} = \ell_{y'y} = \cos f$, and $\ell_{x'y} = -\ell_{y'x} = \sin \phi$. The variation of the shear stress component, $\tau_{x'y'}$, can be found by substituting these conditions into the stress transformation equation (1.8b). Substituting $\ell_{x'z} = -\ell_{y'z} = 0$,

$$\tau_{x'y'} = \cos\phi \sin\phi(-\sigma_{xx} + \sigma_{yy}) + (\cos^2\phi - \sin^2\phi)\tau_{xy}. \tag{1.14a}$$

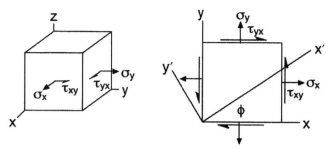

Figure 1.7. Stress state to which Mohr's circle treatment applies. Two shear stresses, τ_{yz} and τ_{zx}, are zero.

Similar substitution into the expressions for $\sigma_{x'}$ and $\sigma_{y'}$ results in

$$\sigma_{x'} = \cos^2\phi\,\sigma_x + \sin^2\phi\,\sigma_y + 2\cos\phi\sin\phi\,\tau_{xy}. \tag{1.14b}$$

and

$$\sigma_{y'} = \sin^2\phi\,\sigma_x + \cos^2\phi\,\sigma_y + 2\cos\phi\sin\phi\,\tau_{xy}. \tag{1.14c}$$

These can be simplified by substituting the trigonometric identities, $\sin 2\phi = 2\sin\phi\cos\phi$ and $\cos 2\phi = \cos^2\phi - \sin^2\phi$,

$$\tau_{x'y'} = -[(\sigma_x - \sigma_y)/2]\sin 2\phi + \tau_{xy}\cos 2\phi \tag{1.15a}$$

$$\sigma_{x'} = (\sigma_x + \sigma_y)/2 + [(\sigma_x - \sigma_y)/2]\cos 2\phi + \tau_{xy}\sin 2\phi. \tag{1.15b}$$

and

$$\sigma_{y'} = (\sigma_x + \sigma_y)/2 - [(\sigma_x - \sigma_y)/2]\cos 2\phi + \tau_{xy}\sin 2\phi. \tag{1.15c}$$

Setting $\tau_{x'y'} = 0$ in equation 1.15a, becomes the angle, θ, between the principal stresses axes and the x and y axes. See Figure 1.8. $\tau_{x'y'} = 0 = \sin 2\theta(\sigma_x - \sigma_y)/2 + \cos 2\theta\,\tau_{xy}$ or

$$\tan 2\theta = \tau_{xy}/[(\sigma_x - \sigma_y)/2]. \tag{1.16}$$

Figure 1.8. Mohr's circles for stresses showing the stresses in the x-y plane. *Note:* The 1-axis is rotated counterclockwise from the x-axis in real space (a), whereas in the Mohr's circle diagram, the 1-axis is rotated clockwise from the x axis (b).

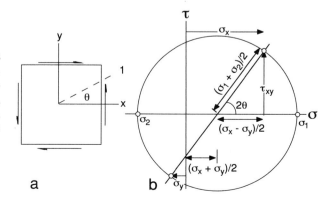

The principal stresses, σ_1 and σ_2, are the values of $\sigma_{x'}$ and $\sigma_{y'}$ for this value of ϕ,

$$
\begin{aligned}
\sigma_{1,2} &= (\sigma_x + \sigma_y)/2 \pm [\sigma_x - \sigma_y)/2]\cos 2\theta + \tau_{xy}\sin 2\theta \quad \text{or} \\
\sigma_{1,2} &= (\sigma_x + \sigma_y)/2 \pm (1/2)[(\sigma_x - \sigma_y)^2 + 4\tau_{xy^2}]^{1/2}
\end{aligned} \tag{1.17}
$$

A Mohr's circle diagram is a graphical representation of equations (1.16) and (1.17). It plots as a circle with a radius $(\sigma_1 - \sigma_2)/2$ centered at

$$
(\sigma_1 + \sigma_2)/2 = (\sigma_x + \sigma_y)/2, \tag{1.17a}
$$

as shown in Figure 1.8. The normal stress components, σ, are represented on the ordinate and the shear stress components, τ, on the abscissa. Consider the triangle in Figure 1.8b. Using the Pythagorean theorem, the hypotenuse,

$$
(\sigma_1 - \sigma_2)/2 = \left\{[(\sigma_x + \sigma_y)/2]^2 + \tau_{xy}^2\right\}^{,1/2} \tag{1.17b}
$$

and

$$
\tan(2\theta) = [\tau_{xy}/[(\sigma_x + \sigma_y)/2]. \tag{1.17c}
$$

The full three-dimensional stress state may be represented by three Mohr's circles (Figure 1.9).

The three principal stresses, σ_1, σ_2, and σ_3, are plotted on the horizontal axis. The circles connecting these represent the stresses in the 1–2, 2–3, and 1–3 planes. The largest shear stress may be either $(\sigma_1 - \sigma_2)/2$, $(\sigma_2 - \sigma_3)/2$, or $(\sigma_1 - \sigma_3)/2$.

EXAMPLE PROBLEM 1.3: A body is loaded under stresses, $\sigma_x = 150$ MPa, $\sigma_y = 60$ MPa, $\tau_{xy} = 20$ MPa, $\sigma_z = \tau_{yz} = \tau_{zx} = 0$. Find the three principal stresses, sketch the three-dimensional Mohr's circle diagram for this stress state, and find the largest shear stress in the body.

Solution: σ_1, $\sigma_2 = (\sigma_x + \sigma_y)/2 \pm \{[(\sigma_x - \sigma_y)/2]^2 + \tau_{xy}^2\}^{1/2} = 154.2$, 55.8 MPa, $\sigma_3 = \sigma_z = 0$. Figure 1.10 is the Mohr's circle diagram. Note that the largest shear stress, $\tau_{max} = (\sigma_1 - \sigma_3)/2 = 77.1$ MPa, is not in the 1–2 plane.

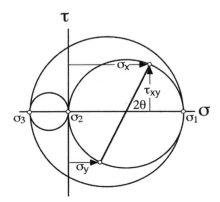

Figure 1.9. Three Mohr's circles representing a stress state in three dimensions. The three circles represent the stress states in the 2–3, 3–1, and 1–2 planes.

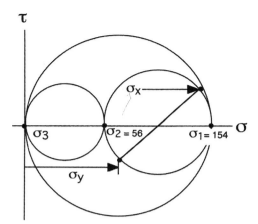

Figure 1.10. Mohr's circles for example problem 1.3.

Strains

An infinitesimal normal strain is defined as strain by the change of length, L, of a line:

$$d\varepsilon = dL/L. \tag{1.18}$$

Integrating from the initial length, L_o, to the current length, L,

$$\varepsilon = \int dL/L = \ln(L/L_o). \tag{1.19}$$

This finite form is called *true strain* (or *natural strain, or logarithmic strain*). Alternatively, *engineering* or *nominal strain, e*, is defined as

$$e = \Delta L/L_o. \tag{1.20}$$

If the strains are small, then engineering and true strains are nearly equal. Expressing $\varepsilon = \ln(L/L_o) = \ln(1+e)$ as a series expansion, $\varepsilon = e - e^2/2 + e^3/3! \ldots$ so as $e \to 0, \varepsilon \to e$. This is illustrated in example problem 1.4.

EXAMPLE PROBLEM 1.4: Calculate the ratio of e/ε for several values of e.

Solution: $e/\varepsilon = e/\ln(1+e)$. Evaluating:

for $e = 0.001$, $e/\varepsilon = 1.0005$;
for $e = 0.01$, $e/\varepsilon = 1.005$;
for $e = 0.02$, $e/\varepsilon = 1.010$;
for $e = 0.05$, $e/\varepsilon = 1.025$;
for $e = 0.10$, $e/\varepsilon = 1.049$;
for $e = 0.20$, $e/\varepsilon = 1.097$;
for $e = 0.50$, $e/\varepsilon = 1.233$.

Note that the difference e and ε between is about 1% for $e < 0.02$.

There are several reasons that true strains are more convenient than engineering strains.

1. True strains for equivalent amounts of deformation in tension and compression are equal except for sign.

2. True strains are additive. For a deformation consisting of several steps, the overall strain is the sum of the strains in each step.
3. The volume change is related to the sum of the three normal strains. For constant volume, $\varepsilon_x + \varepsilon_y + \varepsilon_z = 0$.

These statements are not true for engineering strains, as illustrated in the following examples.

EXAMPLE PROBLEM 1.5: An element 1 cm long is extended to twice its initial length (2 cm) and then compressed to its initial length (1 cm).

A. Find true strains for the extension and compression.
B. Find engineering strains for the extension and compression.

Solution:

A. During the extension, $\varepsilon = \ln(L/L_o) = \ln 2 = 0.693$, and during the compression,

$$\varepsilon = \ln(L/L_o) = \ln(1/2) = -0.693.$$

B. During the extension, $e = \Delta L/L_o = 1/1 = 1.0$, and during the compression,

$$e = \Delta L/L_o = -1/2 = -0.5.$$

Note that with engineering strains, the magnitude of strain to reverse the shape change is different.

EXAMPLE PROBLEM 1.6: A bar 10 cm long is elongated by (1) drawing to 15 cm, and then (2) drawing to 20 cm.

A. Calculate the engineering strains for the two steps, and compare the sum of these with the engineering strain calculated for the overall deformation.
B. Repeat the calculation with true strains.

Solution:

A. For step 1, $e_1 = 5/10 = 0.5$; for step 2, $e_2 = 5/15 = 0.333$. The sum of these is 0.833, which is less than the overall strain, $e_{tot} = 10/10 = 1.00$
B. For step 1, $\varepsilon_1 = \ln(15/10) = 0.4055$; for step 2, $\varepsilon_1 = \ln(20/15) = 0.2877$. The sum is 0.6931, and the overall strain is $\varepsilon_{tot} = \ln(15/10) + \ln(20/15) = \ln(20/10) = 0.6931$.

EXAMPLE PROBLEM 1.7: A block of initial dimensions L_{x0}, L_{y0}, L_{z0} is deformed so that the new dimensions are L_x, L_y, L_z. Express the volume strain, $\ln(V/V_o)$, in terms of the three true strains, ε_x, ε_y, ε_z.

Solution: $V/V_o = L_x L_y L_z/(L_{xo} L_{yo} L_{zo})$, so

$$\ln(V/V_o) = \ln(L_x/L_{xo}) + \ln(L_y/L_{yo}) + \ln(L_z/L_{zo}) = \varepsilon_x + \varepsilon_y + \varepsilon_z.$$

Note that if there is no volume change, $(\ln(V/V_o) = 0)$, the sum of the normal strains

$$\varepsilon_x + \varepsilon_y + \varepsilon_z = 0.$$

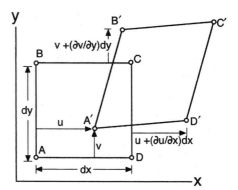

Figure 1.11. Translation, rotation, and distortion of a two-dimensional body.

Small Strains

As bodies deform, they often undergo translations and rotations as well as deformation. Strain must be defined in such a way as to exclude the effects of translation and rotation. Consider a two-dimensional body in Figure 1.11. Normal strains are defined as the fractional extensions (tensile) or contractions (compressive), $\Delta L / L_o$, so $\varepsilon_{xx} = (\overline{AD} - \overline{AD})/\overline{AD} = \overline{AD}/\overline{AD} - 1$. For a small strain, this reduces to

$$\varepsilon_{xx} = (\partial u/\partial x)dx/dx = \partial u/\partial x. \tag{1.21}$$

Similarly,

$$\varepsilon_{yy} = (\partial v/\partial y)/dy/dy = \partial v/\partial y. \tag{1.22}$$

Shear strains are similarly defined in terms of the angles between AD and A′D′ and between AB and A′B′, which are, respectively,

$$(\partial v/\partial x)dx/dx = \partial v/\partial x \quad \text{and} \quad (\partial u/\partial y)dy/dy = \partial u/\partial y.$$

The total engineering shear strain, γ_{yx}, is the sum of these angles,

$$\gamma_{yx} = \partial v/\partial x + \partial u/\partial y = \gamma_{xy}. \tag{1.23}$$

Figure 1.12 shows that this definition excludes the effects of rotation for small strains.

For a three-dimensional body with displacements w in the z-direction,

$$\varepsilon_{yy} = \partial v/\partial y, \quad \gamma_{yz} = \gamma_{zy} = \partial w/\partial y + \partial v/\partial z \quad \text{and} \quad \gamma_{yx} = \gamma_{xy} = \partial v/\partial x + \partial u/\partial y.$$

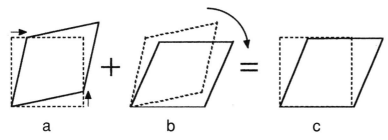

Figure 1.12. Illustration of shear and rotation. With small deformations, (a) differs from (c) only by a rotation, (b).

Small strains can be treated as tensors,

$$\varepsilon_{ij} = \begin{vmatrix} \varepsilon_{xx} & \varepsilon_{yx} & \varepsilon_{zx} \\ \varepsilon_{xy} & \varepsilon_{yy} & \varepsilon_{zy} \\ \varepsilon_{xz} & \varepsilon_{yz} & \varepsilon_{zz} \end{vmatrix}, \tag{1.24}$$

where the mathematical shear strains, ε_{ij}, are one-half of the engineering shear strains, γ_{ij}:

$$\varepsilon_{yz} = \varepsilon_{zy} = (1/2)\gamma_{yz} = (1/2)(dv/dz + dw/dx)$$
$$\varepsilon_{zx} = \varepsilon_{xz} = (1/2)\gamma_{zx} = (1/2)(\partial w/\partial x + \partial u/\partial z) \tag{1.25}$$
$$\varepsilon_{xy} = \varepsilon_{yx} = (1/2)\gamma_{xy} = (1/2)(\partial v/\partial y + \partial v/\partial x).$$

Transformation of Axes

Small strains may be transformed from one set of axes to another in a manner completely analogous to the transformation of stresses (equation (1.9),

$$\varepsilon_{ij} = \ell_{im}\ell_{jn}\varepsilon_{mn}, \tag{1.26}$$

where double summation is implied. For example,

$$\varepsilon_{x'x'} = \ell_{x'x}\ell_{x'x}\varepsilon_{xx} + \ell_{x'y}\ell_{x'x}\varepsilon_{yx} + \ell_{x'z}\ell_{x'x}\varepsilon_{zx}$$
$$+ \ell_{x'x}\ell_{x'y}\varepsilon_{xy} + \ell_{x'y}\ell_{x'y}\varepsilon_{yy} + \ell_{x'z}\ell_{x'y}\varepsilon_{zy}$$
$$+ \ell_{x'x}\ell_{x'z}\varepsilon_{xz} + \ell_{x'y}\ell_{x'z}\varepsilon_{yz} + \ell_{x'z}\ell_{x'z}\varepsilon_{zz} \tag{1.27a}$$

and

$$\varepsilon_{x'y'} = \ell_{x'x}\ell_{y'x}\varepsilon_{xx} + \ell_{x'y}\ell_{y'x}\varepsilon_{yx} + \ell_{x'z}\ell_{y'x}\varepsilon_{zx}$$
$$+ \ell_{x'x}\ell_{y'y}\varepsilon_{xy} + \ell_{x'y}\ell_{y'y}\varepsilon_{yy} + \ell_{x'z}\ell_{y'y}\varepsilon_{zy}$$
$$+ \ell_{x'x}\ell_{y'z}\varepsilon_{xz} + \ell_{x'y}\ell_{y'z}\varepsilon_{yz} + \ell_{x'z}\ell_{y'z}\varepsilon_{zz}. \tag{1.27b}$$

These can be written more simply in terms of the usual shear strains,

$$\varepsilon_{x'} = \ell_{x'x^2}\varepsilon_x + \ell_{x'y^2}\varepsilon_y + \ell_{x'z^2}\varepsilon_z$$
$$+ \ell_{x'y}\ell_{x'z}\gamma_{yz} + \ell_{x'z}\ell_{x'x}\gamma_{zx} + \ell_{x'x}\ell_{x'y}\gamma_{xy} \tag{1.28a}$$

and

$$\gamma_{x'y'} = 2(\ell_{x'x}\ell_{y'x}\varepsilon_x + \ell_{x'y}\ell_{y'y}\varepsilon_y + \ell_{x'z}\ell_{y'z}\varepsilon_z)$$
$$+ (\ell_{x'y}\ell_{y'z} + \ell_{x'z}\ell_{y'y})\gamma_{yz} + (\ell_{x'z}\ell_{y'x} + \ell_{x'x}\ell_{y'z})\gamma_{yz}$$
$$+ (\ell_{x'x}\ell_{y'y} + \ell_{x'y}\ell_{y'z})\gamma_{xy}. \tag{1.28b}$$

Large strains are not tensors and cannot be transformed from one axis system to another by a tensor transformation. The reason is that the angles between material directions are altered by deformation. With small strains, changes of angle are small and can be neglected. Example problem 1.8 illustrates this point.

EXAMPLE PROBLEM 1.8: A two-dimensional square body initially 1.00 cm by 1.00 cm was deformed into a rectangle 0.95 cm by 1.10 cm, as shown in Figure 1.13.

A. Calculate the strain, $e_{x'}$, along the diagonal from its initial and final dimensions. Then calculate the strains, e_x and e_y, along the edges and use the

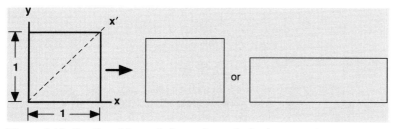

Figure 1.13. Small and large deformations of a body.

transformation equation, $e_{ij} = \sum \sum \ell_{im}\ell_{jn}e_{mn}$, to find the strain along the diagonal and compare with the two values of $e_{x'}$.

B. Repeat A for a 1.00-cm by 1.00-cm square deformed into a 0.50-cm by 2.0-cm rectangle.

Solution:

A. The initial diagonal $= \sqrt{2} = 1.414214$, and for the small deformation, the final diagonal becomes $\sqrt{(\phi.95^2 + 1.1^2)} = 1.4534$, so $e_{x'} = (L - L_o)/L_o = L/L_o - 1 = 1.4534/1.414214 - 1 = 0.0277$. Taking the angle, θ, between the x' and x (or y) axes as 45 degrees, $\varepsilon_{x'} = \ell_{x'x}^2\varepsilon_x + \ell_{x'y}^2\varepsilon_y = (1/2)(1) + (1/2)(-0.5) - 0.0250$, which is very close to 0.0277.

B. For the large deformation, the diagonal becomes $\sqrt{(2^2 + 0.5^2)} = 2.062$, so calculating the strain from this, $e_{x'} = 2.062/1.414214 - 1 = 0.4577$.

The strains on the edges are $e_x = 1$ and $e_y = -0.5$, so $e_{x'} = \ell_{x'x}2e_x + \ell_{x'y}2e_y = (1/2)(1) + (1/2)(-0.5) = 0.0250$, which does not agree $e_{x'} = 0.4577$ calculated from the specimen dimensions. With true strains, the agreement is not much better. Direct calculation of $e_{x'}$ from the diagonal gives $e_{x'} = \ln(2.062/1.414) = 0.377$, and calculation from the strains along the sides gives $e_{x'} = (1/2)\ln(2) + (1/2)\ln(\phi.5) = 0$. The reason is that with large strains, the angle, θ, changes with deformation.

Mohr's Strain Circles

Because small strains are tensor quantities, Mohr's circle diagrams apply if the strains, e_x, e_y, and γ_{xy} are known along two axes and the third axis, z, is a principal strain axis ($\gamma_{yz} = \gamma_{zx} = 0$). The strains may be either engineering, e, or true, ε, because they are equal when small. Care must be taken to remember that the tensor shear–strain terms are only one half of the conventional shear strains. A plot of $\gamma/2$ versus e (or ε) is a circle, as shown in Figure 1.14. The equations, analogous to those for stresses, are

$$(e_1 + e_2)/2 = (e_x + e_y)/2 \tag{1.29}$$

$$(e_1 - e_2)/2 = \left\{[(e_x - e_y)/2]^2 + (\gamma_{xy}/2)\right\}^{1/2} \tag{1.30}$$

$$e_1, e_2 = (e_x + e_y)/2 \pm \left\{[(e_x - e_y)/2]^2 + (\gamma_{xy}/2)^2\right\}^{1/2} \tag{1.31}$$

$$\tan(2\theta) = (\gamma_{xy}/2)/[(e_x - e_y)/2] = \gamma_{xy}/(e_x - e_y). \tag{1.32}$$

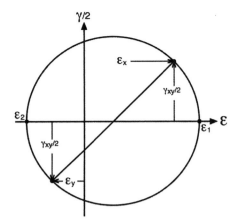

Figure 1.14. Mohr's circle for strain. This is similar to the Mohr's stress circle except that the normal strain, ε (or e), is plotted on the horizontal axis (instead of σ), and $\gamma/2$ is plotted vertically (instead of τ).

As with Mohr's stress circles, a three-dimensional strain state can be represented by three Mohr's circles. It is emphasized that the strain transformation equations, including the Mohr's circle equations, apply to small strains. Errors increase when the strains are large enough to cause rotation of the axes.

EXAMPLE PROBLEM 1.9: Draw the three-dimensional Mohr's circle diagram for an x-direction tension test (assume $e_y = e_z = -e_x/3$), plane strain ($e_y = 0$), and an x-direction compression test (assume $e_y = e_z = -e_x/3$). In each case, determine γ_{max}.

Solution: The three-dimensional Mohr's circle diagrams are shown in Figure 1.15.

Force and Moment Balances

The solutions of many mechanics problems require force and moment balances. The external forces acting on one-half of the body must balance those acting across the cut. Consider force balances to find the stresses in the walls of a capped thin-wall tube loaded by internal pressure.

EXAMPLE PROBLEM 1.10: A capped thin-wall tube having a length, L, a diameter, D, and a wall thickness, t, is loaded by an internal pressure, P. Find the stresses

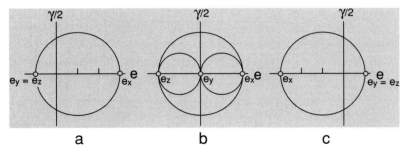

Figure 1.15. Three-dimensional Mohr's strain circles for (a) tension, (b) plane strain, (c) compression. In (a) and (c), the circles between e_y and e_z reduce to a point, and the circles between e_x and e_y coincide with the circles between e_x and e_z.

in the wall assuming that t is much smaller than D and that D is much less than L.

Solution: First make a cut perpendicular to the axis of the tube (Figure 1.16a), and consider the vertical (y-direction) forces. The force from the pressurization is the pressure, P, acting on the end area of the tube, $P\pi D^2/4$. The force in the wall is the stress, σ_y, acting on the cross-section of the wall, $\sigma_y\pi Dt$. Balancing the forces, $P\pi D^2/4 = \sigma_y\pi Dt$, and solving for σ_y,

$$\sigma_y = PD/(4t). \tag{1.33}$$

The hoop stress, σ_x, can be found from a force balance across a vertical cut (Figure 1.16b). The force acting to separate the tube is the pressure, P, acting on the internal area, DL, where L is the tube length. This is balanced by the hoop stress in the two walls, σ_x, acting on the cross-sectional area of the two walls, $2Lt$. (The force in the capped ends is neglected because the tube is long, and we are interested in the stress in regions remote from the ends.) Equating these two forces, $PDL = \sigma_x 2t$,

$$\sigma_x = PD/(2t) = 2\sigma_y. \tag{1.34}$$

The stress in the radial direction though the wall thickness, σ_z, is negligible relative to σ_x and σ_y if $D \gg t$. On the inside surface, $\sigma_z = -P$, and on the outside surface, $\sigma_z = 0$. The average, $-P/2$, is much less than $PD/(2t)$, so for engineering purposes we can take $\sigma_z = 0$.

A moment balance may be made about *any* axis in a body under equilibrium. The moment caused by external forces must equal the moment cause by internal stresses. An example is the torsion of a circular rod.

EXAMPLE PROBLEM 1.11: Relate the internal shear stress, τ_{xy}, in a rod of radius, R, to the torque, T, acting on the rod.

Solution: Consider a differential element of dimensions $(2\pi r)(dr)$ in a tubular element at a radius, r, from the axis and of thickness, dr (Figure 1.17). The shear force acting on this element is the shear stress times the area, $\tau_{xy}(2\pi r)(dr)$. The torque about the central axis caused by this element is the shear force times the

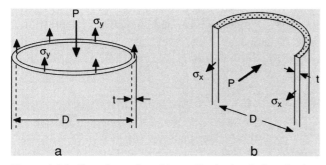

Figure 1.16. Cuts through a thin-wall tube loaded under internal pressure.

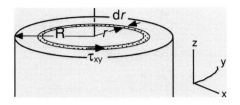

Figure 1.17. A differential element of area $2\pi r\,dr$ in a rod loaded under torsion. The shear stress, τ_{xy}, on this element causes a differential moment, $2\pi\tau_{xy}r^2 dr$.

distance from the axis, r, $dT = \tau_{xy}(2\pi r)(r)\,dr$, so

$$T = 2\pi \int_0^R \tau_{xy}r^2\,dr. \qquad (1.35)$$

An explicit solution requires knowledge of how τ_{xy} varies with r.

Boundary Conditions

It is important, in analyzing mechanical problems, to recognize simple, often unstated, boundary conditions and use them to make simplifying assumptions.

1. Free surfaces: On a free surface, the two shear stress components in the surface vanish (i.e., if z is the normal to a free surface, $\tau_{yz} = \tau_{zx} = 0$). Unless there is a pressure acting on a free surface, the stress normal to it also vanishes (i.e., $\sigma_{zz} = 0$). Likewise, there are no shear stresses, $\tau_{yz} = \tau_{zx}$, acting on surfaces that are assumed to be frictionless.
2. Constraints from neighboring regions: The deformation in a particular region is often controlled by the deformation in a neighboring region. Consider the deformation in a long narrow groove in a plate, as shown in Figure 1.18. The long narrow groove (B) is in close contact with a thicker region (A). As the plate deforms, the deformation in the groove must be compatible with the deformation outside the groove. Its elongation or contraction must be the same as that in the material outside so that $\varepsilon_{xA} = \varepsilon_{xB}$. The y- and z-direction strains in the two regions need not be the same.
3. St. Venant's principle: This principle states that the restraint from any end or edge effect will disappear within one characteristic distance. As an example, the enlarged end of a tensile bar (Figure 1.19) tends to suppress lateral contraction of the gauge section next to it. Here, the characteristic distance is the diameter of the gauge section, so the constraint is almost gone at a distance from the enlarged end equal to the diameter.

Another example of St. Venant's principle is a thin, wide sheet bent to a constant radius of curvature (Figure 1.20). The condition of plane strain ($\varepsilon_y = 0$) prevails over most of the material. This is because the top and bottom surfaces are so close that they restrain one another from contracting or expanding. Appreciable

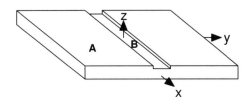

Figure 1.18. Grooved plate. The material outside the groove affects the x-direction flow inside, so $\varepsilon_{xA} = \varepsilon_{xB}$.

Figure 1.19. In the gauge section of a tensile bar, the effect of the ends almost disappears at a distance, d, from the shoulder.

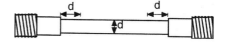

Figure 1.20. Bending of a thin sheet. Plane strain ($\varepsilon_y = 0$) prevails except near the edges, where there is a condition of plane stress ($\sigma_v = 0$).

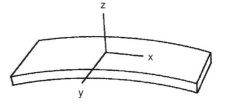

deviation from plane strain occurs only within a distance from the edges of the sheet equal to the sheet thickness. At the edge, $\sigma_y = 0$.

Note

The simplification of the notation for tensor transformation $\sigma_{ij} = \sum_{n=1}^{3} \sum_{m=1}^{3} \ell_{im}\ell_{ij}\sigma_{mn}$ to $\sigma_{ij} = \ell_{im}\ell_{jn}\ell_{mn}$ has been attributed to Albert Einstein.

Problems

1. Consider an aluminum single crystal under a stress state, $\sigma_x = 250$ psi, $\sigma_y = -50$ psi, $\sigma_z = \tau_{yz} = \tau_{zx} = \tau_{xy} = 0$, where $x = [100]$, $y = [010]$, and $z = [001]$.

    A. What is the resolved shear stress, τ_{nd}, on the (111) plane in the $[1\bar{1}0]$ direction (i.e., with n $= [111]$, d $= [1\bar{1}0]$)?

    B. What is the resolved shear stress on the $(11\bar{1})$ plane in the $[101]$ direction?

2. Consider the single crystal in problem 1. Now suppose that slip does occur on the (111) plane in the $[10\bar{1}]$ direction and only on that slip system. Also assume that the resulting strains are small. Calculate the ratios of the resulting strains, $\varepsilon_y/\varepsilon_x$ and $\varepsilon_z/\varepsilon_x$.

3. A body is loaded under a stress state, $\sigma_x = 400$, $\sigma_y = 100$, $\tau_{xy} = 120$, $\tau_{yz} = \tau_{zx} = \sigma_z = 0$.

    A. Sketch the Mohr's circle diagram.

    B. Calculate the principal stresses.

    C. What is the largest shear stress in the body? (Do not neglect the z direction.)

4. Three strain gauges have been pasted on the surface of a piece of steel in the pattern shown in Figure 1.21. While the steel is under load, these gauges indicate the strains parallel to their axes:

    Gauge A 450×10^{-6} Gauge B 300×10^{-6} Gauge C -150×10^{-6}

    A. Calculate the principal strains ε_1 and ε_2.

    B. Find the angle between the 1-axis and the x-axis, where 1 is the axis of the largest principal strain. [*Hint:* Let the direction of gauge B be x', write the strain transformation equation expressing the strain $e_{x'}$ in terms of the strains along the x-y axes, solve for γ_{xy}, and finally use the Mohr's circle equations.]

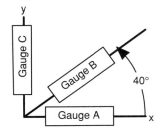

Figure 1.21. Arrangement of strain gauges.

5. Consider a thin-wall tube that is 1 in. in diameter and has a 0.010-in. wall thickness. Let x, y, and z be the axial, tangential (hoop), and radial directions, respectively.

    A. The tube is subjected to an axial tensile force of 80 lb and a torque of 100 in.-lb.

        i. Sketch the Mohr's stress circle diagram showing stresses in the x-y plane.

        ii. What is the magnitude of the largest principal stress?

        iii. At what angles are the principal stress axes, 1 and 2, to the x- and y-directions?

    B. Now let the tube be capped and subject to an internal pressure of 120 psi and a torque of 100 in.-lb.

        i. Sketch the Mohr's stress circle diagram showing stresses in the x-y plane.

        ii. What is the magnitude of the largest principal stress?

        iii. At what angles are the principal stress axes, 1 and 2, to the x- and y-directions?

6. A solid is deformed under plane strain conditions ($\varepsilon_z = 0$). The strains in the x-y plane are $\varepsilon_x = 0.010$, $\varepsilon_y = 0.005$, and $\gamma_{xy} = 0.007$.

    A. Sketch the Mohr's strain circle diagram.

    B. Find the magnitude of ε_1 and ε_2.

    C. What is the angle between the 1 and x axes?

    D. What is the largest shear strain in the body? (Do not neglect the z direction.)

7. A grid of circles, each 10.00 mm in diameter, was etched on the surface of a sheet of steel. (Figure 1.22). When the sheet was deformed, the grid circles were distorted into ellipses. Measurement of one indicated that the major and minor diameters were 11.54 and 10.70 mm, respectively.

    A. What are the principal strains, ε_1 and ε_2?

    B. If the axis of the major diameter of the ellipse makes an angle of 34 degrees to the x-direction, what is the shear strain, γ_{xy}?

    C. Draw the Mohr's strain circle showing ε_1, ε_2, ε_x, and ε_y.

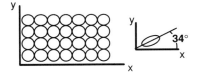

Figure 1.22. Circle grids printed on a metal sheet.

8. Consider an aluminum single crystal under a stress state, $\sigma_x = +75$ psi, $\sigma_y = +25$ psi, $\sigma_z = \tau_{yz} = \tau_{zx} = \tau_{xy} = 0$, where x = [100], y = [010], and z = [001].
What is the resolved shear stress, γnd, on the (111) plane and [10$\bar{1}$] direction?

9. Consider the torsion of a rod that is 1 m long and 50 mm in diameter.

   A. If one end of the rod is twisted by 1.2 degrees relative to the other end, what would be the largest principal strain on the surface?
   B. If the rod were extended by 1.2% and its diameter decreased by 0.4% at the same time it was being twisted, what would be the largest principal strain?

10. Two pieces of rod are glued together along a joint whose normal makes an angle, θ, with the rod axis, x (Figure 1.23). The joint will fail if the shear stress on the joint exceeds its shear strength, τ_{max}. It will also fail if the normal stress across the joint exceeds its normal strength, σ_{max}. The shear strength, τ_{max}, is 80% of the normal strength, σ_{max}. The rod will be loaded in uniaxial tension along its axis, and it is desired that the rod carry as high a tensile force, F_x, as possible. The angle, θ, cannot exceed 65 degrees.

   A. At what angle, θ, should the joint be made so that a maximum force can be carried?
   B. If θ_{max} were limited to 45 degrees, instead of 65 degrees, how would your answer be altered? [*Hint:* Plot σ_x/σ_{max} versus θ for both failure modes.]

Figure 1.23. Glued rod.

11. Consider a tube made by coiling and gluing a strip as shown in Figure 1.24. The diameter is 1.5 in., the length is 6 in., and the wall thickness is 0.030 in. If a tensile

Figure 1.24. Tube formed from a coiled and glued strip.

force of 80 lb. and a torque of 30 in.-lb are applied in the direction shown, what is the stress normal to the glued joint? [*Hint:* Set up a coordinate system.]

2 Elasticity

Introduction

Elastic deformation is reversible. When a body deforms elastically under a load, it will revert to its original shape as soon as the load is removed. A rubber band is a familiar example. Most materials, however, can undergo much less elastic deformation than rubber. In crystalline materials, elastic strain is small, usually less than 1/2%. It is safe for most materials other than rubber to assume that the amount of deformation is proportional to the stress. This assumption is the basis of the following treatment. Because elastic strains are small, it does not matter whether the relations are expressed in terms of engineering strains, e, or true strains, ε.

The treatment in this chapter will start with the elastic behavior of isotropic materials, the temperature dependence of elasticity, and thermal expansion. Then anisotropic elastic behavior and thermal expansion will be covered.

Isotropic Elasticity

An isotropic material is one that has the same properties in all directions. If uniaxial tension is applied in the x-direction, the tensile strain is $\varepsilon_x = \sigma_x/E$, where E is *Young's modulus*. Uniaxial tension also causes lateral strains, $\varepsilon_y = \varepsilon_z = -\upsilon\varepsilon_x$, where υ is *Poisson's ratio*. Consider the strain, ε_x, produced by a general stress state, $\sigma_x, \sigma_y, \sigma_z$. The stress, σ_x, causes a contribution $\varepsilon_x = \sigma_x/E$. The stresses, σ_y, σ_z, cause Poisson contractions, $\varepsilon_x = -\upsilon\sigma_y/E$ and $\varepsilon_x = -\upsilon\sigma_z/E$. Taking into account these Poisson contractions, the general statement of *Hooke's law* is

$$\varepsilon_x = (1/E)[\sigma_x - \upsilon(\sigma_y + \sigma_z)]. \tag{2.1a}$$

Shear strains are affected only by the corresponding shear stress so

$$\gamma_{yz} = \tau_{yz}/G = 2\varepsilon_{yz}, \tag{2.1b}$$

where G is the shear modulus. Similar expressions apply for all directions, so

$$\begin{aligned}
\varepsilon_x &= (1/E)[\sigma_x - \upsilon(\sigma_y + \sigma_z)] & \gamma_{yz} &= \tau_{yz}/G \\
\varepsilon_y &= (1/E)[\sigma_y - \upsilon(\sigma_z + \sigma_x)] & \gamma_{zx} &= \tau_{zx}/G \\
\varepsilon_z &= (1/E)[\sigma_z - \upsilon(\sigma_x + \sigma_y)] & \gamma_{xy} &= \tau_{xy}/G.
\end{aligned} \tag{2.2}$$

Equations (2.1) and (2.2) hold whether or not the x-, y-, and z-directions are directions of principal stress.

For an isotropic material, the shear modulus, G, is not independent of E and v. This can be demonstrated by considering a state of pure shear, τ_{xy}, with $\sigma_x = \sigma_y = \sigma_z = \tau_{yz} = \tau_{zx} = 0$. The Mohr's circle diagram (Figure 2.1) shows that the principal stresses are $\sigma_1 = \tau_{xy}$, $\sigma_2 = -\tau_{xy}$, and $\sigma_3 = 0$.

From Hooke's law, $\varepsilon_1 = (1/E)[\sigma_1 - v(\sigma_2 + \sigma_3)] = (1/E)[\tau_{xy} - v(-\tau_{xy} + 0)] = [(1 + v)/E]\tau$. The Mohr's strain circle diagram (Figure 2.1) shows that $\gamma_{xy}/2 = \varepsilon_1$. Substituting for σ_1, $\gamma_{xy}/2 = [(1 + v)/E]\tau_{xy}$. Now comparing with $\gamma_{xy} = \tau_{xy}/G$,

$$G = E/[2(1 + v)]. \tag{2.3}$$

The bulk modulus, B, is defined by the relation between the volume strain and the mean stress,

$$\Delta V/V - (1/B)\sigma_m, \tag{2.4}$$

where $\sigma_m = (\sigma_x + \sigma_y + \sigma_z)/3$. The bulk modulus, like the shear modulus, is not independent of E and v. This can be demonstrated by considering the volume strain produced by a state of hydrostatic stress, $\sigma_x = \sigma_y = \sigma_z = \sigma_m$. It was shown in example problem 1.7 that, for small deformations, the volume strain, $\Delta V/V = \varepsilon_x + \varepsilon_y + \varepsilon_z$. Substituting $\varepsilon_x = (1/E)[\sigma_x - v(\sigma_y + \sigma_z)] = (1/E)[\sigma_m - v(\sigma_m + \sigma_m)] = [(1 - 2v)/E]\sigma_m$. Then $\Delta V/V = 3\sigma_m(1 - 2v)/E$. Comparing with $\Delta V/V = (1/B)s_m$,

$$B = E/[3(1 - 2v)]. \tag{2.5}$$

For most materials, Poisson's ratio is between 0.2 and 0.4. The value of v for an isotropic material cannot exceed 0.5; if it were, B would be negative, which would imply that an increase of pressure would cause an increase of volume.

Any of the four elastic constants, E, v, G, and B can be expressed in terms of two others:

$$E = 2G(1 + v) = 3B(1 - 2v) = 9BG/(G + 3B), \tag{2.6a}$$

$$v = E/(2G) - 1 = 1/2 - E/(6B) = 1/2 - (3/2)G/(3B + G), \tag{2.6b}$$

$$G = E/[2(1 + v)] = 3EB/(9B - E) = (3/2)E(1 - 2v)/(1 + v), \tag{2.6c}$$

$$B = E/[3(1 - 2v)] = EG/[3(3G - E)] = (2/3)G(1 + v)/(1 - 2v). \tag{2.6d}$$

Only three of the six normal stresses and strain components, e_x, e_y, e_z, σ_x, σ_y, or σ_z, need be known to find the other three. This is illustrated in the following examples.

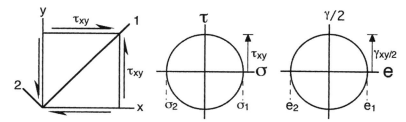

Figure 2.1. Mohr's stress and strain circles for shear.

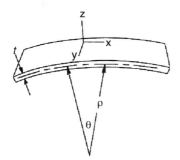

Figure 2.2. Bent sheet. The strain, e_y, and the stress, σ_z, are zero.

EXAMPLE PROBLEM 2.1: A wide sheet (1 mm thick) of steel is bent elastically to a constant radius of curvature, $\rho = 50$ cm, measured from the axis of bending to the center of the sheet, as shown in Figure 2.2. Knowing that $E = 208$ GPa, and $\upsilon = 0.29$ for steel, find the stress in the surface. Assume that there is no net force in the plane of the sheet.

Solution: Designate the directions as shown in Figure 2.2. Inspection shows that $\sigma_z = 0$ because the stress normal to a free surface is zero. Also, $e_y = 0$ because the sheet is wide relative to its thickness, e_y must be the same on top and bottom surfaces, and therefore equals zero. Finally, the strain, e_x, at the outer surface can be found geometrically as $e_x = (t/2)/r = (1/2)/500 = 0.001$. Now substituting into Hooke's laws (equation 2.2), $e_y = (1/E)[\sigma_y - \upsilon(\sigma_x - 0)]$, so $\sigma_y = \upsilon\sigma_x$. Substituting again, $e_x = t/(2r) = (1/E)[\sigma_x - \upsilon(\sigma_y - 0)] = (1/E)(\sigma_x - \upsilon^2\sigma_x) = \sigma_x(1 - \upsilon^2)/E$. $\sigma_x = [t/(2r)]E/(1 - \upsilon^2)] = (0.001)(208 \times 10^9)/(1 - 0.29^2) = 227$ MPa. $\sigma_y = \upsilon\sigma_x = 0.29 \times 227 = 65.8$ MPa.

EXAMPLE PROBLEM 2.2: A body is loaded elastically so that $e_y = 0$ and $\sigma_z = 0$. Both σ_x and σ_y are finite. Derive an expression for e_z in terms of σ_x, E, and υ only.

Solution: Substituting $e_y = 0$ and $\sigma_z = 0$ into Hooke's law (equation 2.2), $0 = (1/E)[\sigma_y - \upsilon(0 + \sigma_x)]$, so $\sigma_y = \upsilon\sigma_x$. Substituting into Hooke's law again, $e_z = (1/E)[0 - \upsilon(\sigma_x + \upsilon\sigma_x)] = -\sigma_x\upsilon(1 + \upsilon)/E$.

EXAMPLE PROBLEM 2.3: Draw the three-dimensional Mohr's strain circle diagram for a body loaded elastically under balanced biaxial tension, $\sigma_1 = \sigma_2$. Assume $\upsilon = 0.25$. What is the largest shear strain, γ?

Solution: $\sigma_1 = \sigma_2$ and $\sigma_3 = 0$. From symmetry, $e_1 = e_2$, so using Hooke's laws, $e_1 = (1/E)(\sigma_1 - \upsilon\sigma_1) = 0.75\sigma_1/E$, $e_3 = (1/E)[0 - \upsilon(\sigma_1 + \sigma_1)] = -0.5\sigma_1$. The maximum shear strain, $\gamma_{max} = (e_1 - e_3) = 1.25\sigma_1/E$. In Figure 2.3, e_1, e_2, and e_3 are plotted on the horizontal axis. The circle has a radius of $0.625\sigma_1/E = (5/6)e_1$ with $e_2 = e_1$ and $e_3 = (-2/3)e_1$.

Variation of Young's Modulus

The binding energy between two neighboring atoms or ions can be represented by a potential well. Figure 2.4 is a schematic plot showing how the binding energy varies with separation of atoms. At absolute zero, the equilibrium separation corresponds

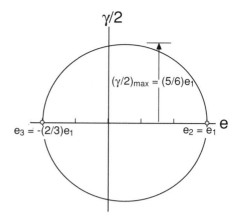

Figure 2.3. Mohr's strain circles for an elastic body under biaxial tension. Because the $(\sigma_1 - \sigma_2)$ circle reduces to a point, the $(\sigma_1 - \sigma_3)$ and $(\sigma_2 - \sigma_3)$ circles are superimposed as a single circle.

to the lowest energy. The Young's modulus and the melting point are both related to the potential well; the modulus depends on the curvature at the bottom of the well, and the melting point depends on the depth of the well. The curvature and the depth tend to be related so the elastic moduli of different elements roughly correlate with their melting points (Figure 2.5). The modulus also correlates with the heat of fusion and the latent heat of melting, which are related to the depth of the potential well. For a given metal, the elastic modulus decreases as the temperature is increased from absolute zero to the melting point. This is illustrated for several metals in Figure 2.6. The value of the elastic modulus at the melting point is in the range of one-third and one-fifth its value at absolute zero.

For crystalline materials, the elastic modulus is generally regarded as being relatively structure insensitive to changes of microstructure. Young's moduli of body-centered cubic (bcc) and face-centered cubic (fcc) forms of iron differ by a relatively small amount. Heat treatments that have large effects on hardness and yield strength have little effect on the elastic properties. Cold working and small alloy additions also have relatively small effects because only a small fraction of the near-neighbor bonds are affected by these changes.

The elastic behavior of polymers is very different from that of metals. The elastic strains are largely a result of straightening polymer chains by rotation of bonds rather than by bond stretching. Consequently, Young's moduli of typical polymers are orders of magnitudes lower than those of metals and ceramics. Only in highly oriented polymers is stretching of covalent bonds the primary source of elastic

Figure 2.4. Schematic plot of the variation of binding energy with atomic separation at absolute zero.

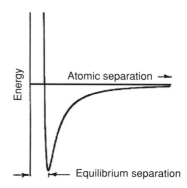

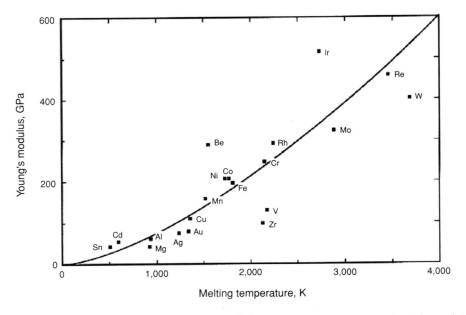

Figure 2.5. Correlation between elastic modulus at room temperature and melting point. Elements with higher melting points have higher elastic moduli.

strains, and the elastic moduli of highly oriented polymers are comparable with metals and ceramics. For polymers, the magnitude of elastic strain is often much greater than for metals and ceramics. Hooke's law does not describe the elasticity of rubber very well. This will be treated in Chapter 20.

Isotropic Thermal Expansion

When the temperature of an isotropic material is changed, its fractional change in length is

$$\Delta L/L = \alpha \Delta T. \tag{2.7}$$

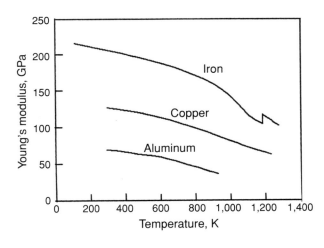

Figure 2.6. Variation of Young's moduli with temperature. Young's moduli decrease with increasing temperature. However, the moduli at the melting point are less than five times that at 0 K. Data from W. Koster, *Zeitschrift fur Metallkunde*, v. 39 (1948).

Such dimension changes can be considered strains, and Hooke's law can be generalized to include thermal strains,

$$e_x = (1/E)[\sigma_x - \upsilon(\sigma_y + \sigma_z)] + \alpha\Delta L. \tag{2.8}$$

This generalization is useful for finding the stresses that arise when constrained bodies are heated or cooled.

> **EXAMPLE PROBLEM 2.4:** A glaze is applied on a ceramic body by heating it above 600°C, which allows it to flow over the surface. On cooling, the glaze becomes rigid at 500°C. See Figure 2.7. The coefficients of thermal expansion of the glaze and the body are $\alpha_g = 4.0 \times 10^{-6}/°C$ and $\alpha_b = 5.5 \times 10^{-6}/°C$. The elastic constants for the glaze are $E_g = 70$ GPa and $\upsilon_g = 0.3$. Calculate the stresses in the glaze when it has cooled to 20°C.
>
> **Solution:** Let the direction normal to the surface be z so the x- and y-directions lie in the surface. The strains in the x and y directions must be the same in the glaze and body. $e_{xg} = (1/E_g)[\sigma_{xg} - \upsilon_g(\sigma_{yg} + \sigma_{zg})] + \alpha_g\Delta T = (1/E_b)[\sigma_{xb} - \upsilon_b(\sigma_{yb} + \sigma_{zb})] + \alpha_b\Delta T$. Let $\sigma_{zg} = \sigma_{zb} = 0$ (free surface) and $\sigma_{xg} = \sigma_{yg}$ (symmetry) and $\sigma_{xb} = \sigma_{yb} = 0$. Because the glaze is very thin and $t_b\sigma_{xb} + t_g\sigma_{bxg} = 0$, the stresses, $\sigma_{xb} = \sigma_{yb}$, in the ceramic body are negligible. Hence, $(1/E_g)[\sigma_{xg}(1 - \upsilon_g)] + \alpha_g\Delta T = \alpha_b\Delta T$, $\sigma_{xg} = [E_g/(1 - \upsilon_g)](\alpha_b - \alpha_g)\Delta T = [(70 \times 10^9$ Pa$)/(1 - 0.3)]](4 - 5.5)(10^{-6}/°C)(-480°C) = -72$ MPa (compression) on cooling.

The elastic constants and thermal expansion coefficients for various materials are given in Table 2.1.

Bimetallic strips used for sensing temperature depend on the differences of the thermal expansion of the two materials.

Elastic Anisotropy

Hooke's law for anisotropic materials can be expressed in terms of compliances, s_{ijmn}, which relate the response of individual strain components to individual stress components. In the most general case, every strain component depends linearly on the stress components,

$$e_{ij} = s_{ijmn}\sigma_{mn}, \tag{2.9}$$

Figure 2.7. Stresses in a glaze.

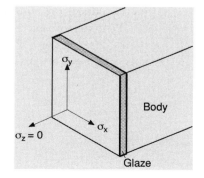

Table 2.1. *Elastic constants and thermal expansion coefficients for various materials*

Material	E (GPa)	Poisson's ratio	Thermal expansion coefficient ($\times 10^{-6}$°C)
Aluminum	62.0	0.24	23.6
Iron	208.2	0.291	11.8
Copper	128.0	0.35	16.5
Magnesium	44.0	—	27.1
Nickel	207.0	0.31	13.3
MgO	205.0	—	9.0
Al$_3$O$_3$	350.0	—	9.0
Glass	70.0	—	9.0
Polystyrene	2.8	—	63.0
PVC	0.35	—	190.0
Nylon	2.8	—	100.0

Here the compliances, s_{ijmn}, form a fourth-order tensor. This is greatly simplified by the relations, $\sigma_{mn} = \sigma_{nm}$ and $\gamma_{ij} = \gamma_{ji} = 2e_{mn} = 2\varepsilon_{nm}$, and by adopting a new subscript convention for the compliances: two subscripts identify the row and column of the compliance in the matrix rather than the stress and strain axes. With these changes, Hooke's law becomes

$$
\begin{aligned}
e_{11} &= s_{11}\sigma_{11} + s_{12}\sigma_{22} + s_{13}\sigma_{33} + s_{14}\sigma_{23} + s_{15}\sigma_{31} + s_{16}\sigma_{12} \\
e_{22} &= s_{21}\sigma_{11} + s_{22}\sigma_{22} + s_{23}\sigma_{33} + s_{24}\sigma_{23} + s_{25}\sigma_{31} + s_{26}\sigma_{12} \\
e_{33} &= s_{31}\sigma_{11} + s_{32}\sigma_{22} + s_{33}\sigma_{33} + s_{34}\sigma_{23} + s_{35}\sigma_{31} + s_{36}\sigma_{12} \\
\gamma_{23} &= s_{41}\sigma_{11} + s_{42}\sigma_{22} + s_{43}\sigma_{33} + s_{44}\sigma_{23} + s_{45}\sigma_{31} + s_{46}\sigma_{12} \\
\gamma_{31} &= s_{51}\sigma_{11} + s_{52}\sigma_{22} + s_{53}\sigma_{33} + s_{54}\sigma_{23} + s_{55}\sigma_{31} + s_{56}\sigma_{12} \\
\gamma_{12} &= s_{61}\sigma_{11} + s_{62}\sigma_{22} + s_{63}\sigma_{33} + s_{64}\sigma_{23} + s_{65}\sigma_{31} + s_{66}\sigma_{12}.
\end{aligned}
\tag{2.10}
$$

Note the subscripts on the strain and stress terms refer to the crystallographic axes. Because it can be shown that $s_{ij} = s_{ji}$, further simplification results in

$$
\begin{aligned}
e_{11} &= s_{11}\sigma_{11} + s_{12}\sigma_{22} + s_{13}\sigma_{33} + s_{14}\sigma_{23} + s_{15}\sigma_{31} + s_{16}\sigma_{12} \\
e_{22} &= s_{21}\sigma_{11} + s_{22}\sigma_{22} + s_{23}\sigma_{33} + s_{24}\sigma_{23} + s_{25}\sigma_{31} + s_{26}\sigma_{12} \\
e_{33} &= s_{13}\sigma_{11} + s_{23}\sigma_{22} + s_{33}\sigma_{33} + s_{34}\sigma_{23} + s_{35}\sigma_{31} + s_{36}\sigma_{12} \\
\gamma_{23} &= s_{14}\sigma_{11} + s_{24}\sigma_{22} + s_{34}\sigma_{33} + s_{44}\sigma_{23} + s_{45}\sigma_{31} + s_{46}\sigma_{12} \\
\gamma_{31} &= s_{15}\sigma_{11} + s_{25}\sigma_{22} + s_{35}\sigma_{33} + s_{45}\sigma_{23} + s_{55}\sigma_{31} + s_{56}\sigma_{12} \\
\gamma_{12} &= s_{16}\sigma_{11} + s_{26}\sigma_{22} + s_{36}\sigma_{33} + s_{46}\sigma_{23} + s_{56}\sigma_{31} + s_{66}\sigma_{12}.
\end{aligned}
\tag{2.11}
$$

So there are at most 21 independent compliances. In the majority of cases, symmetry reduces the number of constants.

With orthotropic symmetry, the 23, 31, and 12 planes are planes of mirror symmetry, making the 1, 2, and 3 axes of twofold symmetry. Paper, wood, and rooled sheets of metal are examples of such materials. Hook's laws simplify to

$$
\begin{aligned}
e_{11} &= s_{11}\sigma_{11} + s_{22}\sigma_{12} + s_{33}\sigma_{13} \\
e_{22} &= s_{12}\sigma_{11} + s_{22}\sigma_{12} + s_{23}\sigma_{13} \\
e_{33} &= s_{13}\sigma_{11} + s_{23}\sigma_{12} + s_{33}\sigma_{13} \\
\gamma_{23} &= s_{44}\sigma_{23} \\
\gamma_{31} &= s_{55}\sigma_{31} \\
\gamma_{12} &= s_{66}\sigma_{12}.
\end{aligned}
\tag{2.12}
$$

For cubic crystals, 1, 2, and 3 are the cubic $<100>$ axes. Because of symmetry, $s_{13} = s_{23} = s_{12}$, $s_{22} = s_{33} = s_{44}$ and $s_{66} = s_{55} = s_{44}$, so the relation simplify to

$$
\begin{aligned}
e_{11} &= s_{11}\sigma_{11} + s_{12}\sigma_{22} + s_{12}\sigma_{33} \\
e_{22} &= s_{12}\sigma_{11} + s_{11}\sigma_{22} + s_{12}\sigma_{33} \\
e_{11} &= s_{12}\sigma_{11} + s_{12}\sigma_{22} + s_{11}\sigma_{33} \\
\gamma_{23} &= s_{44}\sigma_{23} \\
\gamma_{31} &= s_{44}\sigma_{31} \\
\gamma_{12} &= s_{44}\sigma_{11}.
\end{aligned}
\tag{2.13}
$$

For hexagonal and cylindrical symmetry, z is taken as the hexagonal (or cylindrical) axis. The 1 and 2 axes are any axes normal to z. In this case, $s_{13} = s_{23}$, $s_{22} = s_{33}$, and $s_{66} = s_{55}$, and $s_{44} = (s_{11} - s_{12})$. Equations (2.10) simplifies to

$$
\begin{aligned}
e_{11} &= s_{11}\sigma_{11} + s_{12}\sigma_{22} + s_{13}\sigma_{33} \\
e_{22} &= s_{12}\sigma_{11} + s_{21}\sigma_{22} + s_{13}\sigma_{33} \\
e_{11} &= s_{13}\sigma_{11} + s_{13}\sigma_{22} + s_{33}\sigma_{33} \\
\gamma_{23} &= s_{44}\sigma_{23} \\
\gamma_{31} &= s_{55}\sigma_{31} \\
\gamma_{12} &= 2(s_{11} - s_{12})\sigma_{12}.
\end{aligned}
\tag{2.14}
$$

Orientation Dependence of Elastic Response

Equations (2.9) to (2.13) describe the elastic response along the symmetry axes. The elastic response along any other set of axes may be calculated by first resolving the stress state onto the symmetry axes, then using the matrices of elastic constants to find the strains along the symmetry axes, and finally resolving these strains onto the axes of interest.

> **EXAMPLE PROBLEM 2.5:** Derive expressions for the effective Young's modulus and Poisson's ratio for a cubic crystal stressed along the [112] direction.
>
> **Solution:** Let x = [112]. The axes y and z must be normal to x and each other (e.g., $y = [11\bar{1}]$ and $z = [1\bar{1}0]$). The direction cosines between these axes and the cubic axes of the crystal can be found by dot products. For example, $\ell_{1x} = [100][112] = (1.1 + 0.1 + 0.2)/(1^2 + 0^2 + 0^2)(1^2 + 1^2 + 2^2) = 1/\sqrt{6}$. Forming a table of direction cosines,
>
> | | x = 112 | y = 11$\bar{1}$ | z = 1$\bar{1}$0 |
> |---|---|---|---|
> | 1 = 100 | $1/\sqrt{6}$ | $1/\sqrt{3}$ | $1/\sqrt{2}$ |
> | 2 = 010 | $1/\sqrt{6}$ | $1/\sqrt{3}$ | $-1/\sqrt{2}$ |
> | 3 = 001 | $2/\sqrt{6}$ | $-1/\sqrt{3}$ | 0 |
>
> For tension along [112], $\sigma_y = \sigma_z = \sigma_{yz} = \sigma_{zx} = \sigma_{xy} = 0$, so the only finite stress on the x, y, z axis system is σ_x. Transforming to the 1, 2, and 3 axes using equation 1.8:
>
> $$
> \begin{aligned}
> \sigma_1 &= (1/6)\sigma_x & \sigma_{23} &= (1/3)\sigma_x \\
> \sigma_2 &= (1/6)\sigma_x & \sigma_{31} &= (1/3)\sigma_x
> \end{aligned}
> $$
>
> From equation (2.13) for cubic crystal, the resulting elastic strains are:
>
> $$e_1/\sigma_x = (1/6)s_{11} + (1/6)s_{12} + (2/3)s_{12} = (1/6)s_{11} + (5/6)s_{12},$$

$$e_2/\sigma_x = (1/6)s_{12} + (1/6)s_{11} + (2/3)s_{12} = (1/6)s_{11} + (5/6)s_{12},$$
$$\gamma_{23}/\sigma_x = \gamma_{31}/\sigma_x = (1/3)s_{44}, \quad \gamma_{12}/s_x = (1/6)s_{44}.$$

Using equation (1.27), transform these strains onto the x, y, and z axes,

$$
\begin{aligned}
e_x &= (1/6)e_1 + (1/6)e_2 + (2/3)e_3 + (1/3)\gamma_{23} + (1/3)\gamma_{31} + (1/6)\gamma_{12} \\
&= [(1/6)s_{11} + (1/2)s_{12} + (1/4)s_{44}]/\sigma_x, \\
e_y &= (1/3)e_1 + (1/3)e_2 + (1/3)e_3 - (1/3)\gamma_{23} - (1/3)\gamma_{31} + (1/3)\gamma_{12} \\
&= [(1/6)s_{11} + (2/3)s_{12} + (1/6)s_{44}]/\sigma_x, \\
e_z &= (1/2)e_1 + (1/2)e_2 - (1/2)\gamma_{23} \\
&= [(1/6)s_{11} + (5/6)s_{12} - (1/6)s_{44}]/\sigma_x.
\end{aligned}
$$

Young's modulus, $E_{[112]} = (e_x/\sigma_x)^{-1} = [(1/6)s_{11} + (1/2)s_{12} + (1/4)s_{44}]^{-1}$.

$$
\begin{aligned}
\upsilon_y &= -e_y/e_x \\
&= -[(1/3)s_{11} + (2/3)s_{12} + (16)s_{44}]/[(1/6)s_{11} + (1/2)s_{12} + (1/4)s_{44}].
\end{aligned}
$$

There are two values of Poisson's ratio: $\upsilon_y = -e_y/e_x$ and $\upsilon_z = -e_z/e_x$.

$$
\begin{aligned}
\upsilon_y &= -(s_{11} + 8s_{12} + 2s_{44})/(2s_{11} + 6s_{12} + 3s_{44}), \\
\upsilon_z &= -e_z/e_x = -[(1/6)s_{11} + (5/6)s_{12} - (1/6)s_{44}][(1/6)s_{11} + (1/2) + (1/4)s_{44}] \\
&= -(2s_{11} + 10s_{12} - 2s_{44})/(2s_{11} + 6s_{12} + 3s_{44}).
\end{aligned}
$$

Orientation Dependence in Cubic Crystals

A similar treatment can be used to find the effective Young's modulus, E_d, along any direction, d. Consider a cubic crystal loaded under a single tensile stress, σ_d, applied along the d-direction. Let the direction cosines of d with the 1, 2, and 3 axes be α, β, and γ as shown in Figure 2.8. The stresses along the cubic axes are then

$$
\begin{array}{ll}
\sigma_1 = \alpha^2 \sigma_d & \sigma_{23} = \beta\gamma\sigma_d \\
\sigma_2 = \beta^2 \sigma_d & \sigma_{31} = \gamma\alpha\sigma_d \\
\sigma_3 = \gamma^2 \sigma_d & \sigma_{12} = \alpha\beta\sigma_d.
\end{array}
\tag{2.15}
$$

From equation (2.13) for cubic crystals, the corresponding strains are:

$$
\begin{array}{ll}
e_1/\sigma_d = s_{11}\alpha^2 + s_{12}\beta^2 + s_{12}\gamma^2, & \gamma_{23}/\sigma_d = s_{44}\beta\gamma, \\
e_2/\sigma_d = s_{12}\alpha^2 = s_{11}\beta^2 + s_{12}\gamma^2, & \gamma_{31}/\sigma_d = s_{44}\gamma\alpha, \\
e_3/\sigma_d = s_{12}\alpha^2 = s_{12}\beta^2 + s_{11}\gamma^2, & \gamma_{12}/\sigma_d = s_{44}\alpha\beta.
\end{array}
\tag{2.16}
$$

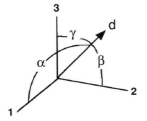

Figure 2.8. Coordinate system for finding E_d. In this figure, α, β, and γ are the direction cosines of the angles indicated.

Table 2.2. *Values of $E_{[111]}/E_{[100]}$ for several cubic crystals*[*]

Material	$E_{[111]}/E_{[100]}$	Material	$E_{[111]}/E_{[100]}$
Cr	0.76	Fe	2.14
Mo	0.92	Nb	0.52
W	1.01	Al	1.20
Cu	2.91	Ni	2.36
Si	1.44	MgO	1.39
KCl	0.44	NaCl	0.74
ZnS	2.17	GaAs	1.66

[*] Calculated from Table 2.1.

Resolving these components onto the d axis using equation 1.27a,

$$e_d/\sigma_d = \alpha^2 e_1 + \beta^2 e_2 + \gamma^2 e_3 + \beta\gamma \cdot \gamma_{23} + \gamma\alpha \cdot \gamma_{31} + \alpha\beta \cdot \gamma_{12}, \qquad (2.17)$$

and substituting equation (2.14) into equation (2.15),

$$e_d/\sigma_d = s_{11}(\alpha^4 + \beta^4 + \gamma^4) + 2s_{12}(\beta^2\gamma^2 + \gamma^2\alpha^2 + \alpha^2\beta^2) + s_{44}(\beta^2\gamma^2 + \gamma^2\alpha^2 + \alpha^2\beta^2)$$
$$= s_{11}(\alpha^4 + \beta^4 + \gamma^4) + (2s_{12} + s_{44})(\beta^2\gamma^2 + \gamma^2\alpha^2 + \alpha^2\beta^2). \qquad (2.18)$$

Substituting the identity, $\alpha^2 + \beta^2 + \gamma^2 = 1$, in the form of $(a^4 + b^4 + g^4) = 1 - 2(\beta^2\gamma^2 + \gamma^2\alpha^2 + \alpha^2\beta^2)$ into equation (2.18),

$$1/E_d = e_d/\sigma_d = s_{11} + (-2s_{11} + 2s_{12} + s_{44})(\beta^2\gamma^2 + \gamma^2\alpha^2 + \alpha^2\beta^2). \qquad (2.19)$$

If d is expressed by direction indices, $[hk\ell]$,

$$\alpha = h/\sqrt{(h^2 + k^2 + \ell^2)}, \quad \beta = k/\sqrt{(h^2 + k^2 + \ell^2)}, \quad \gamma = \ell/\sqrt{(h^2 + k^2 + \ell^2)}.$$

Then equation (2.19) can be rewritten as

$$1/E_d = s_{11} + (2s_{12} - 2s_{11} + s_{44})(k^{22} + \ell^2 h^2 + h^2 k^2)/(h^2 + k^2 + \ell^2)^2. \qquad (2.20)$$

For an isotropic material, E_d is independent of direction. Equations (2.19) and (2.20) indicate that this is possible only if $(2s_{12} - 2s_{11} + s_{44}) = 0$. Therefore, for isotropy,

$$s_{44} = 2(s_{11} - s_{12}). \qquad (2.21)$$

According to equation (2.19), the extreme values of E in a cubic crystal occur for the [100] direction ($\alpha = \beta = \gamma = 0$) and for the [111] direction ($\alpha = \beta = \gamma = 0$ $1/(\alpha = \beta = \gamma = 1/\sqrt{3})$. Because $1/E_{[100]} = s_{11}$ and $1/E_{[111]} = (s_{11} + 2s_{12} + s_{44})/3$, the ratio is given by

$$E_{[111]}/E_{[100]} = 3s_{11}/(s_{11} + 2s_{12} + s_{44}). \qquad (2.22)$$

This ratio represents the largest possible anisotropy in cubic crystals. Table 2.2 gives the $E_{[111]}/E_{[100]}$ ratios for several cubic crystals. For most, $E_{111} > E_{100}$, but there are exceptions. For Mo, Nb, Cr, and the alkali halides, $E_{100} > E_{111}$. Tungsten is almost elastically isotropic.

The orientation dependence in equation (2.19) can be expressed as

$$1/E_d = 1/E_{100} + f[1/E_{111} - 1/E_{100}], \qquad (2.23)$$

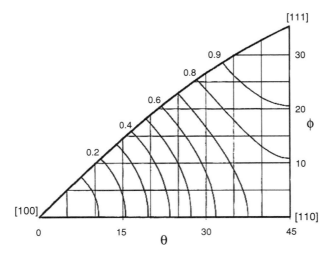

Figure 2.9. Contour plot of f in $1/E_d = 1/E_{100} + [1/E_{111} - 1/E_{100}]$. The extremes of E are found at the [100] and [111] orientations. From W. F. Hosford, *The Mechanics of Crystals and Textured Polycrystals*, used with permission of Oxford University Press (1993).

where $f = 3(\beta^2\gamma^2 + \gamma^2\alpha^2 + \alpha^2\beta^2)$. The orientation dependence of E_d is plotted in Figure 2.9 on the basic orientation triangle. (see Appendix II for plotting of angles.)

Orientation Dependence in Noncubic Crystals

Similar expressions can be derived for other crystal classes. For hexagonal crystals,

$$1/E = (1 - \gamma^2)s_{11} + \gamma^4 s_{33} + \gamma^2(1 - \gamma^2)(2s_{13} + s_{44}), \qquad (2.24)$$

where γ is the cosine of the angle between the direction, d, and the c axis (Figure 2.10). For orthorhombic

$$s_{11}\alpha^4 + s_{22}\beta^4 + s_{33}\gamma^4 + \beta^2\gamma^2(2s_{23} + s_{44}) + \gamma^2\alpha^2(2s_{31} + s_{55}) + \alpha^2\beta^2(2s_{12} + s_{66}).$$
$$(2.25)$$

EXAMPLE PROBLEM 2.6: Pressure, p, affects the c/a ratio of hexagonal crystals. Write an expression for $d(c/a)/dp$ in terms of the elastic constants and c/a. Assume that $\sigma_1 = \sigma_2 = \sigma_3 = -p$.

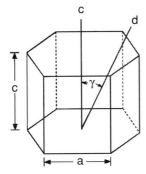

Figure 2.10. Hexagonal cell showing the c axis and a direction, d. The dimensions of the cell are a and c.

Solution: $d(c/a) = (adc - cda)/a^2 = (c/a)(dc/c - da/a) = (c/a)(de_3 - de_1)$. Now substituting $de_1 = s_{11}d\sigma_1 + s_{12}d\sigma_2 + s_{13}d\sigma_3 = -(s_{11} + s_{12} + s_{13})dp$ and $de_3 = s_{13}d\sigma_1 + s_{13}d\sigma_2 + s_{33}d\sigma_3 = (2s_{13} + s_{33})dp$, $d(c/a)/dp = -(c/a)(2s_{13} + s_{33} - s_{11} - s_2 - s_{13}) = -(c/a)(s_{13} + s_{33} - s_{11} - s_{12})$.

Orientation Dependence in Materials Other Than Single Crystals

The basic forms of the elastic constant matrix can be used for materials that are not single crystals, but that have similar symmetries of structure and properties. Such materials may be either natural or artificial composites. Examples include

1. Orthotropic (same as orthorhombic) symmetry (three axes of two-fold symmetry):
   a. Rolled sheets or plates, both before and after recrystallization. The anisotropy is caused by crystallographic texture. The axes are the rolling, transverse, and thickness directions.
   b. Tubes and swaged wire or rod having cylindrical crystallographic textures. The axes are the radial, tangential (hoop), and axial directions.
   c. Wood. The axes are the radial, tangential, and axial directions.
2. Tetragonal symmetry (one axis of four-fold symmetry, two axes of two-fold symmetry):
   a. Plywood with equal number of plies at 0 and 90 degrees. The four-fold axis is normal to the sheet and the two twofold axes are parallel to the wood grain in the two plies. The anisotropy is from oriented fibers.
   b. Woven cloth, window screen, and composites reinforced with woven material if the cloth or screen has the same weft and warp (the two directions of thread in a woven cloth). The four-fold axis is normal to the cloth and the two two-fold axes are parallel to the threads of the weft and warp.
3. Axial symmetry (same as hexagonal symmetry):
   a. Drawn or extruded wire and rod. The anisotropy is caused by a fiber texture, all directions normal to the wire axis being equivalent.
   b. Electroplates and portions of ingots with columnar grains. All directions normal to the growth direction are equivalent.
   c. Uniaxially aligned composites. The axis of symmetry is parallel to the fibers.

In principle, it should be possible to obtain the elastic moduli for a polycrystal from a weighted average of the elastic behavior of all orientations of crystals present in the polycrystal. However, the appropriate way to average is not obvious.

Anisotropic Thermal Expansion

As in the isotropic case, small dimensional changes caused by temperature changes can be treated as strains, $e_{ij} = \alpha_{ij}\Delta T$, where α_{ij} is the linear coefficient of thermal expansion and Hooke's laws can be generalized to include these terms. In the most

Table 2.3. *Coefficients of thermal expansion for some noncubic crystals*

Material	Struct.	Temp. range (°C)	$\alpha_{11}(\mu m/m)/K$	$\alpha_{22}(\mu m/m)/K$	$\alpha_{33}(\mu m/m)/K$
Zirconium	hcp	300–900	5.7	$\alpha_{22} = \alpha_{11}$	11.4
Zinc	hcp	0–100	15.0	$\alpha_{22} = \alpha_{11}$	61.5
Magnesium	hcp	0–35	24.3	$\alpha_{22} = \alpha_{11}$	27.1
Cadmium	hcp	at 0	19.1	$\alpha_{22} = \alpha_{11}$	54.3
Titanium	hcp	0–700	11.0	$\alpha_{22} = \alpha_{11}$	12.8
Tin	Tetr.	at 50	16.6	$\alpha_{22} = \alpha_{11}$	32.9
Calcite	Hex.	0–85	−5.6	$\alpha_{22} = \alpha_{11}$	25.1
Uranium	Ortho.	at 75	20.3	−1.4	22.2

general case, the thermal expansion coefficients form a symmetric tensor

$$\begin{matrix} \alpha_{11} & \alpha_{21} & \alpha_{31} \\ \alpha_{12} & \alpha_{22} & \alpha_{32} \\ \alpha_{13} & \alpha_{23} & \alpha_{33}, \end{matrix} \qquad (2.26)$$

where the subscripts on the a terms refer to the crystal axes, 1, 2, 3, and are the same as on the corresponding strain component. Using a shorter notation system,

$$e_1 = \alpha_1 \Delta T, \quad e_2 = \alpha_2 \Delta T, \quad e_3 = \alpha_3 \Delta T, \quad \text{and}$$
$$\gamma_{23} = \alpha_4 \Delta T, \quad \gamma_{31} = \alpha_5 \Delta T, \quad \gamma_{12} = \alpha_6 \Delta T, \qquad (2.27)$$

where $\alpha_1 = \alpha_{11}$, $\alpha_2 = \alpha_{22}$, $\alpha_3 = \alpha_{33}$, $\alpha_4 = 2\alpha_{23}$, $\alpha_5 = 2\alpha_{31}$, and $\alpha_6 = 2\alpha_{12}$.

There are simplifications for crystals of higher symmetry:

(1) Cubic: $\alpha_1 = \alpha_2 = \alpha_3, \alpha_4 = \alpha_5 = \alpha_6 = 0$
(2) Hexagonal and tetragonal: $\alpha_1 = \alpha_2 \neq \alpha_3, \alpha_4 = \alpha_5 = \alpha_6 = 0$
(3) Orthorhombic crystals: $\alpha_1 \neq \alpha_2 \neq \alpha_3, \alpha_4 = \alpha_5 = \alpha_6 = 0$

The coefficients of thermal expansion for some noncubic crystals are given in Table 2.3.

REFERENCES

H. B. Huntington, *The Elastic Constants of Crystals*, Academic Press (1964).
G. Simmons and H. Wang, *Single Crystal Elastic Constants and Calculated Properties: A Handbook*, MIT Press (1971).
W. Köster and H. Franz, *Metals Review*, v. 6 (1961), pp. 1–55.

Notes

Robert Hooke (1635–1703) was appointed curator of experiments of the Royal Society on the recommendation of Robert Boyle. In that post, he devised apparati to clarify points of discussion and demonstrate his ideas. He noted that when he put weights on a string or a spring, the deflection was proportional to the weight. Hooke was a contemporary of Isaac Newton, whom he envied. Several times he claimed to have discovered things attributed to Newton, but he was never able to substantiate these claims.

Thomas Young (1773–1829) studied medicine, but because of his interest in science, he took a post as lecturer at the Royal Institution. In his lectures, he was the first to define the concept of a modulus of elasticity, although he defined it quite differently than we do today. He described the modulus of elasticity in terms of the deformation at the base of a column due to its own weight.

Navier (1785–1836) wrote a book on the strength of materials in 1826, in which he defined the elastic modulus as the load per unit cross-sectional area and determined E for iron.

Because his family was poor, S. D. Poisson (1781–1840) did not have a chance to learn to read or write until he was 15. After 2 years of visiting mathematics classes, he passed the exams for admission to École Polytechnique with distinction. He realized that axial elongation, e, must be accompanied by lateral contraction, which he took as $(-1/4)e$, not realizing that the ratio of contractile strain to elongation strain is a property that varies from material to material.

Problems

1. Reconsider problem 4 in Chapter 1 and assume that $E = 205$ GPa and $v = 0.29$.
   A. Calculate the principal stresses under load.
   B. Calculate the strain, ε_z.

2. Consider a thin-wall tube, capped at each end and loaded under internal pressure. Calculate the ratio of the axial strain to the hoop strain, assuming that the deformation is elastic. Assume $E = 10^7$ psi and $v = 1/3$. Does the length of the tube increase, decrease, or remain constant?

3. A sheet of metal was deformed elastically under balanced biaxial tension ($\sigma_x = \sigma_y$, $\sigma_z = 0$).
   A. Derive an expression for the ratio of elastic strains, e_z/e_x, in terms of the elastic constants.
   B. If E = 70 GPa and $v = 0.30$, and e_y is measured as 1.00×10^{-3}, what is the value of e_z?

4. A cylindrical plug of gummite* is placed in a cylindrical hole in a rigid block of stiffite.* Then the plug is compressed axially (parallel to the axis of the hole). Assume that the plug exactly fits the hole and that the stiffite does not deform at all. Assume elastic deformation and that Hooke's law holds. Derive an expression for the ratio of the axial strain to the axial stress, ε_a/σ_a, in the gummite in terms of Young's modulus, E, and Poisson's ratio, v, of the gummite.

5. Strain gauges mounted on a free surface of a piece of steel ($E = 205$ GPa, $v = 0.29$) indicate strains of $e_x = 0.00042$, $e_y = 0.00071$, and $\gamma_{xy} = 0.00037$.
   A. Calculate the principal strains.
   B. Use Hooke's laws to find the principal stresses from the principal stresses.
   C. Calculate σ_x, σ_y, and τ_{xy} directly from e_x, e_y, and γ_{xy}.

* "gummite" and "stiffite" are fictitious names.

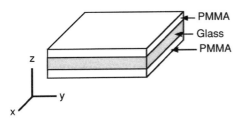

Figure 2.11. Laminated sheets of PMMA and glass.

D. Calculate the principal stresses directly from σ_x, σ_y, and τ_{xy}, and compare your answers with the answers to B.

6. A steel block ($E = 30 \times 10^6$ psi and $\upsilon = 0.29$) is loaded under uniaxial compression along x.

A. Draw the Mohr's strain circle diagram.

B. There is an axis, x′, along which the strain $E_{x'} = 0$. Find the angle between x and x′.

7. Poisson's ratio for rubber is $1/2$. What does this imply about the bulk modulus?

8. A sandwich is made of a plate of glass surrounded by two plates of polymethylmethacrylate, as shown in Figure 2.11. Assume that the composite is free of stresses at 40°C. Find the stresses when the sandwich is cooled to 20°C. The properties of the glass and the polymethylmethacrylate are given as follows. The total thicknesses of the glass and the PMMA are equal. Assume each to be isotropic, and that creep is negligible.

	Glass	PMMA
Thermal expansion coefficient (K^{-1})	9×10^{-6}	90×10^{-6}
Young's modulus (GPa)	69	3.45
Poisson's ratio	0.28	0.38

9. A bronze sleeve, 0.040 in. thick, was mounted on a 2.000-in.-diameter steel shaft by heating it to 100°C, while the temperature of the shaft was maintained at 20°C. Under these conditions, the sleeve just fit on the shaft with zero clearance. Find the principal stresses in the sleeve after it cools to 20°C. Assume that friction between the shaft and the sleeve prevented any sliding at the interface during cooling, and assume that the shaft is so massive and stiff that strains in the shaft itself are negligible. For the bronze, $E = 16 \times 10^6$ psi, $\upsilon = 0.30$, and $\alpha = 18.4 \times 10^{-6}(°C)^{-1}$.

10. Calculate Young's modulus for an iron crystal when tension is applied along a <122> direction.

11. Zinc has the following elastic constants:

$$s_{11} = 0.84 \times 10^{-11} \text{ Pa}^{-1}; \quad s_{33} = 2.87 \times 10^{-11} \text{ Pa}^{-11}; \quad s_{12} = 0.11 \times 10^{-11} \text{ Pa}^{-1}$$
$$s_{13} = -0.78 \times 10^{-11} \text{ Pa}^{-1}; \quad s_{44} = 2.64 \times 10^{-11} \text{ Pa}^{-11}; \quad s_{66} = 2(s_{11} - s_{12})$$

Find the bulk modulus of zinc.

12. Calculate the effective Young's modulus for a cubic crystal loaded in the [110] direction in terms of the constants, s_{11}, s_{12}, and s_{44}. Do this by assuming uniaxial tension along [110] and expressing $\sigma_1, \sigma_2, \ldots \tau_{12}$ in terms of $\sigma_{[110]}$. Then use the matrix of elastic constants to find $e_1, e_2, \ldots \gamma_{12}$, and finally resolve theses strains onto the [110] axis to find $e_{[110]}$.

13. When a polycrystal is elastically strained in tension, it is reasonable to assume that the strains in all grains are the same. Using this assumption, calculate for iron, the ratio of the stress in grains oriented with <111> parallel to the tensile axis to the average stress, $\sigma_{111}/\sigma_{av} = E_{111}/E_{av}$. Calculate σ_{100}/σ_{av}.

14. Take the cardboard back of a pad of paper and cut it into a square. Then support the cardboard horizontally with blocks at each end and apply a weight in the middle. Measure the deflection. Next rotate the cardboard 90 degrees and repeat the experiment using the same weight. By what factor do the two deflections differ? By what factor does the elastic modulus, E, vary with direction? Why was the cardboard used for the backing of the pad placed in the orientation that it was?

3 Mechanical Testing

Introduction

Tensile properties are used in selecting materials for different applications. Material specifications often include minimum tensile properties to ensure quality so tests must be made to guarantee that materials meet these specifications. Tensile properties are also used in research and development to compare new materials or processes. With plasticity theory (Chapter 5), tensile data can be used to predict a material's behavior under forms of loading other than uniaxial tension.

Often the primary concern is strength. The level of stress that causes appreciable plastic deformation is called its *yield stress*. The maximum tensile stress that a material carries is called its *tensile strength* (or *ultimate strength* or *ultimate tensile strength*). Both measures are used, with appropriate caution, in engineering design. A material's ductility may also be of interest. Ductility describes how much the material can deform before it fractures. Rarely, if ever, is the ductility incorporated directly into design. Rather, it is included in specifications to ensure quality and toughness. Elastic properties may be of interest, but these are measured ultrasonically.

Tensile Specimens

Figure 3.1 shows a typical tensile specimen. It has enlarged ends or shoulders for gripping. The important part of the specimen is the gauge section. The cross-sectional area of the gauge section is less than that of the shoulders and grip region, so the deformation will occur here. The gauge section should be long compared to the diameter (typically, four times). The transition between the gauge section and

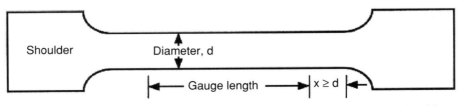

Figure 3.1. Typical tensile specimen with a reduced gauge section and larger shoulders.

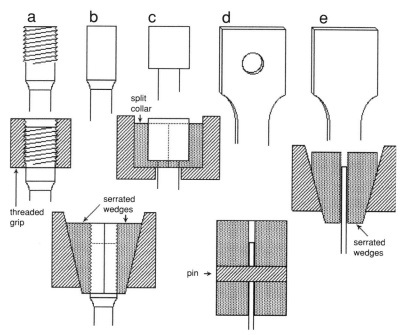

Figure 3.2. Systems for gripping tensile specimens. For round specimens, these include threaded grips (a), serrated wedges (b), and split collars for butt-end specimens (c). Sheet specimens may be gripped by pins (d) or serrated wedges (e). From W. F. Hosford in *Tensile Testing*, ASM Int. (1992).

the shoulders should be gradual to prevent the larger ends from constraining deformation in the gauge section.

There are several ways of gripping specimens, as shown in Figure 3.2. The ends may be screwed into threaded grips, pinned, or held between wedges. Special grips are used for specimens with butt ends. The gripping system should ensure that the slippage and deformation in the grip region are minimized. It should also prevent bending.

Stress–Strain Curves

Figure 3.3 is a typical engineering stress–strain curve for a ductile material. For small strains, the deformation is elastic and reverses if the load is removed. At higher stresses, plastic deformation occurs. This is not recovered when the load is removed. Bending a wire or paper clip with the fingers (Figure 3.4) illustrates the difference. For small strains, the deformation is elastic and reverses if the load is removed. At higher stresses, plastic deformation occurs. This is not recovered when the load is removed. If the wire is bent, a small amount of it will snap back when released. However, if the bend is more severe, it will only partly recover, leaving a permanent bend. The onset of plastic deformation is usually associated with the first deviation of the stress–strain curve from linearity.*

* For some materials, there may be nonlinear elastic deformation.

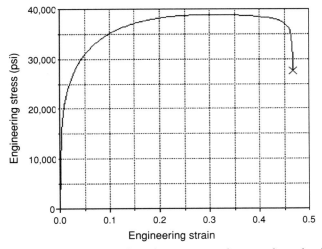

Figure 3.3. Typical engineering stress–strain curve for a ductile material.

It is tempting to define an *elastic limit* as the stress that causes the first plastic deformation and to define a *proportional limit* as the first departure from linearity. However, neither definition is very useful because they both depend on how accurately strain is measured. The more accurate the strain measurement is, the lower is the stress at which plastic deformation and nonlinearity can be detected.

To avoid this problem, the onset of plasticity is usually described by an *offset yield strength*. It is found by constructing a straight line parallel to the initial linear portion of the stress–strain curve, but offset from it by $e = 0.002$ (0.2%). The offset yield strength is taken as the stress level at which this straight line intersects the stress–strain curve (Figure 3.5). The rationale is that if the material had been loaded to this stress and then unloaded, the unloading path would have been along this offset line, resulting in a plastic strain of $e = 0.002$ (0.2%). The advantage of this way of defining yielding is that it is easily reproduced. Occasionally, other offset strains are used, and yielding is defined in terms of the stress necessary to achieve a specified total strain (e.g., 0.5%) instead of a specified plastic strain. In all cases, the criterion should be made clear to the user of the data.

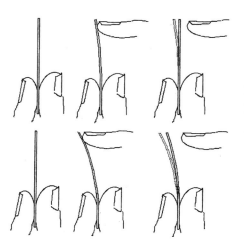

Figure 3.4. Using the fingers to sense the elastic and plastic responses of a wire. With a small force (*top*), all bending is elastic and disappears when the force is released. With a greater force (*bottom*), the elastic part of the bending is recoverable, but the plastic part is not. From W. F. Hosford, *Ibid.*

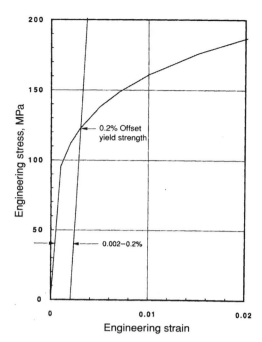

Figure 3.5. Low strain region of the stress–strain curve for a ductile material. From W. F. Hosford, *Ibid*.

Yield points: The stress–strain curves of some materials (e.g., low-carbon steels and linear polymers) have an initial maximum followed by lower stress, as shown in Figures 3.6a and 3.6b. At any given instant after the initial maximum, the deformation occurs within a relatively small region of the specimen. For steels, this deforming region is called a *Lüder's* band. Continued elongation occurs by movement of the Lüder's band along the gauge section, rather than by continued deformation within the band. Only after the band has traversed the entire gauge section, does the stress rise again. In the case of linear polymers, the yield strength is usually defined as the initial maximum stress. For steels, the subsequent lower yield strength is used to describe yielding. Because the initial maximum stress is extremely sensitive to the alignment of the specimen, it is not useful in describing yielding. Even so,

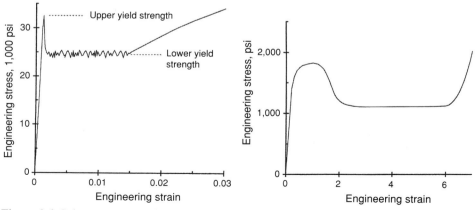

Figure 3.6. Inhomogeneous yielding of low-carbon steel (*left*) and a linear polymer (*right*). After the initial stress maximum, the deformation in both materials occurs within a narrow band that propagates the length of the gauge section before the stress rises again.

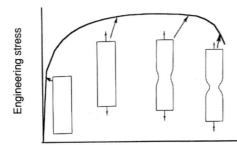

Figure 3.7. After a maximum of the stress–strain curve, deformation localizes to form a neck.

the lower yield strength is sensitive to the strain rate. ASTM standards should be followed. The stress level during Lüder's band propagation fluctuates. Some laboratories report the minimum level as the yield strength, and others use the average level.

As long as the engineering stress–strain curve rises, the deformation will occur uniformly along the gauge length. For a ductile material, the stress will reach a maximum well before fracture. When the maximum is reached, the deformation localizes forming a neck, as shown in Figure 3.7.

The *tensile strength* (or ultimate strength) is defined as the highest value of the engineering stress (Figure 3.8). For ductile materials, the tensile strength corresponds to the point at which necking starts. Less ductile materials fracture before they neck. In this case, the fracture stress is the tensile strength. Very brittle materials (e.g., glass) fracture before they yield. Such materials have tensile strengths, but no yield stresses.

Ductility

Two common parameters are used to describe the ductility of a material. One is the *percent elongation,* which is simply defined as

$$\% \, El = (L_f - L_o)/L_o \times 100\%, \tag{3.1}$$

where L_o is the initial gauge length and L_f is the length of the gauge section at fracture. Measurements may be made on the broken pieces or under load. For most

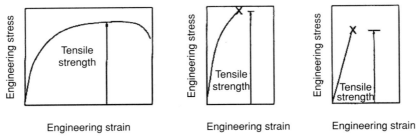

Figure 3.8. The tensile strength is the maximum engineering stress, regardless of whether the specimen necks or fractures before necking. From W. F. Hosford, *Ibid.*

materials, the elastic elongation is so small compared to the plastic elongation that it can be neglected. When this is not so, as with brittle materials or with rubberlike materials, it should be made clear whether or not the percent elongation includes the elastic portion.

The other common measure of ductility is the *percent reduction of area* at fracture, defined as

$$\% RA = (A_o - A_f)/A_o \times 100\%, \tag{3.2}$$

where A_o is the initial cross-sectional area and A_f is the cross-sectional area of the fracture. If the failure occurs before the necking, the %El can be calculated from the %RA by assuming constant volume. In this case,

$$\% El = \% RA/(100 - \% RA) \times 100\%. \tag{3.3}$$

The %El and %RA are no longer directly related after a neck has formed.

Percent elongation, as a measure of ductility, has the disadvantage that it combines the uniform elongation that occurs before necking and the localized elongation that occurs during necking. The uniform elongation depends on how the material strain hardens rather than on the fracture behavior. The necking elongation is sensitive to the specimen shape. With a gauge section that is very long compared to the diameter, the contribution of necking to the total elongation is very small. On the other hand, if the gauge section is very short, the necking elongation accounts for most of the elongation. For round bars, this problem has been remedied by standardizing the ratio of the gauge length to diameter at 4 : 1. However, there is no simple relation between the percent elongation of such standardized round bars and the percent elongation measured on sheet specimens or wires.

Percent reduction of area, as a measure of ductility, does not depend on the ratio of the gauge length to diameter. However, for very ductile materials, it is difficult to measure the final cross-sectional area, especially with sheet specimens.

True Stress and Strain

If the tensile tests are used to predict how the material will behave under other forms of loading, true stress–true strain curves are useful. The true stress, σ, is defined as

$$\sigma = F/A, \tag{3.4}$$

where A is the instantaneous cross-sectional area corresponding to the force, F. Before necking begins, the true strain, ε, is given by

$$\varepsilon = \ln(L/L_o). \tag{3.5}$$

The engineering stress, s, is defined as the force divided by the original area, $s = F/A_o$, and the engineering strain, e, as the change in length divided by the original length, $e = \Delta L/L_o$. As long as the deformation is uniform along the gauge length, the true stress and true strain can be calculated from the engineering quantities. With constant volume, $LA = L_oA_o$,

$$A_o/A = L/L_o, \tag{3.6}$$

so $A_o/A = 1 + e$. Rewriting equation (3.4) as $\sigma = (F/A_o)(A_o/A)$ and substituting $A_o/A = 1 + e$ and $s = F/A_o$,

$$\sigma = s(1 + e). \tag{3.7}$$

Substitution of $L/L_o = 1 + e$ into equation (3.5) gives

$$\varepsilon = \ln(1 + e). \tag{3.8}$$

These expressions are valid only if the deformation is uniformly distributed along the gauge section. After necking starts, equation (3.4) is still valid for true stress, but the cross-sectional area at the base of the neck must be measured independently. Equation (3.5) could still be used if L and L_o were known for a gauge section centered on the middle of the neck and so short that the variations of area along its length are negligible. Then equation (3.6) would be valid over such a gauge section, so the true strain can be calculated as

$$\varepsilon = \ln(A_o/A), \tag{3.9}$$

where A is the area at the base of the neck. Figure 3.9 shows the comparison of engineering and true stress–strain curve for the same material.

> **EXAMPLE PROBLEM 3.1:** In a tensile test, a material fractured before necking. The true stress and strain at fracture were 630 MPa and 0.18, respectively. What is the tensile strength of the material?
>
> **Solution:** The engineering strain at fracture was $e = \exp(0.18) - 1 = 0.197$. Because $s = \sigma/(1 + e)$, the tensile strength $= 630/1.197 = 526$ MPa.

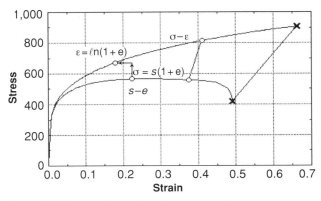

Figure 3.9. Comparison of engineering and true stress–strain curves. Before necking, a point on the true stress–strain curve (σ–ε) can be constructed from a point on the engineering stress–strain curve (s–e) with equations (3.7) and (3.8). After necking, the cross-sectional area at the neck must be measured to find the true stress and strain.

The Bridgman Correction

The state of stress at the center of a neck is not uniaxial tension. As material in the center of the neck is being stretched in the axial direction, it must contract laterally. This contraction is resisted by the adjacent regions immediately above and below that have larger cross-sections and are therefore not deforming. The net effect is that the center of a neck is under triaxial tension. Figure 3.10 shows the stress distribution calculated by Bridgman.*

Only that part of the axial stress, which exceeds the lateral stress, is effective in causing yielding. Bridgman showed that the effective part of the stress, $\overline{\sigma}$, is

$$\overline{\sigma}/\sigma = 1/\{(1 + 2R/a)\ln[1 + a/(2R)]\}, \tag{3.10}$$

where σ is the measured stress, F/A; a is the radius of specimen at the base of the neck; and R is the radius of curvature of the neck profile.

Figure 3.11 is a plot of the Bridgman correction factor, $\overline{\sigma}/\sigma$, as a function of a/R according to equation (3.10). A simple way of measuring the radius of curvature, R, can be measured by sliding a calibrated cone along the neck until it becomes tangent at the base of the neck.

Temperature Rise

Most of the mechanical energy expended by the tensile machine is liberated as heat in the tensile specimen. If the testing is rapid, little heat is lost to the surrounding, and the temperature rise can be surprisingly high. With very slow testing, most of the heat is dissipated to the surroundings so the temperature rise is much less.

> **EXAMPLE PROBLEM 3.2:** Calculate the temperature rise in a tension test of a low-carbon steel after a tensile elongation of 22%. Assume that 95% of the energy goes to heat and remains in the specimen. Also assume that Figure 3.3

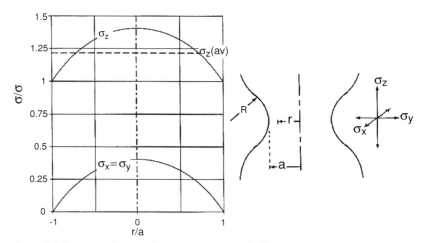

Figure 3.10. Stress distribution across a neck (*left*) and corresponding geometry of the neck (*right*). Both axial and lateral tensile stresses are a maximum at the center of the neck.

* Percy Bridgman, *Studies in Large Plastic Flow and Fracture*, McGraw-Hill, New York (1952).

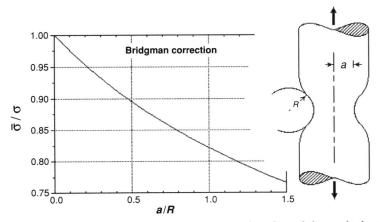

Figure 3.11. Bridgman correction factor as a function of the neck shape. The plot gives the ratio of the effective stress to the axial true stress for measured values of a/R.

represents the stress–strain curve of the material. For steel, the heat capacity is 447 J/kgK, and the density is 7.88 Mg/m³.

Solution: The heat released equals $0.95 \int s\,de$. From Figure 3.3, $\int s\,de = s_{av}e =$ about 34×0.22 ksi $= 51.6$ MPa $= 61.6$ MJ/m³. $\Delta T = Q/C$, where $Q = 0.95\,(61.6 \times 10^6 \text{ J/m}^3)$, and C is heat capacity per volume $= (7.8 \times 10^3 \text{ kg/m}^3)$ (447 J/kg/K). Substituting, $\Delta T = 0.95(61.6 \times 10^6 \text{ J/m}^3)/[(7.8 \times 10^3 \text{ kg/m}^3)$ $(447 \text{ J/kg/K})] = 17°\text{C}$. This is a moderate temperature rise. For high-strength materials, the rise can be much higher.

Sheet Anisotropy

The angular variation of yield strength in many sheet materials is not large. However, such a lack of variation does not indicate that the material is isotropic. The parameter that is commonly used to characterize the anisotropy is the *strain ratio* or *R value** (Figure 3.12). This is defined as the ratio, R, of the contractile strain in the width direction to that in the thickness direction during a tension test,

$$R = \varepsilon_w/\varepsilon_t. \tag{3.11}$$

If a material is isotropic, the width and thickness strains, ε_w and ε_t, are equal, so for an isotropic material, $R = 1$. For most materials, however, R is usually either greater or less than 1 in real sheet materials. Direct measurement of the thickness strain in thin sheets is inaccurate. Instead, ε_t is usually deduced from the width and length strains, ε_w and ε_l assuming constancy of volume, $\varepsilon_t = -\varepsilon_w - \varepsilon_l$, .

$$R = -\varepsilon_w/(\varepsilon_w + \varepsilon_l). \tag{3.12}$$

To avoid constraint from the shoulders, strains should be measured well away from the ends of the gauge section. Some workers suggest that the strains be

---

* Some authors use the symbol, r, instead of R.

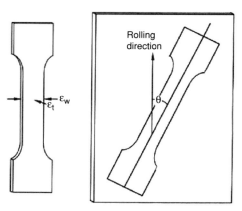

Figure 3.12. Tensile specimen cut from A sheet (*left*). The R value is the ratio of the lateral strains, $\varepsilon_w/\varepsilon_t$, during the extension (*right*).

measured when the total elongation is 15%, if this is less than the necking strain. The change of R during a tensile test is usually quite small, and the lateral strains at 15% elongation are great enough to be measured with accuracy.

EXAMPLE PROBLEM 3.3: Consider the accuracy of R measurement for a material having an R of about 1.00. Assume the width is about 0.5 in. and can be measured to an accuracy of ± 0.001 in., which corresponds to an uncertainty of strain of $\pm 0.001/0.5 = \pm 0.002$. The errors in measuring the length strain are much smaller. Estimate the error in finding R if the total strain was 5%. How would this be reduced if the total strain was 15%?

Solution: At an elongation of 5%, the lateral strain would be about 0.025, so the error in ε_w would be $\pm 0.002/0.025 = 8\%$. If the R value were 1, this would cause an error in R of approximately $0.08 \times 0.5/(1 - 0.5)$ or 8%. At 15% elongation, the percent error should be about one third of this or roughly $\pm 3\%$.

The value of R usually depends on the direction of testing. An average R value is conventionally taken as

$$\overline{R} = (R_0 + R_{90} + 2R_{45})/4. \tag{3.13}$$

The angular variation of R is characterized by ΔR, defined as

$$\Delta R = (R_0 + R_{90} - 2R_{45})/2. \tag{3.14}$$

Both are important in analyzing what happens during sheet metal forming.

Measurement of Force and Strain

In most tensile testing machines, the force is applied through load cells. The load cell is built so that it will deform elastically under the applied loads. The amount of elastic deformation is sensed and converted to an electric signal that in turn may be recorded electronically or used to drive a recording pen.

Several methods may be used to measure strain. With ductile polymers and metals, the deformation is often so great that it may be calculated from the cross-head movement. In this case, no direct measurements on the specimen are required. With screw-driven machines, the cross-head movement can be deduced from time. If it is assumed that all the cross-head displacement corresponds to specimen elongation, the engineering strain is simply the cross-head displacement divided by the gauge length. This procedure ignores the fact that some of the cross-head movement corresponds to elastic distortion of the machine and the gripping system as well as some plastic deformation in the shoulders of the specimen. The machine displacements can be found as a function of load by running a calibration experiment with a known specimen. Subtracting these machine grip displacements from the cross-head displacement will increase the accuracy of the method.

For small strains, much greater accuracy can be achieved with an extensometer attached to the specimen. An extensometer is a device specifically designed to measure small displacements by using resistance strain gauges, differential capacitances, or differential inductances. The change in resistance, capacitance, or inductance is converted to an electric signal that in turn controls a pen or paper drive or is stored in a computer file.

Axial Alignment

For measurement of strains in the elastic region, substantial errors may result with the use of a single extensometer unless the specimen is very straight and axially aligned. Nonaxial alignment and specimen curvature have analogous effects during loading, causing a bending or unbending of the specimen, as sketched in Figure 3.13. The extensometer will respond to the bending as well as the axial elongation. A simple way of compensating for bending and misalignment is to mount two extensometers on opposite sides of the specimen. The averaged response will not be affected by the bending.

Axiality of loading is of particular importance during the testing of brittle materials. In bending, the load on one side of the specimen is higher than on the opposite side, but the average stress deduced from the load will be less than the stress on the more heavily stressed side. Thus, the recorded tensile strength will be lower than the actual stress at the location where the fracture initiates. With low-carbon steels, nonaxial loading may mask the upper yield point. However, with ductile materials, nonaxiality is not a problem after the initial yielding because the specimens straighten during the first plastic deformation.

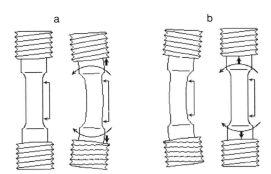

Figure 3.13. Nonaxiality can result from off-center loading (a). With the bending caused by off-center loading, strain measurements on the outside of the bend will be higher than the centerline strain. Straightening of an initially bents pecimen (b) will cause measurements of strain on the outside of the bend to be lower than the strain at the centerline. From W. F. Hosford, *Ibid.*

Special Problems

In testing of wire or rope, there is no simple way of using a reduced gauge section. Instead, the wire or rope is wound around a drum so that friction provides the gripping. When specimens are tested at high or low temperatures, the entire gauge section should be at the test temperature. For low temperature testing, the specimen can be immersed in a constant-temperature liquid bath. For testing at elevated temperatures, the specimen is usually surrounded by a controlled furnace.

Compression Test

Because necking limits the uniform elongation in tension, tension tests are not useful for studying the plastic stress–strain relationships at high strains. Much higher strains can be reached in compression, torsion, and bulge tests. The results from these tests can be used, together with the theory of plasticity (Chapter 5), to predict the stress–strain behavior under other forms of loading.

Much higher strains are achievable in compression tests than in tensile tests. However, two problems limit the usefulness of compression tests: friction and buckling. Friction on the ends of the specimen tends to suppress the lateral spreading of material near the ends (Figure 3.14). A cone-shaped region of *dead metal* (undeforming material) can form at each end, with the result that the specimen becomes barrel shaped. Friction can be reduced by lubrication, and the effect of friction can be lessened by increasing the height-to-diameter ratio, h/d, of the specimen.

If the coefficient of friction, μ, between the specimen and platens is constant, the average pressure to cause deformation is

$$P_{av} = Y(1 + (\mu d/h)/3 + (\mu d/h)^2/12 + \cdots),\tag{3.15}$$

where Y is the true flow stress of the material. If, however, there is a constant shear stress at the interface, such as would be obtained by inserting a thin film of a soft material (e.g., lead, polyethylene, or Teflon), the average pressure is

$$P_{av} = Y + (1/3)k(d/h),\tag{3.16}$$

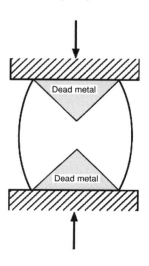

Figure 3.14. Unless the ends of a compression specimen are well lubricated, there will be a conical region of undeforming material (dead metal) at each end of the specimen. As a consequence, the midsection will bulge out or *barrel.*

Figure 3.15. Photograph of the end of a compression specimen. The darker central region was the original end. The lighter region outside was originally part of the cylindrical wall that folded up with the severe barreling. From G. W. Pearsall and W. A. Backofen, *Journal of Engineering for Industry, Trans ASME*, v. 85B (1963), pp. 68–76.

where k is the shear strength of the soft material. However, these equations usually do not accurately describe the effect of friction because neither the coefficient of friction nor the interface shear stress is constant. Friction is usually highest at the edges where liquid lubricants are lost, and thin films may be cut during the test by sharp edges of the specimens. Severe barreling caused by friction may cause the sidewalls to fold up and become part of the ends, as shown in Figure 3.15. Periodic unloading to replace or relubricate the film will help reduce these effects.

Although increasing h/d reduces the effect of friction, the specimen will buckle if it is too long and slender. Buckling is likely if the height-to-diameter ratio is greater than about 3. If the test is so well lubricated that the ends of the specimen can slide relative to the platens, buckling can occur for $h/d \geq 1.5$ (Figure 3.16).

One way to circumvent the effects of friction is to test specimens with different diameter/height ratios. The strains at several levels of stress are plotted against d/h. By the extrapolating the stresses to $d/h = 0$, the stress levels can be found for an infinitely long specimen in which the friction effects would be negligible (Figure 3.17).

During compression, the load-carrying cross-sectional area increases. Therefore, in contrast to the tension test, the absolute value of engineering stress is greater than the true stress (Figure 3.18). The area increase, together with work hardening, can lead to very high forces during compression tests, unless the specimens are very small.

The shape of the engineering stress–strain curve in compression can be predicted from the true stress–strain curve in tension, assuming that absolute values of true stress in tension and compression are the same at the same absolute strain

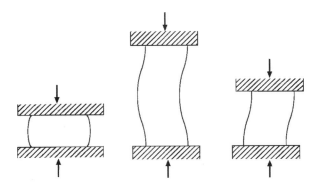

Figure 3.16. Problems with compression testing: (a) friction at the ends prevents spreading, which results in barreling; and (b) buckling of poorly lubricated specimens can occur if the height-to-diameter ratio, h/d, exceeds about 3. Without any friction at the ends (c), buckling can occur if h/d is greater than about 1.5.

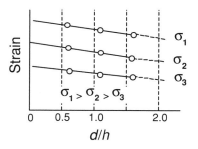

Figure 3.17. Extrapolation scheme for Eliminating frictional effects in compression testing. Strains at different levels of stress $(\sigma_1, \sigma_2, \sigma_3)$ are plotted for specimens of differing heights. The strain for "frictionless" conditions is obtained by extrapolating d/h to 0.

values. Equations (3.7) and (3.8) apply, but it must be remembered that both the stress and strain are negative in compression,

$$e_{\text{comp}} = \exp(\varepsilon) - 1, \tag{3.17}$$

and

$$s_{\text{comp}} = \sigma/(1 + e). \tag{3.18}$$

EXAMPLE PROBLEM 3.3: In a tensile test, the engineering stress $s = 100$ MPa at an engineering strain of $e = 0.20$. Find the corresponding values of σ and ε. At what engineering stress and strain in compression would the values of $|\sigma|$ and $|\varepsilon|$ equal those values of σ and ε?

Solution: In the tensile test, $\sigma = s(1 + e) = 100(1.2) = 120$ MPa, $\varepsilon = \ln(1+e) = \ln(1.2) = 0.182$. At a true strain of -0.182 in compression, the engineering strain would be $e_{\text{comp}} = \exp(-0.18) - 1 = -0.1667$, and the engineering stress would be $s_{\text{comp}} = \sigma/(1 + e) = -120$ MPa$/(1 - 0.1667) = -144$ MPa.

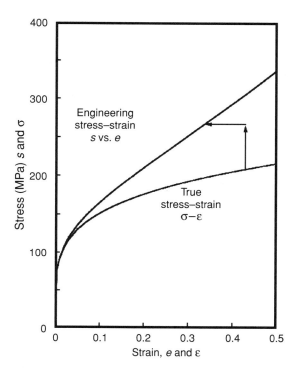

Figure 3.18. Stress–strain relations in compression for a ductile material. Each point, σ, ε, on the true stress–true strain curve corresponds to a point, s, e, on the engineering stress–strain curve. The arrows connect these points.

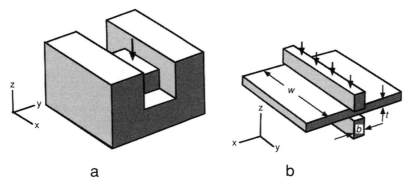

a   b

Figure 3.19. Plane–strain compression tests: (a) compression in a channel with the side walls preventing spreading, and (b) plane–strain compression of a wide sheet with a narrow indenter. Lateral constraint forcing, $\varepsilon_y = 0$, is provided by the adjacent material that is not under the indenter.

Compression failures of brittle materials occur by shear fractures on planes at 45 degrees to the compression axis. In materials of high ductility, cracks may occur on the barreled surface, either at 45 degrees to the compression axis or perpendicular to the hoop direction. In the latter case, secondary tensile stresses are responsible. These occur because the frictional constraint on the ends causes the sidewalls to bow outward. Because of this barreling, the axial compressive stress in the bowed walls is lower than in the center. Therefore, a hoop direction tension must develop to aid in the circumferential expansion.

Plane–Strain Compression

There are two simple ways of making plane–strain compression tests. Small samples can be compressed in a channel that prevents spreading (Figure 3.19a). In this case, there is friction on the sidewalls of the channel as well as on the platens so the effect of friction is even greater than in uniaxial compression. An alternative way of producing plane–strain compression is to use a specimen that is very wide relative to the breadth of the indenter (Figure 3.19b). This eliminates the sidewall friction, but the deformation at and near the edges deviates from plane strain. This departure from plane strain extends inward for a distance approximately equal to the indenter

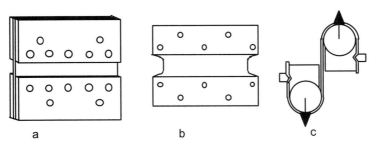

a   b   c

Figure 3.20. Several ways of making plane–strain tension tests on sheet specimens. All have a gauge section that is very short relative to its width: (a) enlarged grips produced by welding to additional material, (b) reduced gauge section cut into edge, and (c) very short gauge section achieved by friction on the cylindrical grips.

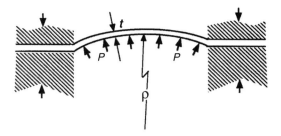

Figure 3.21. Schematic of a hydraulic bulge test of a sheet specimen. Hydraulic pressure causes biaxial stretching of the clamped sheet.

width. To minimize this effect, it is recommended that the ratio of the specimen width to indenter width, w/b, be about 8. It is also recommended that the ratio of the indenter width to sheet thickness, b/t, be about 2. Increasing b/t increases the effect of friction. Both tests simulate the plastic conditions that prevail during flat rolling of sheet and plate. They find their greatest usefulness in exploring the plastic anisotropy of materials.

Plane–Strain Tension

Plane–strain can be achieved in tension with specimens having gauge sections that are much wider than they are long. Figure 3.20 shows several possible specimens and specimen gripping arrangements. Such tests avoid the frictional complications of plane–strain compression. However, the regions near the edges lack the constraint necessary to impose plane strain. At the very edge, the stress preventing contraction disappears so the stress state is uniaxial tension. Corrections must be made for departure from plane–strain flow near the edges.

Biaxial Tension (Hydraulic Bulge Test)

Much higher strains can be reached in bulge tests than in uniaxial tension tests. This allows evaluation of the stress–strain relationships at high strains. A set-up for bulge testing is sketched in Figure 3.21. A sheet specimen is placed over a circular hole, clamped, and bulged outward by oil pressure acting on one side. Consider a force balance on a small circular element of radius ρ near the pole when $\Delta\theta$ is small (Figure 3.22). Using the small angle approximation, the radius of this element be $\rho\Delta\theta$, where ρ is the radius of curvature. The stress, σ, on this circular region acts

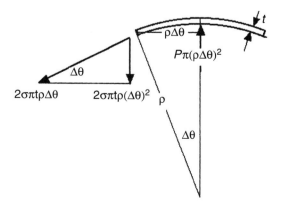

Figure 3.22. Force balance on a small circular region near the dome. Force acting upward is the pressure times the circular area. Total tangential force equals the thickness times the tangential stress. Downward force is the vertical component of this.

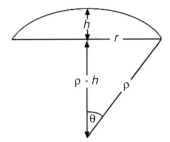

Figure 3.23. Geometry of a bulged surface.

on an area $2\pi\rho\Delta\theta t$ and creates a tangential force equal to $2\pi\sigma\rho\Delta\theta t$. The vertical component of the tangential force is $2\pi\sigma\rho\Delta\theta t$ times $\Delta\theta$, or $2\pi\sigma\rho(\Delta\theta)^2 t$. This is balanced by the pressure, P, acting on an area $\pi(\rho\Delta\theta)^2$ and creating an upward force of $P\pi(\rho\Delta\theta)^2$. Equating the vertical forces,

$$\sigma = Pr/(2t). \tag{3.19}$$

To find the stress, the pressure, radius of curvature, and radial strain must be measured simultaneously. The thickness, t, is then deduced from the original thickness, t_o, and the radial strain, ε_r,

$$t = t_o\exp(-2\varepsilon_r). \tag{3.20}$$

EXAMPLE PROBLEM 3.4: Assume that in a hydraulic bulge test, the bulged surface is a portion of a sphere and that at every point on the thickness of the bulged surface the thickness is the same. (Note: This is not strictly true.) Express the radius of curvature, ρ, in terms of the die radius, r, and the bulged height, h. Also express the thickness strain at the dome in terms of r and h. See Figure 3.23.

Solution: Using the Pythagorean theorem, $(\rho - h)^2 + r^2 = \rho^2$, so $2\rho h = h^2 + r^2$ or $\rho = (r^2 + h^2)/(2h)$, where ρ is radius of the sphere and θ is the internal angle.

The area of the curved surface of a spherical segment is $A = 2\pi rh$. The original area of the circle was $A_o = \pi r^2$, so the average thickness strain is $\varepsilon_t \approx \ln[2\pi rh/(\pi r^2)] = \ln(2h/r)$.

Hydrostatic compression superimposed on the state of biaxial tension at the dome of a bulge is equivalent to a state of through-thickness compression, as shown schematically in Figure 3.24. The hydrostatic part of the stress state has no effect on plastic flow. Therefore, the plastic stress–strain behavior of biaxial tension (in the plane of the sheet) and through-thickness compression are equivalent.

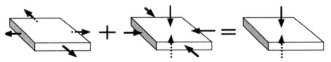

Figure 3.24. Equivalence of biaxial tension and through-thickness compression.

Torsion Test

Very high strains can be reached in torsion. The specimen shape remains constant so there is no necking instability or barreling. There is no friction on the gauge section. Therefore torsion testing can be used to study plastic stress–strain relations to high strains. In a torsion test, each element of the material deforms in pure shear, as shown in Figure 3.25. The shear strain, γ, in an element is given by

$$\gamma = r\theta/L, \tag{3.21}$$

where r is the radial position of the element, θ is the twist angle, and L is the specimen length. The shear stress, τ, cannot be measured directly or even determined unequivocally from the torque. This is because the shear stress, τ, depends on γ, which varies with radial position. Therefore, τ depends on r. Consider an annular element of radius, r, and width, dr, having an area, $2\pi r\,dr$. The contribution of this element to the total torque, T, is the product of the shear force on it, $\tau \cdot 2\pi r\,dr$, times the lever arm, r,

$$dT = 2\pi \tau r^2\,dr \quad \text{and}$$

$$T = 2\pi \int_0^R \tau r^2 dr. \tag{3.22}$$

Equation (3.22) cannot be integrated directly because τ depends on r. Integration requires substitution of the stress–strain (τ–γ) relation. Handbook equations for torque are usually based on assuming elasticity. In this case, $\tau = G\gamma$. Substituting this and equation (3.21) into equation (3.22),

$$T = 2\pi(\theta/L)\int_0^R r^3 dr = (\pi/2)(\theta/L)Gr^4. \tag{3.23}$$

Because $\tau_{yz} = G\gamma_{yz}$ and $\gamma_{yz} = r\theta/L$, the shear stress varies linearly with the radial position and can be expressed as $\tau_{yz} = \tau_s(r/R)$, where τ_s is the shear stress at the surface. The value of τ_s for elastic deformation can be found from the measured torque by substituting $\tau_{yz} = \tau_s(r/R)$ into equation (3.23),

$$T = 2\pi \int_0^r \tau_s(r/R)r^2 dr = (\pi/2)\tau_s R^3, \quad \text{or} \quad \tau_s = 2T/(\pi R^3). \tag{3.24}$$

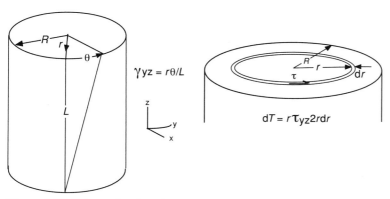

Figure 3.25. Schematic of torsion test.

If the bar is not elastic, Hooke's law cannot be assumed. The other extreme is when the entire bar is plastic and the material does not work harden. In this case, τ is a constant.

EXAMPLE PROBLEM 3.5: Consider a torsion test in which the twist is great enough that the entire cross-section is plastic. Assume that the shear yield stress, τ, is a constant. Express the torque, T, in terms of the bar radius, R, and τ.

Solution: The shear force on a differential annular element at a radius, r, is $\tau \cdot 2\pi r\,dr$. This causes a differential torque of $dT = r(\tau \cdot 2\pi r\,dr)$. Integrating,

$$= 2\pi\tau \int_0^R r^2\,dr = (2/3)\pi R^3 \tau. \tag{3.25}$$

If the torsion test is being used to determine the stress–strain relationship, the form the stress–strain relationship cannot be assumed so one does not know how the stress varies with radial position. One way around this problem might be to test thin-wall tube in which the variation of stress and strain across the wall would be small enough that the variation of τ with r could be neglected. In this case, the integral (equation 3.22) could be approximated as

$$T = 2\pi r^2 \Delta r \tau, \tag{3.26}$$

where Δr is the wall thickness. However, thin-wall tubes tend to buckle and collapse when subjected to torsion. The buckling problem can be circumvented by making separate torsion tests on two bars of slightly different diameter. The difference between the two curves is the torque-twist curve for cylinder whose wall thickness is half of the diameter difference.

The advantage of torsion tests is that very high strains can be reached, even at elevated temperatures. Because of this, torsion tests have been used to simulate the deformation in metal during hot rolling so that the effects of simultaneous hot deformation and recrystallization can be studied. It should be realized that in a torsion test, the material rotates relative to the principal stress axes. Because of this, the strain path in the material is constantly changing.

Bend Tests

Bend tests are used chiefly for materials that are very brittle and difficult to machine into tensile bars. In bending, as in torsion, the stress and strain vary with location. The engineering strain, e, varies linearly with distance, z, from the neutral plane

$$e = z/\rho, \tag{3.27}$$

where ρ is the radius of curvature at the neutral plane as shown in Figure 3.26. Consider bending a plate of width w. The bending moment, dM, caused by the stress on a differential element at z is the force $\sigma w\,dz$ times the lever arm, z, $dM = \sigma w z\,dz$. The total bending moment is twice the integral of dM from the neutral plane to the outside surface,

$$M = 2 \int_0^{t/2} \sigma w z\,dz. \tag{3.28}$$

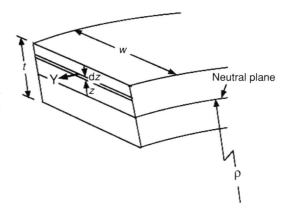

Figure 3.26. Bending moment caused by a differential element of width, w, thickness, dz, at a distance, z, from the neutral plane.

To integrate, σ must be expresses in terms of z. If the entire section is elastic, and $w \ll t$, $\sigma = eE$ (or if $w \gg t$, $\sigma = eE/(1 - v^2)$). Substituting for e, $\sigma = (z/r)E$. Integrating

$$M = 2w(E/\rho) \int_0^{t/2} z^2 \, dz,$$
$$M = 2w(E/\rho)(t/2)^3/3 = w(E/\rho)t^3/12. \tag{3.29}$$

The surface stress $\sigma_s = E(t/2)/r$. Substituting $r = wEt^3/(12M)$ from equation (3.29),

$$\sigma_s = 6M/(wt^2). \tag{3.30}$$

In three-point bending, $M = FL/4$, and in four point-bending, $M = FL/2$ as shown in Figure 3.27. The relation between the moment and the stress is different if plastic deformation occurs.

EXAMPLE PROBLEM 3.6: Consider a four-point bend test on a flat sheet of width, w, of a material for which the stress to cause plastic deformation is a constant, Y. Assume also that the bend is sharp enough so that the entire thickness is plastic. Derive an expression that can be used to determine Y from the force, F.

Solution: The differential force on an element of thickness, dz, at a distance z from the center is $Y(w \, dz)$. This force causes a differential moment,

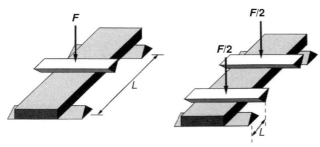

Figure 3.27. Three-point bending (*left*) and four-point bending (*right*) tests. Note that L is defined differently for each test.

$dM = wYz\,dz$. The net moment is twice the integral of dM from the center to the surface ($z = t/2$). Then

$$M = 2wY(t/2)^2/2 = wt^2Y/4.$$

Substituting $M = FL/2$ (equation 3.30 for four-point bending),

$$Y = 2FL/wt^2.$$

Measurements of fracture strengths of brittle materials are usually characterized by a large amount of scatter because of preexisting flaws, so many duplicate tests are usually required. The fracture stress is taken as the value of the surface stress, σ_s, at fracture. This assumes that no plastic deformation has occurred. See Chapters 14 and 19.

Hardness Tests

Hardness tests are simple to make, and they can be made on production parts as quality control checks without destroying the part. They depend on measuring the amount of deformation caused when a hard indenter is pressed into the surface with a fixed force. The disadvantage is that although hardness of a material depends on the plastic properties, the stress–strain relation cannot be obtained. Figure 3.28 shows the indenters used for various tests. The Rockwell tests involve measuring the depth of indentation. There are several different Rockwell scales, each of which uses different shapes and sizes of indenters and different loads. Conversion from one scale to another is approximate and empirical.

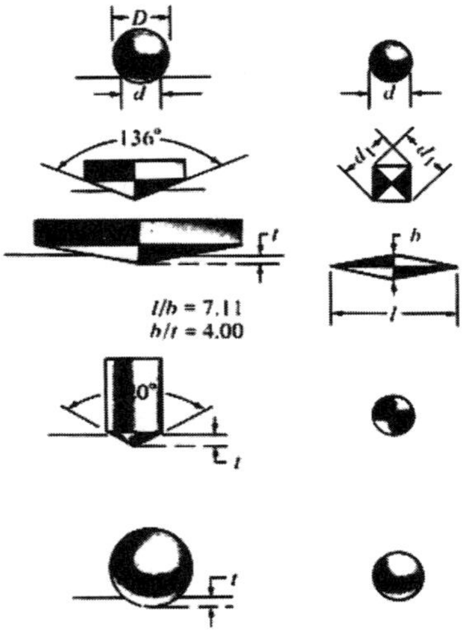

Test	Indenter	Shape of Indentation Side View	Shape of Indentation Top View	Load	Formula for Hardness Number
Brinell	10-mm sphere of steel or tungsten carbide			P	$BHN = \dfrac{2P}{\pi D(D - \sqrt{D^2 - d^2})}$
Vickers	Diamond pyramid	136°		P	$VHN = 1.72\,P/d_1^2$
Knoop microhardness	Diamond pyramid	$l/b = 7.11$ $b/t = 4.00$		P	$KHN = 14.2\,P/l^2$
Rockwell A C D	Diamond cone			60 kg 150 kg 100 kg	$\left.\begin{array}{l} R_A = \\ R_C = \\ R_D = \end{array}\right\} 100 - 500t$
B F G	$\frac{1}{16}$-in.-diameter steel sphere			100 kg 60 kg 150 kg	$\left.\begin{array}{l} R_B = \\ R_F = \\ R_G = \end{array}\right\} 130 - 500t$
E	$\frac{1}{8}$-in.-diameter steel sphere			100 kg	$R_E =$

Figure 3.28. Various hardness tests. From H. W. Hayden, William G. Moffat, and John Wulff, *Structure and Properties of Materials, Vol. III. Mechanical Behavior*, Wiley (1965).

Brinell hardness is determined by pressing a ball, 1 cm in diameter, into the surface under a fixed load (500 or 3,000 kg). The diameter of the impression is measured with an eyepiece and converted to hardness. The scale is such that the Brinell hardness number, H_B, is given by

$$H_B = F/A_s \qquad (3.31)$$

where F is the force expressed in kg and A_s is the spherical surface area of the impression in mm^2. This area can be calculated from

$$A_s = \pi D[D - (D^2 - d^2)^{1/2}]/2 \qquad (3.32)$$

where D is the ball diameter and d is the diameter of the impression, but it is more commonly found from tables.

Brinell data can be used to determine the Meyer hardness, H_M, which is defined as the force divided by the projected area of the indentation, $\pi d^2/4$. The Meyer hardness has greater fundamental significance because it is relatively insensitive to load. Vickers and Knoop hardnesses are also defined as the indentation load divided by the projected area. They are nearly equal for the same material. Indentations under different loads are geometrically similar unless the indentation is so shallow that the indenter tip radius is not negligible compared to the indentation size. For this reason, the hardness does not depend on the load. For Brinell, Knoop, and Vickers hardness, a rule of thumb is that the hardness is about three times the yield strength when expressed in the same units. (*Note*: Hardness is conventionally expressed in kg/mm^2 and strength in MPa.) To convert kg/mm^2 to MPa, multiply by 9.807. Figure 3.29 shows approximate conversions between several hardness scales.

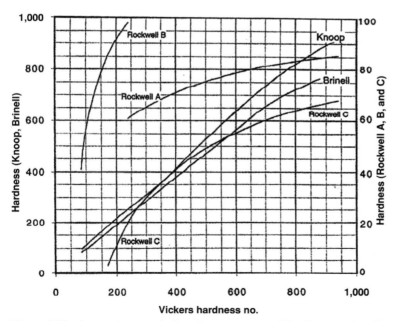

Figure 3.29. Approximate relations between several hardness scales. For Brinell (3,000 kg) and Knoop, scales read left. For Rockwell A, B, and C, scales, read right. Data from *Metals Handbook*, vol. 8, 9th ed., American Society for Metals (1985).

The Brinell, Meyers, Vickers, and Knoop hardness numbers are defined as the force on the indenter divided by area. (This is the projected area, except in the case of the Brinell number.) The hardness number is related to the yield strength measured at the strain characteristic of the indentation. For a nonwork hardening material under plane–strain indentation, theoretical analysis predicts

$$H = (2/\sqrt{3})(1 + \pi/2)Y = 2.97Y \approx 3Y. \qquad (3.33)$$

There are several approximate relations between the different scales.

$$B \approx 0.95\,V, \qquad (3.34)$$

$$K \approx 1.05\,V, \qquad (3.35)$$

$$R_C \approx 100 - 1{,}480/\sqrt{B}, \qquad (3.36)$$

and

$$R_B \approx 134 - 6{,}700/B, \qquad (3.37)$$

where B, V, K, R_C, and R_B are the Brinell, Vickers, Knoop, Rockwell C, and Rockwell B hardness numbers. Another useful approximation for steels is that the tensile strength (expressed in psi) is about 500 times the Brinell hardness number (expressed in kg/mm^2).

Possible errors in making hardness tests include:

1. Making indentations too close to an edge of the material. If the plastic zone around an indenter extends to an edge, the reading will be too low.
2. Making an indentation too close to a prior indentation. If the plastic zone around an indenter overlaps that of a prior indentation, the strain hardening during the prior test will cause the new reading to be too high.
3. Making too large an indentation on a thin specimen. If the plastic zone penetrates to the bottom surface, the reading will be in error.

With Rockwell tests, it is important that the bottom of the specimen be flat and supported on a rigid anvil. This is necessary because the dial reading, which indicates the hardness by measuring the depth of penetration, is influenced by bending or settling of the specimen, with the result of recording too low a hardness.

Among the other hardness tests is the Shore seleroscope test, which measures the rebound of a ball dropped from a fixed height onto the surface to be measured. This works on the principle that the energy absorbed when the ball hits the surface is a measure of hardness. With hard materials, little energy is absorbed, so the rebounds are high.

Mineralogists frequently classify minerals by the Moh's scratch hardness. The system is based on ranking minerals on a scale of 1 to 10. The scale is such that a mineral higher on the scale will scratch a mineral lower on the scale. The scale is arbitrary and not well suited to metals because most metals tend to fall in the range between 4 and 8. Actual values vary somewhat with how the test is made (e.g., the angle of inclination of the scratching edge). Figure 3.30 shows the approximate relationship between Vickers and Moh's hardnesses.

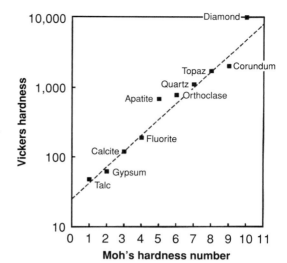

Figure 3.30. Relation between Vickers and Moh's hardness scales.

Mutual Indentation Hardness

Hardness tests are normally made with indenters that do not deform. However, it is possible to measure hardness when the indenters do deform. This is useful at very high temperatures, where it is impossible to find a suitable material for the indenter. Defining the hardness as the indentation force/area of indentation, the hardness, H, of a material is proportional to its yield strength,

$$H = cY, \tag{3.38}$$

where c depends mainly on the geometry of the test. For example, in a Brinell test where the indenting ball is much harder than the test material, $c \approx 2.8$. For mutual indentation of crossed wedges, $c \approx 3.4$ (Figure 3.31) and for crossed cylinders, $c \approx 2.4$. The constant c has been determined for other geometries and for cases where the indenters have different hardnesses. The forces in automobile accidents have been estimated from examining the depths of mutual indentations.

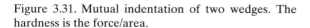

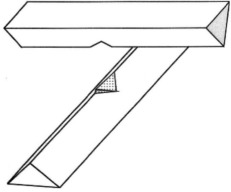

Figure 3.31. Mutual indentation of two wedges. The hardness is the force/area.

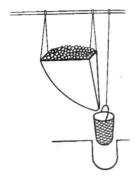

Figure 3.32. Leonardo's sketch of a system for tensile testing wire. The load was applied by pouring sand into a basket suspended from the wire.

REFERENCES

Tensile Testing, ASM International, Materials Park, OH (1992).
Percy Bridgman, *Studies in Large Plastic Flow and Fracture,* McGraw-Hill, New York (1952).
Metals Handbook, 9th ed., vol. 8, Mechanical Testing, ASM (1985).
D. K. Felbeck and A. G. Atkins, *Strength and Fracture of Engineering Solids,* Prentice Hall (1984).

Notes

Leonardo da Vinci (1452–1519) described a tensile test of a wire. Fine sand was fed through a small hole into a basket attached to the lower end of the wire until the wire broke. Figure 3.32 shows his sketch of the apparatus.

Galileo (1564–1642) stated that the strength of a bar in tension was proportional to its cross-sectional area and independent of its length. Petrus von Musschenbrök (1692–1761) devised the tensile testing machine and grip system illustrated in Figure 3.33.

Percy W. Bridgman (1882–1961) was born in Cambridge, Massachusetts, and attended Harvard, where he graduated in 1904 and received his PhD in 1908. He discovered the high-pressure forms of ice, is reputed to have discovered "dry ice," and wrote a classic book, *Physics of High Pressures,* in 1931. In 1946, he was awarded the Nobel prize in physics for his work at high pressure. Another book, *Studies in Large Plastic Flow and Fracture* (1952), summarizes his work with the mechanics of solids.

The yield point effect in linear polymers may be experienced by pulling the piece of plastic sheet that holds a six-pack of carbonated beverage cans together.

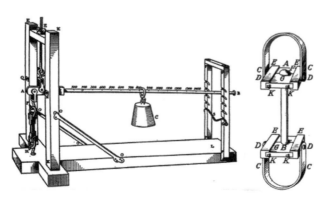

Figure 3.33. Musschenbrök's design (1729) of a tensile testing machine that uses a lever to increase the applied force. His system for gripping the specimen is shown at the right.

Figure 3.34. Galileo's sketch of a bending test.

When one pulls hard enough, the plastic will yield and the force drop. A small thinned region develops. As the force is continued, this region will grow. The yield point effect in low-carbon steel may be experienced by bending a low-carbon steel wire. (Florist's wire works well.) First, heat the wire in a flame to anneal it. Then bend a 6-in. length by holding it only at the ends. Instead of bending uniformly, the deformation localizes to form several sharp kinks. Why? Bend an annealed copper wire for comparison.

An apparatus suggested by Galileo for bending tests is illustrated in Figure 3.34.

Problems

1. The results of a tensile test on a steel test bar are given. The initial gauge length was 25.0 mm, and the initial diameter was 5.00 mm. The diameter at the fracture was 2.6 mm. The engineering strain and engineering stress in MPa are

Strain	Stress	Strain	Stress	Strain	Stress
0.0	0.0	0.06	319.8	0.32	388.4
0.0002	42.0	0.08	337.9	0.34	388.0
0.0004	83.0	0.10	351.1	0.38	386.5
0.0006	125.0	0.15	371.7	0.40	384.5
0.0015	155.0	0.20	382.2	0.42	382.5
0.005	185.0	0.22	384.7	0.44	378.0
0.02	249.7	0.24	386.4	0.46	362.0
0.03	274.9	0.26	387.6	0.47	250.0
0.04	293.5	0.28	388.3		
0.05	308.0	0.30	388.9		

A. Plot the engineering stress–strain curve.
B. Determine (i) Young's modulus, (ii) the 0.2% offset yield strength, (iii) the tensile strength, (iv) the percent elongation, and (iv) the percent reduction of area.

2. Construct the true stress–true strain curve for the material in problem 1. Note that necking starts at maximum load, so the construction should be stopped at this point.

3. Determine the engineering strains, e, and the true strains, ε, for each of the following:

    A. Extension from $L = 1.0$ to $L = 1.1$.
    B. Compression from $h = 1$ to $h = 0.9$.
    C. Extension from $L = 1$ to $L = 2$.
    D. Compression from $h = 1$ to $h = 1/2$.

4. The ASM *Metals Handbook* (Vol. 1, 8th ed., p. 1008) gives the percent elongation in a 2-in. gauge section for annealed electrolytic tough-pitch copper as

    55% for a 0.505-in.-diameter bar
    45% for a 0.030-in.-thick sheet
    38.5% for a 0.010-in.-diameter wire.

Suggest a reason for the differences.

5. The tensile strength of iron-carbon alloys increases as the percent carbon increases up to contents of about 1.5% to 2%. Above this, the tensile strength drops rapidly with increased percent carbon. Speculate about the nature of this abrupt change.

6. The area under an engineering stress–strain curve up to fracture is the energy/volume. The area under a true stress–strain curve up to fracture is also the energy/volume. If the specimen necks, these two areas are not equal. What is the difference? Explain.

7. Suppose it is impossible to use an extensometer on the gauge section of a test specimen. Instead a button head specimen (Figure 3.2c) is used, and the strain is computed from the cross-head movement. There are two possible sources of error with this procedure. One is that the gripping system may deform elastically, and the other is that the button head may be drawn partly through the collar. How would each error affect the calculated true stress and true strain?

8. Equation (3.10) relates the average axial stress in the neck, σ, to the effective stress, $\bar{\sigma}$. The variation of the local stresses with distance, r, from the center is given by

$$\sigma_z = \bar{\sigma}\{1 + \ln[1 + a/(2R) - r^2/(2aR)]\}, \quad \text{with } \sigma_x = \sigma_y = \sigma_z - \bar{\sigma}.$$

Derive an expression for the level of hydrostatic stress, $\sigma_H = (\sigma_x + \sigma_y + \sigma_z)/3$, at the center in terms of a, R, and $\bar{\sigma}$.

9. The tensile strengths of brazed joints between two pieces of steel are often considerably higher than the tensile strength of the braze material itself. Furthermore, the strengths of thin joints are higher than those of thick joints. Explain.

10. Two strain gauges were mounted on opposite sides of a tensile specimen. Strains measured as the bar was pulled in tension were used to compute Young's modulus. Readings from one gauge gave a modulus much higher than those from the other gauge. What was the probable cause of this discrepancy?

11. Engineering stress–strain curve from a tension test on a low-carbon steel is plotted in Figure 3.35. From this construct, the engineering stress–strain curve in compression, neglecting friction.

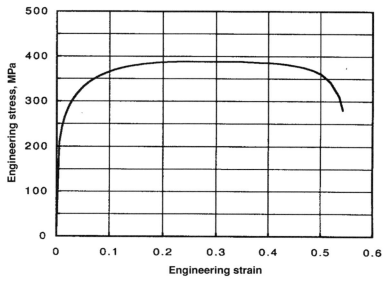

Figure 3.35. Engineering stress–strain curve for a low-carbon steel

12. Discuss the how friction and inhomogeneous deformation affect the results from

    A. The two types of plane–strain compression tests illustrated in Figure 3.19.
    B. The plane–strain tension tests illustrated in Figure 3.20.

13. Sketch the three-dimensional Mohr's stress and strain diagrams for a plane–strain compression test.

14. Draw a Mohr's circle diagram for the surface stresses in a torsion test, showing all three principal stresses. At what angle to the axis of the bar are the tensile stresses the greatest?

15. For a torsion test, derive equations relating the angle, ψ, between the axis of the largest principal stress and the axial direction and the angle, ψ, between the axis of the largest principal strain and the axial direction in terms of L, r, and the twist angle, θ. Note that for finite strains, these two angles are not the same.

16. Derive an expression relating the torque, T, in a tension test to the shear stress at the surface, τs, in terms of the bar diameter, D, assuming that the bar is

    A. Entirely elastic so τ varies linearly with the shear strain, γ.
    B. Entirely plastic and does not work harden so the shear stress, τ, is constant.

17. The principal strains in a circular bulge test are the thickness strain, ε_t; the circumferential (hoop) strain, ε_c; and the radial strain, ε_r. Describe how the ratio, $\varepsilon_c/\varepsilon_r$, varies over the surface of the bulge. Assume that the sheet is locked at the opening.

18. Derive an expression for the fracture stress, S_f, in bending as a function of F_f, L, w, and t for three-point bending of a specimen having a rectangular cross-section, where F_f is the force at fracture, L is the distance between supports, w is the specimen width, and t is the specimen thickness. Assume the deformation is elastic, the deflection, y, in bending is given by $y = \alpha FL^3/(EI)$, where α is a constant and E is Young's modulus. How would you expect the value of α to depend on the ratio of t/w?

19. Equation (3.30) gives the stresses at the surface of bend specimens. The derivation of this equation is based on the assumption of elastic behavior. If there is plastic deformation during the bend test, will the stress predicted by this equation be (a) too low, (b) too high, (c) either too high or too low depending on where the plastic deformation occurs, or (d) correct?

20. By convention, Brinell, Meyer, Vickers, and Knoop hardness numbers are stresses expressed in units of kg/mm^2, which is not an SI unit. To what stress, in MPa, does a Vickers hardness of 100 correspond?

21. In making Rockwell hardness tests, it is important that the bottom of the specimen is flat so that the load does not cause any bending of the specimen. On the other hand, this is not important in making Vickers or Brinell hardness tests. Explain.

Strain Hardening of Metals

Introduction

With elastic deformation, the strains are proportional to the stress, so every level of stress causes some elastic deformation. However, a definite level of stress must be applied before any plastic deformation occurs. As the stress is further increased, the amount of deformation increases, but not linearly. After plastic deformation starts, the total strain is the sum of the elastic strain (which still obeys Hooke's law) and the plastic strain. Because the elastic part of the strain is usually much less than the plastic part, it will be neglected in this chapter and the symbol ε will signify the true plastic strain.

The terms *strain hardening* and *work hardening* are used interchangeably to describe the increase of the stress level necessary to continue plastic deformation. The term *flow stress* is used to describe the stress necessary to continue deformation at any stage of plastic strain. Mathematical descriptions of true stress–strain curves are needed in engineering analyses that involve plastic deformation, such as predicting energy absorption in automobile crashes, designing of dies for stamping parts, and analyzing the stresses around cracks. Various approximations are possible. Which approximation is best depends on the material, the nature of the problem, and the need for accuracy. This chapter will consider several approximations and their applications.

Mathematical Approximations

The simplest model is one with no work hardening. The flow stress, σ, is independent of strain, so

$$\sigma = Y, \tag{4.1}$$

where Y is the tensile yield strength (Figure 4.1a). For linear work hardening (Figure 4.1b),

$$\sigma = Y + A\varepsilon. \tag{4.2}$$

It is more common for materials to work harden with a hardening rate that decreases with strain. For many metals, a log-log plot of true stress versus true strain

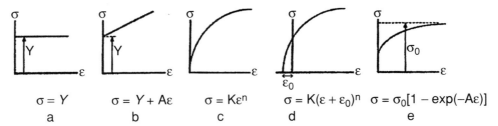

Figure 4.1. Mathematical approximations of the true stress–strain curve.

is nearly linear. In this case, a power law,

$$\sigma = K\varepsilon^n, \tag{4.3}$$

is a reasonable approximation (Figure 4.1c). A better fit is often obtained with

$$\sigma = K(\varepsilon + \varepsilon_0)^n. \tag{4.4}$$

(Figure 4.1d). This expression is useful where the material has undergone a prestrain of ε_0.

Still another model is a saturation model suggested by Voce (Figure 4.1e) is

$$\sigma = \sigma_0[1 - \exp(-A\varepsilon)]. \tag{4.5}$$

Equation (4.5) predicts that the flow stress approaches an asymptote, σ_0 at high strains. This model seems to be reasonable for a number of aluminum alloys.

EXAMPLE PROBLEM 4.1: Two points on a true stress–strain curve are $\sigma = 222$ MPa at $\varepsilon = 0.05$ and $\sigma = 303$ MPa at $\varepsilon = 0.15$.

A. Find the values of K and n in equation (5.3) that best fit the data. Then using these values of K and n, predict the true stress at a strain of $\varepsilon = 0.30$.
B. Find the values of σ_0 and A in equation (5.5) that best fit the data. Then using these values of K and n, predict the true stress at a strain of $\varepsilon = 0.30$.

Solution:

A. Appling the power law at two levels of stress, σ_2 and σ_1. Then taking the ratios, $\sigma_2/\sigma_1 = (\varepsilon_2/\varepsilon_1)^n$, and finally taking the natural logs of both sides, $n = \ln(\sigma_2/\sigma_1)/\ln(\varepsilon_2/\varepsilon_1) = \ln(303/222)/\ln(0.15/0.05) = 0.283$. $K = \sigma/\varepsilon^n = 303/0.15^{0.283} = 518.5$ MPa.

$$\text{At e} = 0.30, \quad \sigma = 518.4(0.30)^{0.283} = 368.8 \text{ MPa}.$$

B. Appling the saturation model two levels of stress, σ_2 and σ_1, $\sigma_1 = \sigma_0[1 - \exp(-A\varepsilon_1)]$ and $\sigma_2 = \sigma_0[1 - \exp(A\varepsilon_2)]$. Then taking the ratios, $\sigma_2/\sigma_1 = [1 - \exp(-A\varepsilon_2)]/[1 - \exp(-A\varepsilon_1)]$. Substituting $\sigma_2/\sigma_1 = 303/222 = 1.365$, $\varepsilon_1 = 0.05$, and $\varepsilon_2 = 0.15$ and solving by trial and error, $A = 25.2$. $\sigma_0 = 303[1 - \exp(-25.2 \times 0.15)] = 296$ MPa.

$$\text{At e} = 0.30, \quad \sigma = 310[1 - \exp(-25.2 \times 0.30)] = 309.8 \text{ MPa}.$$

Table 4.1. *Typical values of n and K**

Material	Strength coefficient, K (MPa)	Strain hardening exponent, n
Low-carbon steels	525–575	0.20–0.23
HSLA steels	650–900	0.15–0.18
Austenitic stainless	400–500	0.40–0.55
Copper	420–480	0.35–0.50
70/30 brass	525–750	0.45–0.60
Aluminum alloys	400–550	0.20–0.30

* From various sources, including Hosford and Caddell, *Metal Forming; Mechanics and Metallurgy*, 3rd ed., Cambridge 2007.

Power Law Approximation

The most commonly used expression is the simple power law (equation 4.3). Typical values of the exponent n are in the range of 0.1 to 0.6. Table 4.1 lists K and n for various materials. As a rule, high-strength materials have lower n values than low-strength materials. Figure 4.2 shows that the exponent, n, is a measure of the persistence of hardening. If n is low, the work hardening rate is initially high, but the

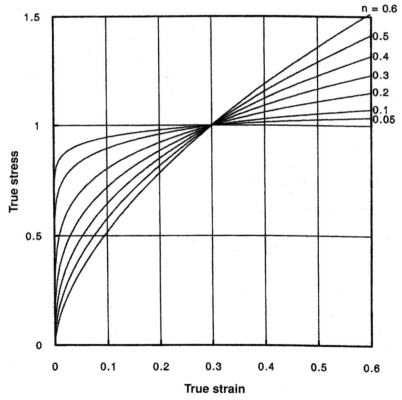

Figure 4.2. True stress–strain curves for $\sigma = K\varepsilon^n$, with several values of n. To make the effect of n on the shape of the curves apparent, the value of K for each curve has been adjusted so that it passes through $\sigma = 1$ at $\varepsilon = 0.3$.

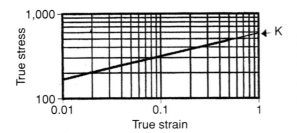

Figure 4.3. A plot of the true stress–strain curve on logarithmic scales. Because $\sigma = K\varepsilon^n$, $\ln \sigma = \ln K + n \ln \varepsilon$. The straight line indicates that $\sigma = K\varepsilon^n$ holds. The slope is equal to n, and K equals the intercept at $\varepsilon = 1$.

rate decreases rapidly with strain. In contrast, with a high n, the initial work hardening is less rapid but continues to high strains. If $\sigma = K\varepsilon^n$, $\ln \sigma = \ln K + n \ln \varepsilon$, so the true stress strain relation plots as a straight line on log-log coordinates as shown in Figure 4.3. The exponent, n, is the slope of the line. The preexponential, K, can be found by extrapolating to $\varepsilon = 1.0$. K is the value of σ at this point. The level of n is particularly significant in stretch forming because it indicates the ability of a metal to distribute the straining over a wide region. Often a log-log plot of the true stress–strain curve deviates from linearity at low or high strains. In such cases, it is still convenient to use equation (4.3) over the strain range of concern. The value of n is then taken as the slope of the linear portion of the curve.

$$n = d(\ln \sigma)/d(\ln \varepsilon) = (\varepsilon/\sigma)d\sigma/d\varepsilon. \tag{4.6}$$

Necking

As a tensile specimen is extended, the level of true stress, σ, rises, but the cross-sectional area carrying the load decreases. The maximum load-carrying capacity is reached when $dF = 0$. The force equals the true stress times the actual area, $F = \sigma A$. Differentiating gives

$$Ad\sigma + \sigma dA = 0. \tag{4.7}$$

Because the volume, AL, is constant, $dA/A = -dL/L = -d\varepsilon$. Rearranging terms $d\sigma = -\sigma dA/A = \sigma d\varepsilon$, or

$$d\sigma/d\varepsilon = \sigma \tag{4.8}$$

Equation (4.8) simply states that the maximum load is reached when the rate of work hardening is numerically equal to the stress level.

As long as $d\sigma/d\varepsilon > \sigma$, deformation will occur uniformly along the test bar. However, once the maximum load is reached ($d\sigma/d\varepsilon = \sigma$), the deformation will localize. Any region that deforms even slightly more than the others will have a lower load-carrying capacity, and the load will drop to that level. Other regions will cease to deform so deformation will localize into a neck. Figure 4.4 is a graphical illustration.

If a mathematical expression is assumed for the stress–strain relationship, the limit of uniform elongation can be found analytically. For example, with power law hardening, equation (4.3), $\sigma = K\varepsilon^n$, and $d\sigma/d\varepsilon = nK\varepsilon^{n-1}$. Substituting into equation (4.8) gives $K\varepsilon^n = nK\varepsilon^{n-1}$, which simplifies to

$$\varepsilon = n, \tag{4.9}$$

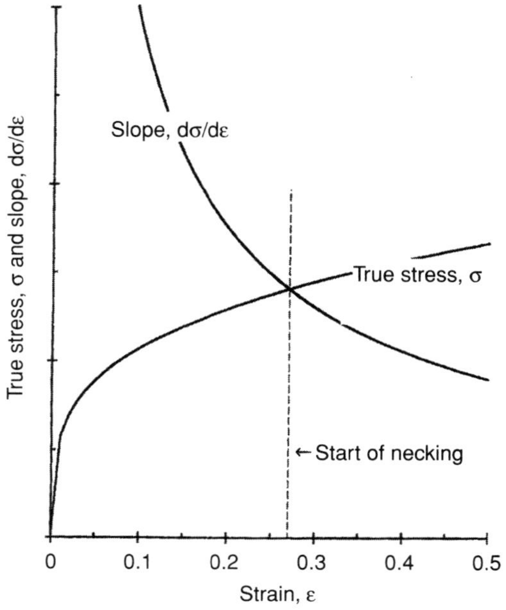

Figure 4.4. The condition for necking in a tension test is met when the true stress, σ, equals the slope, $d\sigma/d\varepsilon$, of the true stress–strain curve.

so the strain at the start of necking equals n. Uniform elongation in a tension test occurs before necking. Therefore, materials with a high n value have large uniform elongations.

Because the ultimate tensile strength is simply the engineering stress at maximum load, the power law hardening rule can be used to predict it. First, find the true stress at maximum load by substituting the strain, n, into, into equation (4.3).

$$\sigma_{\text{max load}} = Kn^n. \tag{4.10}$$

Then substitute $\sigma = \sigma/(1+\varepsilon) = \sigma\exp(-\varepsilon) = \sigma\exp(-n)$ into equation (4.10); the tensile strength is expressed as

$$\sigma_{\text{max}} = Kn^n\exp(-n) = K(n/e)^n, \tag{4.11}$$

where, here, e is the base of natural logarithms.

Similarly, the uniform elongation and tensile strength may be found for other approximations to the true stress–strain curve.

EXAMPLE PROBLEM 4.2: A material has a true stress–strain curve that can be approximated by $\sigma = Y + A\varepsilon$. Express the uniform elongation, ε, in terms of the constants, A and Y.

Solution: The maximum load is reached when $d\sigma/d\varepsilon = \sigma$. Equating $d\sigma/d\varepsilon = A$ and $\sigma = Y + A\varepsilon$, the maximum load is reached when $A = Y + A\varepsilon$ or $\varepsilon = 1 - Y/A$. The engineering strain, $e = \exp(\varepsilon) - 1 = \exp(1 - Y/A) - 1$.

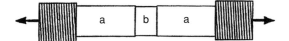

Figure 4.5. Stepped tensile specimen. The initial cross-sectional area of region b is f times the cross-sectional area of A.

Work Per Volume

The area under the true stress–strain curve is the work per volume, w, expended in the deforming a material. That is, $w = \int \sigma \, d\varepsilon$. With power law hardening,

$$w = K\varepsilon^{n+1}/(n+1). \tag{4.12}$$

This is sometimes called the *tensile toughness*.

Localization of Strain at Defects

If the stresses that cause deformation in a body are not uniform, the deformation will be greatest where the stress is highest, and least where the stress is lowest. The differences between the strains in the different regions depend on the value of n. If n is high, the difference will be less than if n is low. For example, consider a tensile bar in which the cross-sectional area of one region is a little less than in the rest of the bar (Figure 4.5). Let these areas be A_{ao} and A_{bo}.

The tensile force, F, is the same in both, $F_a = F_b$, regions, so

$$\sigma_a A_a = \sigma_b A_b. \tag{4.13}$$

Because $\varepsilon = \ln(L/L_o)$ and $(L/L_o) = (A_o/A)$, the instantaneous area, A, may be expressed as $A = A_o \exp(-\varepsilon)$, so $A_a = A_{ao} \exp(-\varepsilon_a)$, $A_b = A_{bo} \exp(-\varepsilon_b)$. Assuming power law hardening (equation 5.3), $\sigma_a = K\varepsilon_a^n$ and $\sigma_b = K\varepsilon_b^n$. Substituting into equation (4.13),

$$K\varepsilon_a^n A_{ao} \exp(-\varepsilon_a) = K\varepsilon_b^n A_{bo} \exp(-\varepsilon_b). \tag{4.14}$$

Now simplify by substituting $f = A_{bo}/A_{ao}$,

$$\varepsilon_a^n \exp(-\varepsilon_a) = f\varepsilon_b^n \exp(-\varepsilon_b). \tag{4.15}$$

Equation (4.15) can be numerically evaluated to find ε_a as a function of ε_b. Figure 4.6 shows that with low values of n, the region with the larger cross-section deforms very little. In contrast, if n is high, there is appreciable deformation in the thicker region, so more overall stretching will have occurred when the thinner region fails. This leads to greater formability.

EXAMPLE PROBLEM 4.3: To ensure that the neck in a tensile bar would occur at the middle of the gauge section, the machinist made the bar with a 0.500-in. diameter in the middle of the gauge section and machined the rest of it to a diameter of 0.505 in. After testing, the diameter away from the neck was 0.470 in. Assume that the stress–strain relation follows the power law, equation (5.3). What was the value of n?

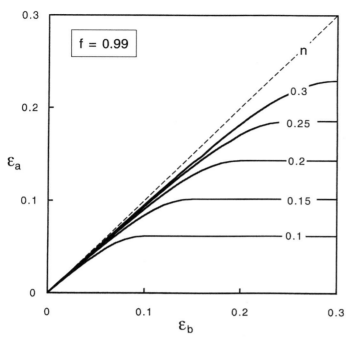

Figure 4.6. The relative strains in two regions of a tensile bar having different initial cross-sectional areas.

Solution: Let the regions with the larger and smaller diameters be designated a and b. When the maximum load is reached, the strain in b is $\varepsilon_b = n$, and the strain in region a is $\varepsilon_a = \ln(A_o/A) = \ln(d_o/d)^2 = 2\ln(0.505/0.470) = 0.1473$. $f = (0.500/0.505)^2 = 0.9803$. Substituting $\varepsilon_b = n$, $e_a = 0.1473$, and $f = 0.98$ into equation (5.15), $0.1437^n\exp(-0.1437) = 0.9803n^n\exp(-n)$. Solving by trial and error, $n = 0.238$. Note that there would have been a large error if the effect of the different initial cross-sections had been ignored and n had been taken as $2\ln(0.505/0.470) = 0.1473$.

Notes

Professor Zdzislaw Marciniak of the Technical University of Warsaw was the first to analyze the effect of a small defect or inhomogeneity on the localization of plastic flow.* His analysis of necking forms the basis for understanding superplastic elongations (Chapter 6) and for calculating forming limit diagrams (Chapter 22). He has been Acting Rector, Senior Professor, and Director of the Institute of Metal Forming at the Technical University of Warsaw and has published a number of books in Polish on metal forming and many papers in the international literature.

One can experience work hardening by holding the ends of a wire in one's fingers and bending it, straightening it, and bending it again. The second time, the bend will occur in a different spot than the first bend. Why?

* Z. Marciniak and K. Kuczynski, *Int. J. Mech. Sci.*, v. 9 (1967), p. 609.

Problems

1. What are the values of K and n in Figure 5.3?

2. The following true stress–true strain data were obtained from a tension test.

Strain	Stress (MPa)	Strain	Stress (MPa)
0.00	0.00	0.10	250.7
0.01	188.8	0.15	270.6
0.02	199.9	0.20	286.5
0.05	223.5		

    A. Plot true stress versus true strain on a logarithmic plot.
    B. What does your plot suggest about n in equation (5.3)?
    C. What does your plot suggest about a better approximation?

3. The true stress–strain curve of a material obeys the power hardening law with $n = 0.18$. A piece of this material was given a tensile strain of $\varepsilon = 0.03$ before being sent to a laboratory for tension testing. The lab workers were unaware of the pre-strain and tried to fit their data to equation (5.3).

    A. What value of n would they report if they determined n from the elongation at maximum load?
    B. What value of n would they report if they determined n from the loads at $\varepsilon = 0.05$ and 0.15?

4. In a tension test, the following values of engineering stress and strain were found: $s = 133.3$ MPa at $e = 0.05$, $s = 155.2$ MPa at $e = 0.10$, and $s = 166.3$ MPa at $e = 0.15$.

    A. Determine whether the data fit equation 4.3.
    B. Predict the strain at necking.

5. Two points on a stress–strain curve for a material are
    $\sigma = 278$ MPa at $\varepsilon = 0.08$ and $\sigma = 322$ MPa at $\varepsilon = 0.16$.

    A. Find K and n in the power law approximation and predict σ at $\varepsilon = 0.20$.
    B. For the approximation $\sigma = K(\varepsilon_0 + \varepsilon)^n$ (equation 5.4) with $\varepsilon_0 = 0.01$, find K and n and predict σ at $\varepsilon = 0.20$.

6. The tensile stress–strain curve of a certain material is best represented by a saturation model, $\sigma = \sigma_0[1 - \exp(-A\varepsilon)]$.

    A. Derive an expression for the true strain at maximum load in terms of the constants A and σ_0.
    B. In a tension test, the maximum load occurred at an engineering strain of $e = 21\%$, and the tensile strength was 350 MPa.

Determine the values of the constants A and σ_0 for the material.
(Remember that the tensile strength is the maximum *engineering* stress.)

7. A material has a stress–strain relation that can be approximated by $\sigma = 150 + 185\varepsilon$. For such a material,

    A. What percent uniform elongation should be expected in a tension test?
    B. What is the material's tensile strength?

8.

    A. Derive expressions for the true strain at the onset of necking if the stress–strain curve is given by $\sigma = K(\varepsilon_o + \varepsilon)^n$ (equation 4.4).

    B. Write an expression for the tensile strength.

9. Consider a tensile specimen made from a material that obeys the power hardening law with $K = 400$ MPa and $n = 0.20$. Assume K is not sensitive to strain rate. One part of the gauge section has an initial cross-sectional area that is 0.99 times that of the rest of the gauge section. What will be the true strain in the larger area after the smaller area necks and reduces to 50% of its original area?

10. Consider a tensile bar that was machined so that most of the gauge section was 1.00 cm in diameter. One short region in the gauge section has a diameter 1/2% less (0.995 cm). Assume the stress–strain curve of the material is described by the power law with $K = 330$ MPa and $n = 0.23$, and the flow stress is not sensitive to strain rate. The bar was pulled in tension well beyond maximum load, and it necked in the reduced section.

    A. Calculate the diameter away from the reduced section.

    B. Suppose that an investigator had not known that the bar initially had a reduced section and had assumed that the bar had a uniform initial diameter of 1.00 cm. Suppose that she measured the diameter away from the neck and had used that to calculate n. What value of n would she have calculated?

11. A tensile bar was machined so that most of the gauge section had a diameter of 0.500 cm. One small part of the gauge section had a diameter 1% smaller (0.495 cm). Assume power law hardening with $n = 0.17$. The bar was pulled until necking occurred.

    A. Calculate the uniform elongation (%) away from the neck.

    B. Compare this with the uniform elongation that would have been found if there were no initial reduced section.

12. Repeated cycles of freezing of water and thawing of ice will cause copper pipes to burst. Water expands about 8.3% when it freezes.

    A. Consider a copper tube as a capped cylinder that cannot lengthen or shorten. If it were filled with water, what would be the circumferential strain in the wall when the water freezes?

    B. How many cycles of freezing/thawing would it take to cause the tube walls to neck? Assume the tube is filled after each thawing. Assume $n = 0.55$.

Plasticity Theory

Introduction

Plasticity theory deals with yielding of materials under complex stress states. It allows one to decide whether a material will yield under a stress state and to determine the shape change that will occur if it does yield. It also allows tensile test data to be used to predict the work hardening during deformation under such complex stress states. These relations are a vital part of computer codes for predicting crashworthiness of automobiles and for designing forming dies.

Yield Criteria

The concern here is to describe mathematically the conditions for yielding under complex stresses. A *yield criterion* is a mathematical expression of the stress states that will cause yielding or plastic flow. The most general form of a yield criterion is

$$f(\sigma_x, \sigma_y, \sigma_z, \tau_{yz}, \tau_{zx}, \tau_{xy}) = C, \tag{5.1}$$

where C is a material constant. For an isotropic material, this can be expressed in terms of principal stresses,

$$f(\sigma_1, \sigma_2, \sigma_3) = C. \tag{5.2}$$

The yielding of most solids is independent of the sign of the stress state. Reversing the signs of the stresses has no effect on whether a material yields. This is consistent with the observation that for most materials, the yield strengths in tension and compression are equal.* Also with most solid materials, it is reasonable to assume that yielding is independent of the level of mean normal stress, σ_m,

$$\sigma_m = (\sigma_1 + \sigma_2 + \sigma_3)/3. \tag{5.3}$$

---

* This may not be true when the loading path is changed during deformation. Directional differences in yielding behavior after prior straining are called the *Bauschinger effect*. This subject is treated in Chapter 18. It is also not true if mechanical twinning is an important deformation mechanism (see Chapter 10). The compressive yield strengths of polymers are generally higher than the tensile yield strengths (see Chapter 20).

It will be shown later that this is equivalent to assuming that plastic deformation causes no volume change. This assumption of constancy of volume is certainly reasonable for crystalline materials that deform by slip and twinning because these mechanisms involve only shear. With slip and twinning, only the shear stresses are important. With this simplification, the yield criteria must be of the form

$$f[(\sigma_2 - \sigma_3), (\sigma_3 - \sigma_1), (\sigma_1 - \sigma_2)] = C. \tag{5.4}$$

In terms of the Mohr's stress circle diagrams, only the sizes of the circles (not their positions) are of importance in determining whether yielding will occur. In three-dimensional stress space (σ_1 vs. σ_2 vs. σ_3), the locus can be represented by a cylinder parallel to the line $\sigma_1 = \sigma_2 = \sigma_3$, as shown in Figure 5.1.

Tresca (maximum shear stress criterion)

The simplest yield criterion is one first proposed by Tresca. It states that yielding will occur when the largest shear stress reaches a critical value. The largest shear stress is $\tau_{max} = (\sigma_{max} - \sigma_{min})/2$, so the Tresca criterion can be expressed as

$$\sigma_{max} - \sigma_{min} = C. \tag{5.5}$$

If the convention is maintained that $\sigma_1 \geq \sigma_2 \geq \sigma_3$, this can be written as

$$\sigma_1 - \sigma_3 = C. \tag{5.6}$$

The constant, C, can be found by considering uniaxial tension. In a tension test, $\sigma_2 = \sigma_3 = 0$, and at yielding $\sigma_1 = Y$, where Y is the yield strength. Substituting into equation (5.6), $C = Y$. Therefore, the Tresca criterion may be expressed as

$$\sigma_1 - \sigma_3 = Y. \tag{5.7}$$

For pure shear, $\sigma_1 = -\sigma_3 = k$, where k is the shear yield strength. Substituting in equation (5.7), $k = Y/2$, so

$$\sigma_1 - \sigma_3 = 2k = C. \tag{5.8}$$

EXAMPLE PROBLEM 5.1: Consider an isotropic material, loaded so that the principal stresses coincide with the x, y, and z axes of the material. Assume that the Tresca yield criterion applies. Make a plot of the combinations of σ_y versus σ_x that will cause yielding with $\sigma_z = 0$.

Figure 5.1. A yield locus is the surface of a body in three-dimensional stress space. Stress states on the locus will cause yielding. Those inside the locus will not cause yielding.

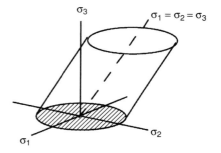

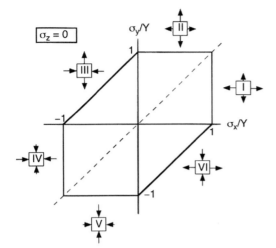

Figure 5.2. Plot of the yield locus for the Tresca criterion for $\sigma_z = 0$. The Tresca criterion predicts that the intermediate principal stress has no effect on yielding. For example, in sector I, the value of σ_y has no effect on the value of σ_x required for yielding. Only if σ_y is negative or if it is higher than σ_x does it have an influence. In these cases, it is no longer the intermediate principal stress.

Solution: Divide the σ_y versus σ_x stress space into six sectors, as shown in Figure 5.2. The following conditions are appropriate:

I $\sigma_x > \sigma_y > \sigma_z = 0$, so $\sigma_1 = \sigma_x, \sigma_3 = \sigma_z = 0$, so $\sigma_x = Y$

II $\sigma_y > \sigma_x > \sigma_z = 0$, so $\sigma_1 = \sigma_y, \sigma_3 = \sigma_z = 0$, so $\sigma_y = Y$

III $\sigma_y > \sigma_z = 0 > \sigma_x$, so $\sigma_1 = \sigma_y, \sigma_3 = \sigma_x$, so $\sigma_y - \sigma_x = Y$

IV $\sigma_z = 0 > \sigma_y > \sigma_x$, so $\sigma_1 = 0, \sigma_3 = \sigma_x, 0 - \sigma_x = Y$

V $\sigma_z = 0 > \sigma_x > \sigma_y$, so $\sigma_1 = 0, \sigma_3 = \sigma_y, 0 - \sigma_y = Y$

VI $\sigma_x > \sigma_z = 0 > \sigma_y$, so $\sigma_1 = \sigma_x, \sigma_3 = \sigma_y$, so $\sigma_x - \sigma_y = Y.$

These are plotted in Figure 5.2.

It seems reasonable to incorporate the effect of the intermediate principal stress into the yield criterion. One might try this by assuming that yielding depends on the average of the diameters of the three Mohr's circles, $[(\sigma_1 - \sigma_2) + (\sigma_2 - \sigma_3) + (\sigma_1 - \sigma_3)]/3$, but the intermediate stress term, σ_2, drops out of the average, $[(\sigma_1 - \sigma_2) + (\sigma_2 - \sigma_3) + (\sigma_1 - \sigma_3)]/3 = (2/3)(\sigma_1 - \sigma_3)$. Therefore, an average diameter criterion reduces to the Tresca criterion.

Von Mises Criterion

The effect of the intermediate principal stress can be included by assuming that yielding depends on the root-mean-square diameter of the three Mohr's circles.[*] This is the von Mises criterion, which can be expressed as

$$\{[(\sigma_2 - \sigma_3)^2 + (\sigma_3 - \sigma_1)^2 + (\sigma_1 - \sigma_2)^2]/3\}^{1/2} = C. \qquad (5.9)$$

Note that each term is squared so the convention, $\sigma_1 \geq \sigma_2 \geq \sigma_3$, is not necessary. Again, the material constant, C, can be evaluated by considering a uniaxial tension test. At yielding, $\sigma_1 = Y$ and $\sigma_2 = \sigma_3 = 0$. If $[0^2 + (-Y)^2 + Y^2]/3 = C^2$, or $C = (2/3)^{1/3}Y$ is substituted, equation (5.9) becomes

$$(\sigma_2 - \sigma_3)^2 + (\sigma_3 - \sigma_1)^2 + (\sigma_1 - \sigma_2)^2 = 2Y^2. \qquad (5.10)$$

---

[*] This is equivalent to assuming that yielding occurs when the elastic distortional strain energy reaches a critical value.

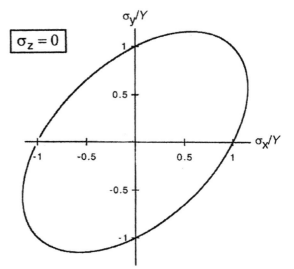

Figure 5.3. The von Mises criterion with $\sigma_z = 0$ plots as an ellipse.

For a state of pure shear, $\sigma_1 = -\sigma_3 = k$ and $\sigma_2 = 0$. Substituting in equation (5.10), $(-k)^2 + [(-k) - k]^2 + k^2 = 2Y^2$, so

$$k = Y/\sqrt{3}. \tag{5.11}$$

If one of the principal stresses is zero (plane–stress conditions, $\sigma_z = 0$), equation (5.10) simplifies to $\sigma_1^2 + \sigma_2^2 - \sigma_1\sigma_2 = Y^2$, which is an ellipse. With further substitution of $\alpha = \sigma_2/\sigma_1$,

$$\sigma_1 = Y/(1 - \alpha + \alpha^2)^{1/2}. \tag{5.12}$$

EXAMPLE PROBLEM 5.2: Consider an isotropic material loaded so that the principal stresses coincide with the x, y, and z axes. Assuming the von Mises yield criterion applies, make a plot of σ_y versus σ_x yield locus with $\sigma_z = 0$.

Solution: Let $\sigma_x = \sigma_1, \sigma_y = \sigma_2$, and $\sigma_z = 0$. Now $a = \sigma_2/\sigma_1$. Figure 5.3 results from substituting several values of a into equation (5.12), solving for σ_x/Y and $\sigma_y/Y = \alpha\sigma_x/Y$, and then plotting.

EXAMPLE PROBLEM 5.3: For plane stress ($\sigma_3 = 0$), what is the largest possible ratio of σ_1/Y at yielding and at what stress ratio, α, does this occur?

Solution: Inspecting equation (5.12), it is clear that maximum ratio of σ_1/Y corresponds to the minimum value $1 - \alpha + \alpha^2$. Differentiating and setting to zero, $d(1 - \alpha + \alpha^2)/d\alpha = -1 + 2\alpha; \alpha = -1/2$. Substituting into equation (5.12), $\sigma_1/Y = [1 - 1/2 + (1/2)^2]^{-1/2} = \sqrt{(4/3)} = 1.155$.

The von Mises yield criterion can also be expressed in terms of stresses that are not principal stresses. In this case, it is necessary to include shear terms,

$$(\sigma_y - \sigma_z)^2 + (\sigma_z - \sigma_x)^2 + (\sigma_x - \sigma_y)^2 + 6\left(\tau_{yz}^2 + \tau_{zx}^2 + \tau_{xy}^2\right) = 2Y^2, \tag{5.13}$$

where x, y, and z are not principal stress axes.

Flow Rules

When a material yields, the ratio of the resulting strains depends on the stress state that causes yielding. The general relations between plastic strains and the stress states are called the *flow rules*. They may be expressed as

$$d\varepsilon_{ij} = d\lambda(\partial f/\partial\sigma_{ij}), \tag{5.14}$$

where f is the yield function corresponding to the yield criterion of concern and $d\lambda$ is a constant that depends on the shape of the stress–strain curve.* For the von Mises criterion, we can write $f = [(\sigma_2 - \sigma_3)^2 + (\sigma_3 - \sigma_1)^2 + (\sigma_1 - \sigma_2)^2]/4$. In this case, equation (5.14) results in

$$d\varepsilon_1 = d\lambda[2(\sigma_1 - \sigma_2) - 2(\sigma_3 - \sigma_1)]/4 = dl \cdot [\sigma_1 - (\sigma_2 + \sigma_3)/2]$$
$$d\varepsilon_2 = d\lambda[\sigma_2 - (\sigma_3 + \sigma_1)/2] \tag{5.15}$$
$$d\varepsilon_3 = d\lambda[\sigma_3 - (\sigma_1 + \sigma_2)/2]^{**}$$

These are known as the Levy-Mises equations. Even though dl is not usually known, these equations are useful for finding the ratio of strains that result from a known stress state or the ratio of stresses that correspond to a known strain state.

> **EXAMPLE PROBLEM 5.4:** Find the ratio of the principal strains that result from yielding if the principal stresses are $\sigma_y = \sigma_x/4, \sigma_z = 0$. Assume the von Mises criterion.
>
> **Solution:** According to equation (5.15),
>
> $$d\varepsilon_1 : d\varepsilon_2 : d\varepsilon_3 = [\sigma_1 - (\sigma_2 + \sigma_3)/2] : [\sigma_2 - (\sigma_3 + \sigma_1)/2] : [\sigma_3 - (\sigma_1 + \sigma_2)/2]$$
> $$= (7/8)\sigma_1 : (-1/4)\sigma_1 : (-5/8)\sigma_1 = 7 : -2 : -5.$$

The flow rules for the Tresca criterion can be found by applying equation (5.14) with $f = \sigma_1 - \sigma_3$. Then $de_1 = d\lambda$, $de_2 = 0$, and $de_3 = -d\lambda$, or

$$d\varepsilon_1 : d\varepsilon_2 : d\varepsilon_3 = 1 : 0 : -1 \tag{5.16}$$

> **EXAMPLE PROBLEM 5.5:** Find the ratio of the principal strains that result from yielding if the principal stresses are $\sigma_y = \sigma_x/4, \sigma_z = 0$, assuming the Tresca criterion.
>
> **Solution:** Here $d\varepsilon_1 = d\varepsilon_x$, $d\varepsilon_2 = d\varepsilon_y$, and $d\varepsilon_3 = d\varepsilon_z$, so $d\varepsilon_x : d\varepsilon_y : d\varepsilon_z = 1 : 0 : -1$

> **EXAMPLE PROBLEM 5.6:** Circles were printed on the surface of a part before it was deformed. Examination after deformation revealed that the principal strains in the sheet are $\varepsilon_1 = 0.18$ and $\varepsilon_2 = 0.078$. Assume that the tools did not touch the surface of concern and that the ratio of stresses remained constant during the deformation. Using the von Mises criterion, find the ratio of the principal stresses.

---

* The constant, $d\lambda$, can be expressed as $d\lambda = d\bar{\varepsilon}/d\bar{\sigma}$, which is the inverse slope of the effective stress–strain curve at the point where the strains are being evaluated.

** Equation (5.15) parallels Hooke's law equations (equation 2.2), where $d\lambda = d\bar{\varepsilon}/d\bar{\sigma}$ replaces $1/E$ and 1/2 replaces Poisson's ratio, υ. For this reason, it is sometimes said that the "plastic Poisson's ratio" is 1/2.

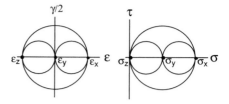

Figure 5.4. Mohr's strain and stress circles for plane–strain.

Solution: First, realize that with a constant ratio of stresses, $d\varepsilon_2 : d\varepsilon_1 = \varepsilon_2 : \varepsilon_1$. Also, if the tools did not touch the surface, it can be assumed that $\sigma_3 = 0$. Then according to equation (5.15), $\varepsilon_2/\varepsilon_1 = d\varepsilon_2/d\varepsilon_1 = (\sigma_2 - \sigma_1/2)/(\sigma_1 - \sigma_2/2)$. Substituting, $\alpha = \sigma_2/\sigma_1$, and

$$\rho = \varepsilon_2/\varepsilon_1, \quad (\alpha - 1/2)/(1 - \alpha/2) = \rho. \text{ Solving for } \alpha,$$

$$\alpha = (\rho + 1/2)/(\rho/2 + 1). \tag{5.17}$$

Now substituting $\rho = 0.078/0.18 = 0.4333$, $\alpha = 0.7671$.

EXAMPLE PROBLEM 5.7: Draw the three-dimensional Mohr's circle diagrams for the stresses and plastic strains in a body loaded under plane–strain tension, $\varepsilon_y = 0$, with $\sigma_z = 0$. Assume the von Mises yield criterion.

Solution: From the flow rules, $\varepsilon_y = 0 = l[\sigma_y - (1/2)(\sigma_x + \sigma_z)]$ with $e_y = 0$ and $\sigma_z = 0$, $\sigma_y = 1/2(\sigma_x)$. The Mohr's stress circles are determined by the principal stresses, σ_x, $\sigma_y = 1/2(\sigma_x)$, and $\sigma_z = 0$. For plastic flow, $\varepsilon_x + \varepsilon_y + \varepsilon_z = 0$, so with $\varepsilon_y = 0$, $\varepsilon_z = -\varepsilon_x$. The Mohr's strain circles (Figure 5.4) are determined by the principal strains, ε_x, $\varepsilon_y = 0$, and $\varepsilon_z = -\varepsilon_x$.

Principle of Normality

The flow rules may be represented by the *principle of normality*. According to this principle, if a normal is constructed to the yield locus at the point of yielding, the strains that result from yielding are in the same ratio as the stress components of the normal. This is illustrated in Figure 5.5. A corollary is that for a σ_1 versus σ_2 yield locus with $\sigma_3 = 0$,

$$d\varepsilon_1/d\varepsilon_2 = -\partial\sigma_2/\partial\sigma_1, \tag{5.18}$$

where $\partial\sigma_2/\partial\sigma_1$ is the slope of the yield locus at the point of yielding. It should be noted that equations (5.14) and (5.17) are general and can be used with other yield criteria, including ones formulated to account for anisotropy and pressure-dependent yielding.

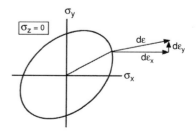

Figure 5.5. The ratios of the strains resulting from yielding are in the same proportion as the components of a vector normal to the yield surface at the point of yielding.

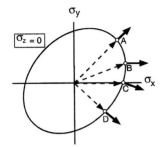

Figure 5.6. The ratio of strains resulting from yielding along several loading paths. At A, $\varepsilon_y = \varepsilon_x$; at B, $\varepsilon_y = 0$; at C (uniaxial tension) $\varepsilon_y = -1/2e_x$; and at D, $\varepsilon_y = -\varepsilon_x$.

Figure 5.6 shows how different shape changes result from different loading paths. The components of the normal at A are $d\sigma_y/d\sigma_x = 1$, so $\varepsilon_y = \varepsilon_x$. At B, the normal has a slope of $d\sigma_y = 0$, so $\varepsilon_y = 0$. For uniaxial tension at C, the slope of the normal is $d\sigma_y/d\sigma_x = -1/2$, which corresponds to $\varepsilon_y = (-1/2)\varepsilon_x$. At D, $d\sigma_y/d\sigma_x = -1$, so $\varepsilon_y = -\varepsilon_x$.

Figure 5.7 is a representation of the normality principle applied to the Tresca criterion. All stress states along a straight edge cause the same ratio of plastic strains. The shape changes corresponding to the corners are ambiguous because it is ambiguous which stress component is $\sigma_{\max}$ and which stress component is $\sigma_{\min}$. For example, with yielding under biaxial tension, $\sigma_x = \sigma_y, 0 \leq \varepsilon_y/\varepsilon_x \leq \infty$.

Effective Stress and Effective Strain

The concepts of effective stress and effective strain are necessary for analyzing the strain hardening that occurs on loading paths other than uniaxial tension. Effective stress, $\bar{\sigma}$, and effective strain, $\bar{\varepsilon}$, are defined so that

1. $\bar{\sigma}$ and $\bar{\varepsilon}$ reduce to σ_x and ε_x in an x-direction tension test.
2. The incremental work per volume done in deforming a material plastically is $dw = \bar{\sigma}d\bar{\varepsilon}$.
3. Furthermore it is usually assumed that the $\bar{\sigma}$ versus $\bar{\varepsilon}$ curve describes the strain hardening for loading under a constant stress ratio, α, regardless of α.*

The effective stress, $\bar{\sigma}$, is the function of the applied stresses that determine whether yielding occurs. When $\bar{\sigma}$ reaches the current flow stress, plastic deformation will occur. For the von Mises criterion,

$$\bar{\sigma} = (1/\sqrt{2})[(\sigma_2 - \sigma_3)^2 + (\sigma_3 - \sigma_1)^2 + (\sigma_1 - \sigma_2)^2]^{1/2}, \qquad (5.19)$$

and for the Tresca criterion,

$$\bar{\sigma} = (\sigma_1 - \sigma_3). \qquad (5.20)$$

Note that in a tension test the effective stress reduces to the tensile stress, so both criteria predict yielding when $\bar{\sigma}$ equals the current flow stress.

---

* It should be understood that in large strains, the $\bar{\sigma}$ versus $\bar{\varepsilon}$ curves actually do depend somewhat on the loading path because of orientation changes within the material. These orientation changes (texture development) depend on the loading path. However, the dependence of strain hardening on loading path is significant only at large strains.

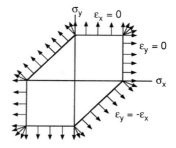

Figure 5.7. Normality principle applied to the Tresca yield criterion. All stress states on the same side of the locus cause the same shape change. The shape changes at the corners are ambiguous. Biaxial tension can produce any shape change from $\varepsilon_y = 0$ to $\varepsilon_x = 0$.

The effective strain, $\bar{\varepsilon}$, is a mathematical function of the strain components defined in such a way that $\bar{\varepsilon}$ reduces to the tensile strain in a tension test and the plastic work per volume is

$$dw = \bar{\sigma}d\bar{\varepsilon} = \sigma_1 d\varepsilon_1 + \sigma_2 d\varepsilon_2 + \sigma_3 d\varepsilon_3. \tag{5.21}$$

For the von Mises criterion, $d\bar{\varepsilon}$, can be expressed as

$$d\bar{\varepsilon} = (\sqrt{2}/3)[(d\varepsilon_2 - d\varepsilon_3)^2 + (d\varepsilon_3 - d\varepsilon_1)^2 + (d\varepsilon_1 - d\varepsilon_2)^2]^{1/2}, \tag{5.22}$$

or as

$$d\bar{\varepsilon} = (2/3)^{1/3}(d\varepsilon_1^2 + d\varepsilon_2^2 + d\varepsilon_3^2)^{1/2}, \tag{5.23}$$

which is completely equivalent.

EXAMPLE PROBLEM 5.8: Show that $d\bar{\varepsilon}$ in equations (5.22) and (5.23) reduce to $d\varepsilon_1$ in a one-direction tension test.

Solution: In a one-direction tension test, $d\varepsilon_2 = d\varepsilon_3 = -(1/2)d\varepsilon_1$. Substituting into equation (5.22),

$$d\bar{\varepsilon} = (\sqrt{2}/3)\{(0)^2 + [-(1/2)d\varepsilon_1 - d\varepsilon_1]^2 + [d\varepsilon_1 - (-1/2)d\varepsilon_1]^2\}^{1/2}$$
$$= (\sqrt{2}/3)[(9/4)d\varepsilon_1^2 + (9/4)d\varepsilon_1^2]^{1/2} = d\varepsilon_1$$

Substituting $d\varepsilon_2 = d\varepsilon_3 = -(1/2)d\varepsilon_1$ into equation (5.22),

$$d\bar{\varepsilon} = \sqrt{(2/3)}[d\varepsilon_1^2 + (-1/2d\varepsilon_1)^2 + (-1/2d\varepsilon_1)^2]^{1/2}$$
$$= \sqrt{(2/3)}[d\varepsilon_1^2 + d\varepsilon_1^2/4 + d\varepsilon_1^2/4]^{1/2} = d\varepsilon_1.$$

If the straining is proportional (constant ratios of $d\varepsilon_1 : d\varepsilon_2 : d\varepsilon_3$), the total effective strain can be expressed as

$$\bar{\varepsilon} = [(2/3)(\varepsilon_1^2 + \varepsilon_2^2 + \varepsilon_3^2)]^{1/2}. \tag{5.24}$$

If the straining is not proportional, $\bar{\varepsilon}$ must be found by integrating $d \gg \varepsilon$ along the strain path.

The effective strain (and stress) may also be expressed in terms of nonprincipal strains (and stresses). For von Mises,

$$\bar{\varepsilon} = [(2/3)(e_x^2 + \varepsilon_y^2 + \varepsilon_z^2) + (1/3)(\gamma_{yz}^2 + \gamma_{zx}^2 + \gamma_{xy}^2)]^{1/2}, \tag{5.25}$$

and

$$\bar{\sigma} = (1/\sqrt{2})[(\sigma_y - \sigma_z)^2 + (\sigma_z - \sigma_x)^2 + (\sigma_x - \sigma_y)^2 + 6(\tau_{yz}^2 + \tau_{zx}^2 + \tau_{xy}^2)]^{1/2}. \tag{5.26}$$

For the Tresca criterion, the effective strain is the absolutely largest principal strain,

$$d\bar{\varepsilon} = |d\varepsilon_i|_{max}. \tag{5.27}$$

Although the Tresca effective strain is not widely used, it is of value because it is extremely simple to find. Furthermore, it is worth noting that the value of the von Mises effective strain can never differ greatly from it:

$$|d\varepsilon_i|_{max} \leq d\bar{\varepsilon}_{Mises} \leq 1.15|d\varepsilon_i|_{max}. \tag{5.28}$$

Equation (5.27) provides a simple check when calculating the von Mises effective strain. If one calculates a value for $d\bar{\varepsilon}_{Mises}$ that does not fall within the limits of equation (5.27), a mistake has been made.

As a material deforms plastically, the level of stress necessary to continue deformation increases. It is postulated that the strain hardening depends only on $\bar{\varepsilon}$. In that case, there is a unique relation,

$$\bar{\sigma} = f(\bar{\varepsilon}). \tag{5.29}$$

Because for a tension test, $\bar{\varepsilon}$ is the tensile strain and $\bar{\sigma}$ is the tensile stress, the $\sigma - \varepsilon$ curve in a tension test is the $\bar{\sigma} - \bar{\varepsilon}$ curve. Therefore, the stress–strain curve in a tension test can be used to predict the stress–strain behavior under other forms of loading.

> **EXAMPLE PROBLEM 5.9:** The strains measured on the surface of a piece of sheet metal after deformation are $\varepsilon_1 = 0.182$, $\varepsilon_2 = -0.035$. The stress–strain curve in tension can be approximated by $\sigma = 30 + 40\varepsilon$. Assume the von Mises criterion, and assume that the loading was such that the ratio of $\varepsilon_2/\varepsilon_1$ was constant. Calculate the levels of ε_1 and ε_2 reached before unloading.
>
> **Solution:** First, find the effective strain. With constant volume, $\varepsilon_3 = -\varepsilon_1 - \varepsilon_2 = -0.182 + 0.035 = -0.147$. The Mises effective strain, $\bar{\varepsilon} = [(2/3)(\varepsilon_1^2 + \varepsilon_2^2 + \varepsilon_3^2)]^{1/2} = [(2/3)(0.182^2 + 0.035^2 + 0.147^2)]^{1/2} = 0.193$. (Note that this is larger than 0.182 and smaller than 1.15×0.182.) Because the tensile stress–strain curve is the effective stress-effective strain relation, $\bar{\sigma} = 30 + 40\bar{\varepsilon} = 30 + 40 \times 0.193 = 37.7$. At the surface, $\sigma_z = 0$, so the effective stress function can be written as $\bar{\sigma}/\sigma_1 = (1/\sqrt{2})[\alpha^2 + 1 + (1 - \alpha)2]^{1/2} = (1 - \alpha + \alpha^2)^{1/2}$. From the flow rules with $\sigma_z = 0$, $\rho = d\varepsilon_2/d\varepsilon_1 = (\alpha - 1/2)/(1 - \alpha/2)$. Solving for α, $\bar{\varepsilon} = (\rho + 1/2)/(1 + \rho/2)$. Substituting $\rho = -0.035/0.182 = -0.192$, $\alpha = (-0.192 + 1/2)/(1 - 0.192/2) = 0.341$, $\bar{\sigma}/\sigma_1 = [1 - \alpha + \alpha^2]^{1/2} = [1 - 0.341 + 0.341^2]^{1/2} = 0.881$. $\sigma_1 = 0.881/\bar{\sigma} = 37.7/0.881 = 42.8$ and $\sigma_2 = \alpha\sigma_1 = 0.341 \times 42.8 = 14.6$.

Other Isotropic Yield Criteria

Von Mises and Tresca are not the only possible isotropic criteria. Both experimental data and theoretical analysis based on a crystallographic model tend to lie between

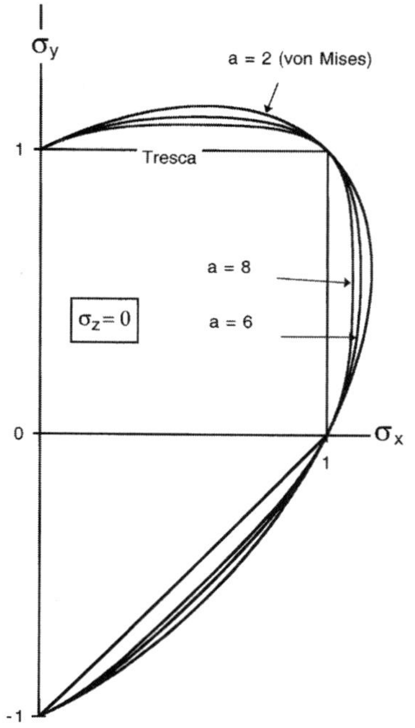

Figure 5.8. Yield loci for $(\sigma_2 - \sigma_3)^a + (\sigma_3 - \sigma_1)^a + (\sigma_1 - \sigma_2)^a = 2Y^a$ with several values of a. Note that the von Mises criterion corresponds to a = 2 and the Tresca criterion to a = 1.

the two and can be represented by[*]

$$|\sigma_2 - \sigma_3|^a + |\sigma_3 - \sigma_1|^a + |\sigma_1 - \sigma_2|^a = 2Y^a. \tag{5.30}$$

This criterion reduces to von Mises for a = 2 and a = 4, and to Tresca for a = 1 and a → ∞. For values of the exponent greater than 4, this criterion predicts yield loci between Tresca and von Mises, as shown in Figure 5.8. If the exponent, a, is an even integer, this criterion can be written without the absolute magnitude signs as

$$(\sigma_2 - \sigma_3)^a + (\sigma_3 - \sigma_1)^a + (\sigma_1 - \sigma_2)^a = 2Y^a. \tag{5.31}$$

Theoretical calculations based on {111}<110> slip suggest an exponent of a = 8 for fcc metals. Similar calculations suggest that a = 6 for bcc. These values fit experimental data as well.

EXAMPLE PROBLEM 5.10: Derive the flow rules for the high exponent yield criterion, using equation (5.29).

Solution: Expressing the yield function as $f = (\sigma_y - \sigma_z)^a + (\sigma_z - \sigma_x)^a + (\sigma_x - \sigma_y)^a$, the flow rules can be found from $de_{ij} = d\lambda(df/d\sigma_{ij})$.

$$d\varepsilon_x = d\lambda[a(\sigma_x - \sigma_z)^{a-1} + a(\sigma_x - \sigma_y)^{a-1}],$$
$$d\varepsilon_y = d\lambda[a(\sigma_y - \sigma_x)^{a-1} + a(\sigma_y - \sigma_z)^{a-1}],$$

and

$$d\varepsilon_z = d\lambda[a(\sigma_z - \sigma_x)^{a-1} + a(\sigma_z - \sigma_y)^{a-1}].$$

* W. F. Hosford, *J. Appl. Mech. (Trans. ASME ser E.)*, v. 39E (1972).

These can be expressed as

$$d\varepsilon_x : d\varepsilon_y : d\varepsilon_z = [(\sigma_x - \sigma_z)^{a-1} + (\sigma_x - \sigma_y)^{a-1}] : [(\sigma_y - \sigma_x)^{a-1} + (\sigma_y - \sigma_z)^{a-1}] :$$
$$[(\sigma_z - \sigma_x)^{a-1} + (\sigma_z - \sigma_y)^{a-1}]. \tag{5.32}$$

Anisotropic Plasticity

Although it is frequently assumed that materials are isotropic (have the same properties in all directions), they rarely are. There are two main causes of anisotropy. One cause is *preferred orientations* of grains, or *crystallographic texture*. The second is *mechanical fibering*, which is the elongation and alignment of microstructural features such as inclusions and grain boundaries. Anisotropy of plastic behavior is almost entirely caused by the presence of preferred orientations.

The first complete quantitative treatment of plastic anisotropy was in 1948 by Hill[*] who proposed an anisotropic yield criterion to accommodate such materials. It is a generalization of the von Mises criterion:

$$F(\sigma_y - \sigma_z)^2 + G(\sigma_z - \sigma_x)^2 + H(\sigma_x - \sigma_y)^2 + 2L\tau_{yz}^2 + 2M\tau_{zx}^2 + 2N\tau_{xy}^2 = 1, \tag{5.33}$$

where the axes x, y, and z are the symmetry axes of the material (e.g., the rolling, transverse, and through-thickness directions of a rolled sheet).

If the loading is such that the directions of principal stress coincide with the symmetry axes and if there is planar isotropy (properties do not vary with direction in the x-y plane), equation (5.32) can be simplified to

$$(\sigma_y - \sigma_z)^2 + (\sigma_z - \sigma_x)^2 + R(\sigma_x - \sigma_y)^2 = (R+1)X^2, \tag{5.34}$$

where X is the yield strength in uniaxial tension.

For plane stress ($\sigma_z = 0$), equation (5.34) plots as an ellipse, as shown in Figure 5.9. The higher the value of R, the more the ellipse extends into the first quadrant. Thus, the strength under biaxial tension increases with R as suggested earlier.

The corresponding flow rules are obtained by applying the general equation, $d\varepsilon_{ij} = dl(\partial f/\partial \sigma_{ij})$ (equation 5.14), where now $f(\sigma_{ij})$ is given by equation (5.33),

$$d\varepsilon_x : d\varepsilon_y : d\varepsilon_z = [(R+1)\sigma_x - R\sigma_y - \sigma_z] : [(R+1)\sigma_y - R\sigma_x - \sigma_z] : [2\sigma_z - \sigma_y - \sigma_x]. \tag{5.35}$$

This means that $\rho = \varepsilon_y/\varepsilon_x$ for $\sigma_z = 0$ and $\alpha = \sigma_y/\sigma_x$ for $\sigma_z = 0$ are related by

$$\rho = [(R+1)\alpha - R]/[(R+1) - R\alpha] \tag{5.36}$$

and

$$\alpha = [(R+1)\rho + R]/[(R+1) + R\rho]. \tag{5.37}$$

Note that if $R = 1$, these reduce to equation (5.15), and the effective stress is

$$\bar{\sigma} = \{[(\sigma_y - \sigma_z)^2 + (\sigma_z - \sigma_x)^2 + R(\sigma_x - \sigma_y)^2]/(R+1\}^{1/2}. \tag{5.38}$$

For $\sigma_z = 0$,

$$\bar{\sigma}/\sigma_x = \{[\alpha^2 + 1 + R(1 - \alpha)^2]/(R+1\}^{1/2}. \tag{5.39}$$

---

[*] R. Hill, *Proc Roy. Soc.*, v. 193A (1948).

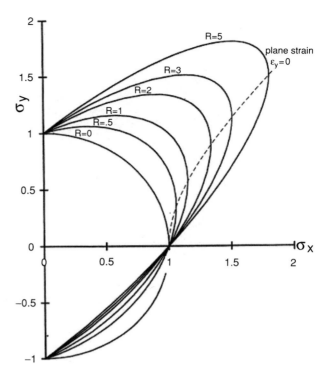

Figure 5.9. Plane stress ($\sigma_z = 0$) yield locus predicted by the Hill 1948 yield criterion for planar isotropy (equation 5.34) with several values of R. The dashed line is the locus of stress states that produce plane strain ($\varepsilon_y = 0$). Note that the strength under biaxial tension increases with R. A high R indicates a resistance to thinning in a tension test, which is consistent with high strength under biaxial tension where thinning must occur. From W. F. Hosford, *The Mechanics of Crystals and Textured Polycrystals.* Used by permission of Oxford University Press (1993).

The effective strain function is given

$$\bar{\varepsilon}/\varepsilon_x = (\sigma_1/\bar{\sigma})(1 + \alpha\rho). \tag{5.40}$$

The Hill criterion often overestimates the effect of R value on the flow stress. A modification of equation (5.34), referred to as the high-exponent criterion, was suggested to overcome this difficulty,[*]

$$(\sigma_y - \sigma_z)^a + (\sigma_z - \sigma_x)^a + R(\sigma_x - \sigma_y)^a = (R+1)X^a, \tag{5.41}$$

where a is an even exponent much higher than 2. Calculations based on crystallographic slip have suggested that a $= 6$ is appropriate for bcc metals and a $= 8$ for fcc metals. Figure 5.10 compares the yield loci predicted by this criterion and the Hill criterion for several levels of R.

With this criterion, the flow rules are

$$d\varepsilon_x : d\varepsilon_y : d\varepsilon_z = [R(\sigma_x - \sigma_y)^{a-1} + (\sigma_x - \sigma_z)^{a-1}] : [(\sigma_y - \sigma_z)^{a-1} + R(\sigma_y - \sigma_x)^{a-1}] :$$
$$[(\sigma_z - \sigma_x)^{a-1} + (\sigma_z - \sigma_y)^{a-1}]. \tag{5.42}$$

The effective stress function corresponding to equation (5.29) is

$$\bar{\sigma} = [(\sigma_y - \sigma_z)^a + (\sigma_z - \sigma_x)^a + R(\sigma_x - \sigma_y)^a]/(R+1)\}^{1/a}. \tag{5.43}$$

The effective strain function is

$$\bar{\varepsilon} = (\sigma_x/\bar{\sigma})(1 + \alpha\rho), \tag{5.44}$$

where $\alpha = (\sigma_y - \sigma_z)/(\sigma_x - \sigma_y)$ and $\rho = \varepsilon_y/\varepsilon_x$.

[*] W. F. Hosford, *7th North Amer. Metalworking Conf., SME*, Dearborn, MI (1979), and R. W. Logan and W. F. Hosford, *Int. J. Mech. Sci.*, v. 22 (1980).

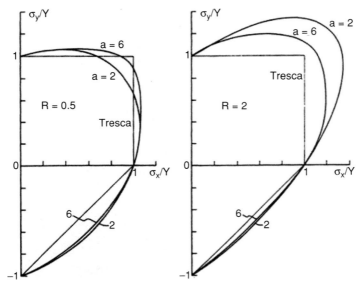

Figure 5.10. Yield loci predicted by the high-exponent criterion for planar isotropy. The loci for the Hill criterion correspond to a = 2. Note that with a high exponent, there is much less effect of R on strength under biaxial tension. From W. F. Hosford, *Ibid*.

Although the nonlinear flow rules for the high-exponent criterion (equation 5.42) cannot be explicitly solved for the stresses, iterative solutions with calculators or personal computers are simple. The nonquadratic yield criterion and accompanying flow rules (equations 5.41 and 5.42) have been shown to give better fit to experimental data than the quadratic form (equations 5.34 and 5.35).

EXAMPLE PROBLEM 5.11: A steel sheet that has an R value of 1.75 in all directions in the sheet was stretched in biaxial tension with $\sigma_z = 0$. Strain measurements indicate that throughout the deformation, $\varepsilon_y = 0$. Find the stress ratio, $\alpha = \sigma_y/\sigma_x$, that prevailed according to the Hill yield criterion (equation 5.33) and and with the nonquadratic yield criterion (equation 5.40) with a = 8.

Solution: For the Hill criterion, equation (5.36) gives $\alpha = [(R+1)\rho + R]/[R+1+R\rho]$. With $\rho = 0$, $\alpha = R/(R+1) = 1.75/2.75 = 0.636$.

For the nonquadratic criterion, the flow rules (equation (5.41)) with $\sigma_z = 0$ and $\rho = 0$ can be expressed as $0 = d\varepsilon_y/d\varepsilon_x = [\alpha^{a-1} + R(\alpha-1)^{a-1}]/[R(1-\alpha)^{a-1} + 1] = [\alpha^7 - 1.75(1-\alpha)^7]/[1.75(1-\alpha)^7 + 1]$. Trial and error solution gives $\alpha = 0.520$.

Effect of Strain Hardening on the Yield Locus

According to the *isotropic hardening* model, the effect of strain hardening is simply to expand the yield locus without changing its shape. The stresses for yielding are increased by the same factor along all loading paths. This is the basic assumption that $\bar{\sigma} = f(\bar{\varepsilon})$ (equation 5.28). The isotropic hardening model can be applied to anisotropic materials. It does not imply that the material is isotropic.

An alternative model is *kinematic hardening*. According to this model, plastic deformation simply shifts the yield locus in the direction of the loading path

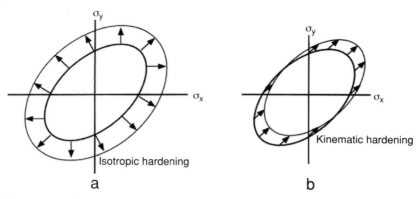

Figure 5.11. Effect of strain hardening on the yield locus. The isotropic model (a) predicts an expansion of the locus. The kinematic hardening model (b) predicts a translation of the locus in the direction of the loading path.

without changing its shape or size. If the shift is large enough, unloading may actually cause plastic deformation. The kinematic model is probably better for describing small strains after a change in load path. However, the isotropic model is better for describing behavior during large strains after a change of strain path. Figure 5.11 illustrates both models.

REFERENCES

W. F. Hosford and R. M. Caddell, *Metal Forming: Mechanics and Metallurgy*, 2nd ed., Prentice Hall (1993).
R. Hill, *The Mathematical Theory of Plasticity*, Oxford (1950).
W. A. Backofen, *Deformation Processing*, Addison-Wesley (1972).

Notes

Otto Z. Mohr (1835–1918) worked as a civil engineer, designing bridges. At 32, he was appointed a professor of engineering mechanics at Stuttgart Polytecknium. Among other contributions, he devised the graphical method of analyzing the stress at a point. He then extended Coulomb's idea that failure is caused by shear stresses into a failure criterion based on maximum shear stress, or diameter of the largest circle. For cast iron, he proposed that the different fracture stresses in tension, shear, and compression could be combined into a single diagram in which the tangents form an envelope of safe stress combinations.

This is essentially the Tresca yield criterion. It may be noted that early workers used the term *failure criteria*, which failed to distinguish between fracture and yielding.

Lamé assumed a maximum stress theory of failure. However, later a maximum strain theory of which Poncelet and Saint-Venant were proponents became generally accepted. They proposed that failure would occur when a critical strain was reached, regardless of the stress state.

In letters to William Thompson, John Clerk Maxwell (1831–1879) proposed that *strain energy of distortion* was critical, but he never published this idea, and it was forgotten. M. T. Huber, in 1904, first formulated the expression for "distortional strain energy," $U = [1/(12G)][(\sigma_2 - \sigma_3)^2 + (\sigma_3 - \sigma_1)^2 + (\sigma_1 - \sigma_2)^2]$, where $U = \sigma_{yp}^2/(6G)$. The same idea was independently developed by von Mises (*Göttinger. Nachr.*, [1913], p. 582) for whom the criterion is generally called. It is also referred to by the names of several people who independently proposed it: Huber, Hencky, as well as Maxwell. It is also known as the *maximum distortional energy* theory and the *octahedral shear stress* theory. The first name reflects that the elastic energy in an isotropic material, associated with shear (in contrast to dilatation) is proportional to the left-hand side of equation (5.10). The second name reflects that the shear terms, $(\sigma_2 - \sigma_3)$, $(\sigma_3 - \sigma_1)$, and $(\sigma_1 - \sigma_2)$, can be represented as the edges of an octahedron in principal stress space.

In 1868, Tresca presented two notes to the French Academy. From these, Saint-Venant established the first theory of plasticity based on the assumptions that

1. Plastic deformation does not change the volume of a material.
2. Directions of principal stresses and principal strains coincide.
3. The maximum shear stress at a point is a constant.

The Tresca criterion is also called the *Guest* or the *maximum shear stress* criterion.

Problems

1. For the von Mises yield criterion with $\sigma_z = 0$, calculate the values of σ_x/Y at yielding if
   a. $\alpha = 1/2$ b. $\alpha = 1$ c. $\alpha = -1$ d. $\alpha = 0$,   where $\alpha = \sigma_y/\sigma_x$.

2. For each of the values of α in problem 1, calculate the ratio $\rho = d\varepsilon_y d\varepsilon_x$.

3. Repeat problems 1 and 2 assuming the Tresca criterion instead of the Mises criterion.

4. Repeat problems 1 and 2 assuming the following yield criterion:

$$(\sigma_2 - \sigma_3)^a + (\sigma_3 - \sigma_1)^a + (\sigma_1 - \sigma_2)^a = 2Y^a, \text{ where a } = 8.$$

5. Consider a plane–strain tension test (Figure 5.12) in which the tensile stress is applied along the x-direction. The strain, ε_y, is zero along the transverse direction, and the stress in the z-direction vanishes. Assuming the von Mises criterion, write expressions for

   A. $\bar{\sigma}$ as a function of σ_x, and
   B. $d\bar{\varepsilon}$ as a function of $d\varepsilon_x$.

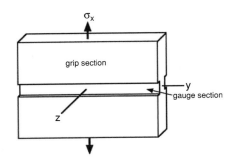

Figure 5.12. Plane–strain tensile specimen. Lateral contraction of material in the groove is constrained by material outside the groove.

C. Using the results of parts A and B, write an expression for the incremental work per volume, dw, in terms of σ_x and $d\varepsilon_x$.

D. Derive an expression for σ_x as a function of ε_x in such a test, assuming that the strain hardening can be expressed by $\bar{\sigma} = K\bar{\varepsilon}^n$.

6. A 1.00-cm-diameter circle was printed onto the surface of a sheet of steel before forming. After forming, the circle was found to be an ellipse with major and minor diameters of 1.18 cm and 1.03 cm, respectively. Assume that both sets of measurements were made when the sheet was unloaded; that during forming the stress perpendicular to the sheet surface was zero; and that the ratio, α, of the stresses in the plane of the sheet remained constant during forming. The tensile stress–strain curve for this steel is shown in Figure 5.13. Assume the von Mises yield criterion.

A. What were the principal strains, ε_1, ε_2, and ε_3?
B. What was the effective strain, $\bar{\varepsilon}$?
C. What was the effective stress, $\bar{\sigma}$?
D. Calculate the ratio, $\rho = \varepsilon_2/\varepsilon_1$, and use this to find the ratio, $\alpha = \sigma_2/\sigma_1$. (Take σ_1 and σ_2, respectively, as the larger and smaller of the principal stresses in the plane of the sheet.)
E. What was the level of σ_1?

Figure 5.13. True tensile stress–strain curve for the steel in problem 6.

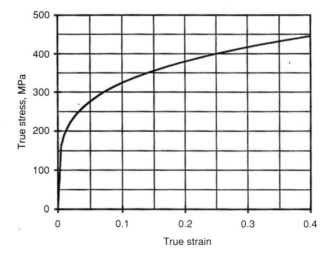

7. Measurements on the surface of a deformed sheet after unloading indicate that $e_1 = 0.154$ and $e_2 = 0.070$. Assume that the von Mises criterion is appropriate and that the loading was proportional (i.e., the ratio, $\alpha = \sigma_y/\sigma_x$, remained constant during loading). It has been found that the tensile stress–strain relationship for this alloy can be approximated by $\sigma = 150 + 185\varepsilon$ (Figure 5.14), where σ is the true stress in MPa and ε is the true strain.

A. What was the effective strain?
B. What was the effective stress?
C. What was the value of the largest principal stress?

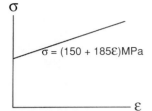

$\sigma = (150 + 185\varepsilon)$MPa   Figure 5.14. True tensile stress–strain curve for the steel in problem 7.

8. The following yield criterion has been proposed for an isotropic material: "yielding will occur when the sum of the diameters of the largest and second largest Mohr's circles reaches a critical value." Defining $\sigma_1 \geq \sigma_2 \geq \sigma_3$, this can be expressed mathematically as

$$\text{if } (\sigma_1 - \sigma_2) \geq (\sigma_2 - \sigma_3), (\sigma_1 - \sigma_2) + (\sigma s_1 - \sigma_3) = C \text{ or } 2\sigma_1 - \sigma_2 - \sigma_3 = C, \quad (1)$$

$$\text{but if } (\sigma_2 - \sigma_3) \geq (\sigma_1 - \sigma_2), (\sigma_1 - \sigma_3) + (\sigma_2 - \sigma_3) = C \text{ or } \sigma_1 + \sigma_2 - 2\sigma_3 = C. \quad (2)$$

Evaluate C in terms of the yield strength, Y, in uniaxial tension or the yield strength, $-Y$, in compression.

Plot the yield locus as σ_x versus σ_y for $\sigma_z = 0$, where σ_x, σ_y, and σ_z are principal stresses. (*Hint*: For each region in σ_x versus σ_y stress space, determine whether (1) or (2) applies.)

9. Consider a long thin-wall tube, capped at both ends. It is made from a steel with a yield strength of 40,000 psi. Its length is 60 in., its diameter is 2.0 in., and the wall thickness is 0.015 in. The tube is loaded under an internal pressure, P, and a torque of 1,500 in.-lb is applied.

    A. What internal pressure can it withstand without yielding according to Tresca?

    B. What internal pressure can it withstand without yielding according to von Mises?

10. In flat rolling of a sheet or plate, the width does not appreciably change. A sheet of aluminum is rolled from 0.050 in. to 0.025 in. thickness. Assume the von Mises criterion.

    A. What is the effective strain, $\bar{\varepsilon}$, caused by the rolling?

    B. What strain in a tension test (if it were possible) would cause the same amount of strain hardening?

11. A piece of ontarium (which has a tensile yield strength of $Y = 700$ MPa) was loaded in such a way that the principal stresses, σ_x, σ_y, and σ_z, were in the ratio of $1 : 0 : -0.25$. The stresses were increased until plastic deformation occurred.

    A. Predict the ratio of the principal strains, $\rho = \varepsilon_y / \varepsilon_x$, resulting from yielding according to von Mises.

    B. Predict the value of $\rho = \varepsilon_y / \varepsilon_x$ resulting from yielding according to Tresca.

    C. Predict the value of σ_x when yielding occurred according to von Mises.

    D. Predict the value of σ_x when yielding occurred according to Tresca.

12. A new yield criterion has been proposed for isotropic materials. It states that yielding will occur when the diameter of Mohr's largest circle plus half of the diameter of Mohr's second largest circle equals a critical value. This criterion can be expressed mathematically, following the convention that $\sigma_1 \geq \sigma_2 \geq \sigma_3$, as

$$(\sigma_1 - \sigma_3) + 1/2(\sigma_1 - \sigma_2) = C \quad \text{if } (\sigma_1 - \sigma_2) \geq (\sigma_2 - \sigma_3) \text{ and}$$
$$(\sigma_1 - \sigma_3) + 1/2(\sigma_2 - \sigma_3) = C \quad \text{if } (\sigma_2 - \sigma_3) \geq (\sigma_1 - \sigma_2).$$

A. Evaluate C in terms of the tensile (or compressive) yield strength, Y.
B. Let x, y, and z be directions of principal stress, and let $\sigma_z = 0$.

Plot the σ_y versus σ_x yield locus. (That is, plot the values of σ_y/Y and σ_x/Y that will lead to yielding according to this criterion.)
(*Hint:* Consider different loading paths (ratios of σ_y/σ_x), and for each, decide which stress ($\sigma_1, \sigma_2,$ or σ_3) corresponds to σ_x, σ_y or $\sigma_z = 0$, then determine whether ($\sigma_1 - \sigma_2$) $\geq (\sigma_2 - \sigma_3)$, substitute s_x, s_y, and 0 into the appropriate expression, solve, and finally plot.)

13. The tensile yield strength of an aluminum alloy is 14,500 psi. A sheet of this alloy is loaded under plane–stress conditions ($\sigma_3 = 0$) until it yields. On unloading, it is observed that $\varepsilon_1 = 2\varepsilon_2$ and both ε_1 and ε_2 are positive.

A. Assuming the von Mises yield criterion, determine the values of σ_1 and σ_2 at yielding.
B. Sketch the yield locus, and show where the stress state is located on the locus.

14. Consider a capped thin-wall cylindrical pressure vessel, made from a material with planar isotropy and loaded to yielding under internal pressure. Predict the ratio of axial to hoop strains, $\rho = \varepsilon_a/\varepsilon_h$, as a function of R, using

A. The Hill criterion and its flow rules (equations 5.34 and 5.35);
B. The high exponent criterion and its flow rules (equations 5.41 and 5.42).

15. In a tension test of an anisotropic sheet, the ratio of the width strain to the thickness strain, $\varepsilon_w/\varepsilon_t$, is R.

A. Express the ratio, $\varepsilon_2/\varepsilon_1$, of the strains in the plane of the sheet in terms of R. Take the 1-direction as the rolling direction, the 2-direction as the width direction in the tension test, and the 3-direction as the thickness direction.
B. There is a direction, x, in the plane of the sheet along which $\varepsilon_x = 0$. Find the angle, θ, between x and the tensile axis.
C. How accurately would this angle have to be measured to distinguish between two materials having R values of 1.6 and 1.4?

16. Redo problem 6, assuming the Tresca criterion instead of the von Mises criterion.

17. The total volume of a foamed material decreases when it plastically deforms in tension.

A. What does this imply about the effect of $\sigma_H = (\sigma_1 + \sigma_2 + \sigma_3)/3$ on the shape of the yield surface in $\sigma_1, \sigma_2, \sigma_3$ space?
B. Would the absolute magnitude of the yield stress in compression be greater, smaller, or the same as the yield strength in tension?
C. When it yields in compression, would the volume increase, decrease, or remain constant?

6 Strain Rate and Temperature Dependence of Flow Stress

Introduction

For most materials, an increase of strain rate raises the flow stress. The amount of the effect depends on the material and the temperature. In most metallic materials, the effect near room temperature is small and often neglected. A factor of 10 increase of strain rate may raise the level of the stress–strain curve by only 1% or 2%. However, at elevated temperatures, the effect of strain rate on flow stress is much greater. Increasing the strain rate by a factor of 10 may raise the stress–strain curve by 50% or more.

Strain localization occurs very slowly in materials that have a high strain rate dependence because less strained regions continue to deform. Under certain conditions, the rate dependence is high enough for materials to behave *superplastically*. Tensile elongations of 1,000% are possible.

There is a close coupling of the effects of temperature and strain rate on the flow stress. Increased temperatures have the same effects as deceased temperatures. This coupling can be understood in terms of the Arrhenius rate equation.

Strain Rate Dependence of Flow Stress

The average strain rate during most tensile tests is in the range of 10^{-3} to 10^{-2}/s. If it takes 5 minutes during the tensile test to reach a strain of 0.3, the average strain rate is $\dot{\varepsilon} = 0.3/(5 \times 60) = 10^{-3}$/s. At a strain rate of $\dot{\varepsilon} = 10^{-2}$/s, a strain of 0.3 will occur in 30 seconds. For many materials, the effect of the strain rate on the flow stress, σ, at a fixed strain and temperature can be described by a power-law expression,

$$\sigma = C\dot{\varepsilon}^{\mathrm{m}}. \tag{6.1}$$

The exponent, m, is called the *strain-rate sensitivity*. The relative levels of stress at two strain rates (measured at the same total strain) is given by

$$\sigma_2/\sigma_1 = (\dot{\varepsilon}_2/\dot{\varepsilon}_1)^{\mathrm{m}}, \tag{6.2}$$

or $\ln(\sigma_2/\sigma_1) = m \ln(\dot{\varepsilon}_2/\dot{\varepsilon}_1)$. If σ_2 is not much greater than σ_1,

$$\ln(\sigma_2/\sigma_1) \approx \Delta\sigma/\sigma. \tag{6.3}$$

Table 6.1. *Typical values of the strain rate exponent, m, at room temperature**

Material	m
Low-carbon steels	0.010–0.015
HSLA steels	0.005–0.010
Austenitic stainless steels	−0.005–0.005
Ferritic stainless steels	0.010–0.015
Copper	0.005
70/30 brass	−0.005–0
Aluminum alloys	−0.005–0.005
A-titanium alloys	0.01–0.02
Zinc alloys	0.05–0.08

* From various sources.

Equation (6.2) can be simplified to

$$\Delta\sigma/\sigma \approx m\ln(\dot{\varepsilon}_2/\dot{\varepsilon}_1) = 2.3\,m\log(\dot{\varepsilon}_2/\dot{\varepsilon}_1). \qquad (6.4)$$

At room temperature, the values of m for most engineering metals are between −0.005 and 0.015, as shown in Table 6.1.

Consider the effect of a ten-fold increase in strain rate, $(\dot{\varepsilon}_2/\dot{\varepsilon}_1 = 10)$, with m = 0.01. Equation (6.4) predicts that the level of the stress increases by only $\Delta\sigma/\sigma = 2.3(0.01)(1) = 2.3\%$. This increase is typical of room temperature tensile testing. It is so small that the effect of strain rate is often ignored. A plot of equation (6.2) in Figure 6.1 shows how the relative flow stress depends on strain rate for several levels of m. The increase of flow stress, $\Delta\sigma/\sigma$, is small unless either m or $(\dot{\varepsilon}_2/\dot{\varepsilon}_1)$ is high.

Figure 6.1. The dependence of flow stress on strain rate for several values of the strain rate sensitivity, m, according to equation (6.2). From W. F. Hosford and R. M. Caddell, *Metal Forming: Mechanics and Metallurgy*, 3nd ed., Cambridge University Press (2007).

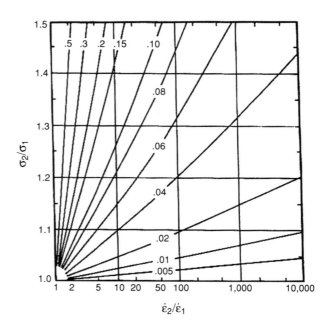

EXAMPLE PROBLEM 6.1: The strain rate dependence of a zinc alloy can be represented by equation (6.1) with m = 0.07. What is the ratio of the flow stress at $\varepsilon = 0.10$ for a strain rate of 10^3/s to that at $\varepsilon = 0.10$ for a strain rate of 10^{-3}/s? Repeat for a low-carbon steel with m = 0.01.

Solution: For zinc (m = 0.07), $\sigma_2/\sigma_1 = (C\dot{\varepsilon}_2{}^m)/(C'\dot{\varepsilon}^m) = (\dot{\varepsilon}_2/\dot{\varepsilon}_1)^m = (10^3/10^{-3})^{0.07} = (10^6)^{0.07} = 2.63$. For steel (m = 0.01), $\sigma_2/\sigma_1 = (10^6)^{0.01} = 1.15$.

EXAMPLE PROBLEM 6.2: The tensile stress in one region of an HSLA steel sheet (m = 0.005) is 1% greater than in another region. What is the ratio of the strain rates in the two regions? Neglect strain hardening. What would be the ratio of the strain rates in the two regions for a titanium alloy (m = 0.02)?

Solution: Using equation (6.2), $\dot{\varepsilon}_2/\dot{\varepsilon}_1 = (\sigma_2/\sigma_1)^{1/m} = (1.01)^{1/0.005} = 7.3$. If m = 0.02, $\dot{\varepsilon}_2/\dot{\varepsilon}_1 = 1.64$. The difference between the strain rates differently stressed locations decreases with increasing values of m.

Figure 6.2 illustrates two ways of determining the value of m. One method is to run two continuous tensile tests at different strain rates and compare the levels of stress at the same fixed strain. The other way is to change the strain rate suddenly during a test and compare the levels of stress immediately before and after the change. The latter method is easier and therefore more common. The two methods may give somewhat different values for m. In both cases, m is found from after the change compared (right), and equation (6.5) can be used to find m:

$$m = \ln(\sigma_2/\sigma_1)/\ln(\dot{\varepsilon}_2/\dot{\varepsilon}_1). \tag{6.5}$$

In rate change tests, $(\dot{\varepsilon}_2/\dot{\varepsilon}_1)$ is typically 10 or 100.

For most metals, the value of the rate sensitivity, m, is low near room temperature but increases with temperature. The increase of m with temperature is quite rapid above half of the melting point ($T > T_m/2$) on an absolute temperature scale. In some cases, m may be 0.5 or higher. Figures 6.3 and 6.4 show the temperature dependence of m for several metals. For some alloys, there is a minimum between $0.2T_m$ and $0.3T_m$. For aluminum alloy 2024, the rate sensitivity is slightly negative in this temperature range.

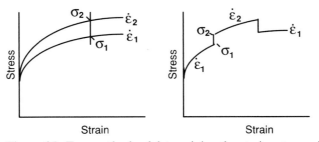

Figure 6.2. Two methods of determining the strain rate sensitivity. Either continuous stress–strain curves at different strain rates can be compared at the same strain (*left*) or sudden changes of strain rate can be made, and the stress levels just before and just after the change compared (*right*). In both cases, equation (6.5) can be used to find m. In rate change tests, $(\dot{\varepsilon}_2/\dot{\varepsilon}_1)$ is typically 10 or 100.

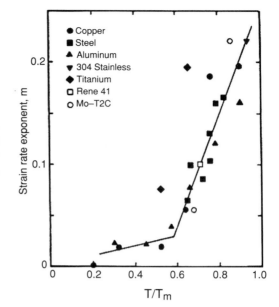

Figure 6.3. Variation of the strain rate sensitivity, m, with temperature for several metals. Above about half of the melting point, m rises rapidly with temperature. From W. F. Hosford and R. M. Caddell, *Ibid*.

Superplasticity

If the rate sensitivity, m, of a material is 0.5 or higher, the material will behave *superplastically*, exhibiting very high tensile elongations. The high elongations occur because the necks are extremely gradual (like those that one observes when chewing gum is stretched). Superplasticity permits forming of parts that require very high strains. The conditions for superplasticity are:

1. Temperatures equal to or above the half of the absolute melting point ($T \geq 0.5T_m$).

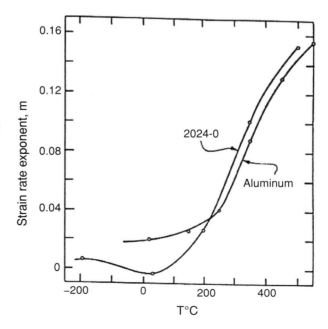

Figure 6.4. Temperature dependence of m for pure aluminum and aluminum alloy 2024. Note that at room temperature m is negative for 2024. From D. S. Fields and W. A. Backofen, *Trans ASM*, v. 51 (1959).

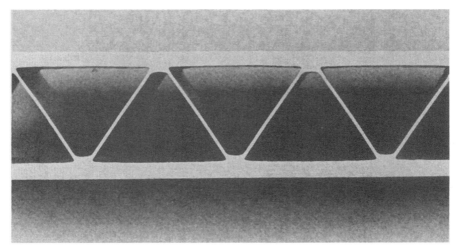

Figure 6.5. A titanium-alloy aircraft panel made by diffusion bonding and superplastic deformation. Three sheets were diffusion bonded at a few locations, and then internal pressurization of the unbonded channels caused the middle sheet to stretch. Superplastic behavior
is required because of the very high uniform extension required in the middle sheet. From
W. F. Hosford and R. M. Caddell, *Ibid.*

  2. Slow strain rates* (usually 10^{-3}/s or slower).
  3. Very fine grain size (grain diameters of a few micrometers or less).

Long times are needed to form useful shapes at low strain rates. With long times
and high temperatures, grain growth may occur, negating an initially fine grain size.
For this reason, most superplastic alloys either have two-phase structures or have a
very fine dispersion of insoluble particles. Both minimize grain growth.

Under superplastic conditions, flow stresses are very low, and extremely high
elongations (1,000% or more) are observed in tension tests. Both the low flow stress
and high elongations can be useful in metal forming. The very low flow stresses
permit slow forging of large, intricate parts with fine detail. The very high tensile
elongation makes it possible to form very deep parts with simple tooling. Examples
are shown in Figures 6.5 and 6.6.

The reason that very high tensile elongations are possible can be understood in
terms of the behavior of a tensile bar having a region with a slightly reduced cross-
sectional area (Figure 6.7). The following treatment is similar to the treatment in
Chapter 4 of strain localization at defects. Assume now, however, that work hardening can be neglected at the high temperatures and that equation (6.1) describes
the strain rate dependence. Let A_a and A_b be the cross-sectional areas of the thicker
and thinner regions. The force carried by both sections is the same, so

$$F_b = \sigma_b A_b = F_a = \sigma_a A_a. \tag{6.6}$$

Substituting $\sigma_a = C\dot{\varepsilon}_a^m$, $\sigma_b = C\dot{\varepsilon}_b^m$, $A_a = A_{ao}\exp(-\varepsilon_a)$, and $A_b = A_{bo}\exp(-\varepsilon_b)$,
equation (6.6) becomes

$$A_{bo}\exp(-\varepsilon_b)C\dot{\varepsilon}_b^m = A_{ao}\exp(-\varepsilon_a)C\dot{\varepsilon}^m. \tag{6.7}$$

---

  * Note that m is somewhat rate dependent, so equation (6.1) is not a complete description of the rate
    dependence. Nevertheless, m, defined by $m = \partial \ln\sigma / \partial \ln\dot{\varepsilon}$, is a useful index for describing the rate
    sensitivity of a material and analyzing rate effects.

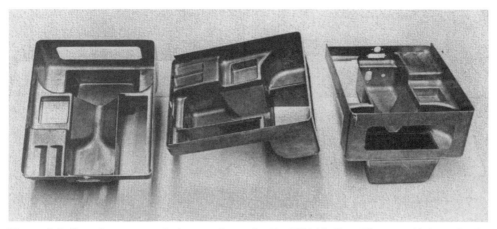

Figure 6.6. Complex part made from a sheet of a Zn-22%Al alloy. The very high strains in the walls require superplasticity. From W. F. Hosford and R. M. Caddell, *Ibid*.

Designating the ratio of areas of the initial cross-sectional areas by $A_{bo}/A_{ao} = f$, $f \exp(-\varepsilon_b)\dot{\varepsilon}_b^m = \exp(-\varepsilon_a)\dot{\varepsilon}_a^m$. Now raising to the (1/m)th power and recognizing that $\dot{\varepsilon} = d\varepsilon/dt$, $f^{1/m} \exp(-\varepsilon_b/m)d\varepsilon_b = \exp(-\varepsilon_a/m)d\varepsilon_a$. Integration from zero to their current values, ε_b and e_a, results in

$$\exp(-\varepsilon_a/m) - 1 = f^{1/m}[\exp(-\varepsilon_b/m) - 1]. \tag{6.8}$$

Under superplastic conditions, the reduction in area is often quite large. The deformation in the unnecked region can be approximated by letting $\varepsilon_b \to \infty$ in equation (6.8). In this case, the limiting strain in the unnecked region, ε_a^*, is

$$\varepsilon_a^* = -m \ln(1 - f^{1/m}). \tag{6.9}$$

Figure 6.8 shows how ε_a^* varies with m and f. Also shown are measured tensile elongations, plotted assuming that $\varepsilon_a^* = \ln(\%El/100 + 1)$. This plot suggests that values of f in the range from 0.99 to 0.995 are not unreasonable.

The high values of m are a result of the contribution of diffusional processes to the deformation. Two possible mechanisms are illustrated in Figure 6.9. One is a net diffusional flux under stress of atoms from grain boundaries parallel to the tensile axis to grain boundaries normal to the tensile axis, causing a tensile elongation. This amounts to a net diffusion of vacancies from grain boundaries normal to the tensile axis to boundaries parallel to it. The other possible mechanism is grain boundary sliding. With grain boundary sliding, compatibility where three grains meet must be accommodated by another mechanism. Such accommodation could occur either by slip or by diffusion, but in both cases the overall m would be below 1. Both

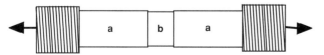

Figure 6.7. Schematic of a tensile bar with one region (b) is slightly smaller in cross-section than the other (a). Both must carry the same tensile force.

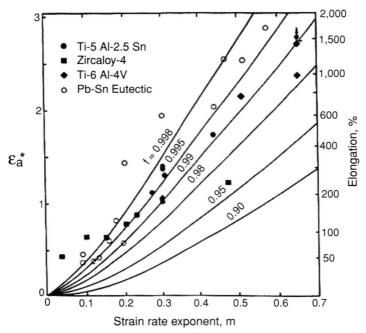

Figure 6.8. Limiting strain, ε_a^*, in the unnecked portion of a tensile specimen as a function of f and m. Reported values of total elongations are also shown as a function of m. These are plotted with the elongation converted to true strain. From W. F. Hosford and R. M. Caddell, *Ibid.*

diffusional creep and grain boundary sliding need a very fine grain size, high temperatures, and low strain rates. These mechanisms are discussed in more detail in Chapter 16.

EXAMPLE PROBLEM 6.3: A tensile bar is machined so that the diameter at one location is 1% smaller than the rest of the bar. The bar is tested at high temperature, so strain hardening is negligible, but the strain rate exponent is 0.25. When the strain in the reduced region is 0.20, what is the strain in the larger region?

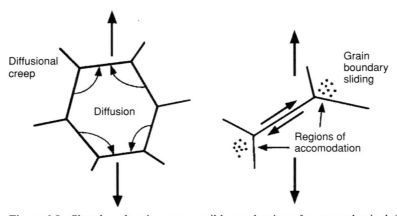

Figure 6.9. Sketches showing two possible mechanisms for superplastic deformation.

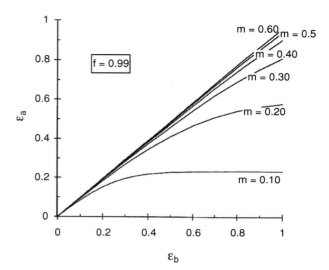

Figure 6.10. Relative strains in unnecked and necked regions for several values of rate sensitivity, m.

Solution: $f = (\pi D_1^2/4)/(\pi D_2^2/4) = (D_1/D_2)^2 = (0.99/1)^2 = 0.98$. Substituting $\varepsilon_b = 0.20$ and $f = 0.98$ into equation (6.8), $\exp(-\varepsilon_a/m) - 1 = f^{1/m}[\exp(-\varepsilon_a/m) - 1]$, $\exp(-\varepsilon_a/0.25) = (0.98)^{1/0.25}[\exp(-0.2/0.25) - 1] + 1 = 0.492$, $\varepsilon_a = -0.25\ln(0.492) = 0.177$.

Figure 6.10 shows that for high values of m, the strain, ε_a, in the region outside of the neck continues to grow even when the neck strain, ε_b, is large. With low values of m, deformation outside the neck ceases early.

Combined Strain and Strain Rate Effects

Rate sensitivity may have appreciable effects at low temperatures, even when m is relatively small. If both strain and strain rate hardening are considered, the true stress may be approximated by

$$\sigma = C'\varepsilon^n\dot\varepsilon^m. \tag{6.10}$$

Reconsider the tension test on an inhomogeneous tensile bar, with initial cross-sections of A_{ao} and $A_{bo} = fA_{oa}$. Substituting $A_a = A_{ao}\exp(-\varepsilon_a)$, $A_b = A_{bo}\exp(-\varepsilon_b)$, $\sigma_a = C\varepsilon_a^n e_a\dot\varepsilon^m$, and $\sigma_b = C\varepsilon_b^n\dot\varepsilon_b e^m$, into a force balance (equation 6.6) $F_a = A_a\sigma_a = F_b = A_b\sigma_b$,

$$A_{ao}\exp(-\varepsilon_a)C'\varepsilon_a^n\dot\varepsilon_a^m = A_{bo}\exp(-\varepsilon_b)C'\varepsilon_b^n\dot\varepsilon_b^m. \tag{6.11}$$

Following the procedure leading to equation (6.8),

$$\int_o^{\varepsilon a}\exp(-\varepsilon_a/m)\varepsilon_a^{n/m}d\varepsilon_a = f^{1/m}\int_o^{\varepsilon b}\exp(-\varepsilon_b/m)\varepsilon_b^{n/m}d\varepsilon_b \tag{6.12}$$

This equation must be integrated numerically. The results are shown in Figure 6.11 for $n = 0.2$, $f = 0.98$, and several levels of m.

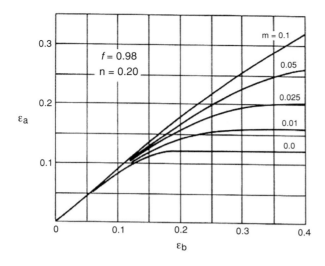

Figure 6.11. Comparison of the strains in the reduced and unreduced sections of a tensile bar for a material that strain hardens and is strain rate sensitive. From W. F. Hosford and R. M. Caddell, *Ibid.*

EXAMPLE PROBLEM 6.4: Reconsider the tension test in example problem 6.3. Let $n = 0.226$ and $m = 0.012$ (typical values for low-carbon steel), and let $f = 0.98$ as in the previous problem. Calculate the strain, ε_a, in the region with the larger diameter if the bar necks to a 25% reduction of area ($\varepsilon_b = \ln(4/3) = 0.288$.)

Solution: Substituting into equation (6.12),

$$\int_o^{\varepsilon_a} \exp(-\varepsilon_a/m)\varepsilon_a^{n/m}d\varepsilon_a = f^{1/m}\int_o^{\varepsilon_b} \exp(-\varepsilon_b/m)\varepsilon_b^{n/m}d\varepsilon_b \quad \text{or}$$

$$\int_o^{\varepsilon_a} \exp(-83.33\varepsilon_a)\varepsilon_a^{18.83}d\varepsilon_a = 0.1587\int_o^{\varepsilon_b} \exp(-83.33\varepsilon_b\varepsilon_b^{18.83}d\varepsilon_b)$$

Numerical integration of the right-hand side gives 0.02185. The left-hand side has the same value when $\varepsilon_a = 0.2625$. The loss of elongation in the thicker section is much less than in example problem 6.3, where it was found that $\varepsilon_a = 0.195$, with $n = 0.226$ and $m = 0$.

Strain Rate Sensitivity of bcc Metals

For bcc metals, including steels, a better description of the strain rate dependence of flow stress is given by

$$\sigma = C + m' \ln \dot{\varepsilon}, \tag{6.13}$$

where C is the flow stress at a reference strain rate and m' is the rate sensitivity. This expression indicates that a change of strain rate from $\dot{\varepsilon}_1$ to $\dot{\varepsilon}_2$ raises the flow stress by an amount

$$\Delta\sigma = m' \ln(\dot{\varepsilon}_2/\dot{\varepsilon}_1), \tag{6.14}$$

which is independent of the stress level. This is in contrast to equation (6.3), which predicts $\Delta\sigma \approx \sigma m \ln(\dot{\varepsilon}_2/\dot{\varepsilon}_1)$. The difference between the predictions of equations (6.3) and (6.14) are illustrated in Figures 6.12a and 6.12b. This difference is also apparent in comparing Figures 6.13 and 6.14. For copper (fcc) at 25°C, the difference

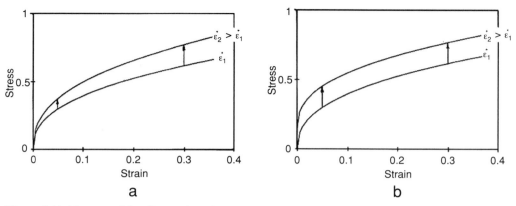

Figure 6.12. Two possible effects of strain rate on stress–strain curves. The stain rates for the two stress–strain curves in each figure differ by a factor of 100. (a) Equation (6.3) predicts that the difference in flow stress, $\Delta\sigma$, is proportional to the flow stress. (b) Equation (6.14) predicts that the difference in flow stress, $\Delta\sigma$, is independent of the flow stress.

Figure 6.13. Stress–strain curves for copper (fcc) at 25°C. Note that the difference, $\Delta\sigma$, between the stress–strain curves at different rates is proportional to the stress level. From P. S. Follansbee and U. F. Kocks, *Acta Met.*, v. 36 (1988).

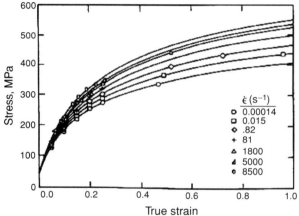

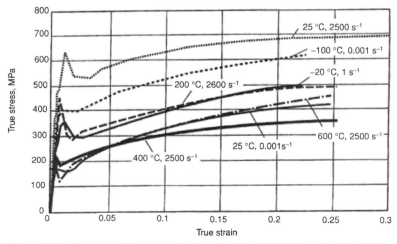

Figure 6.14. Stress–strain curves for iron (bcc) at 25°C. Note that the difference, $\Delta\sigma$, between the stress–strain curves at different rates is independent of the stress level. From G. T. Gray, *ASM Metals Handbook*, v. 8 (2000).

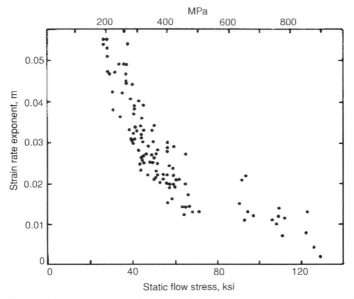

Figure 6.15. Correlation between the strain rate exponent, m, and the static flow stresses of steels. Each point represents a different steel. From W. F. Hosford and R. M. Caddell, *Ibid.* Data from A. Saxena and D. A. Chatfield, *SAE Paper 760209* (1976).

between the stress–strain curves at different rates, $\Delta\sigma$ is proportional to the stress level. In contrast, the difference between the stress–strain curves at different rates, $\Delta\sigma$, for pure iron (bcc) at 25°C is independent of the stress level.

EXAMPLE PROBLEM 6.5: Figure 6.15 is a plot of measured strain rate exponents, m, for a large number of steels as a function of the flow stress at which they were measured. Find the value of m′ in equation (6.14) that best fits the data in Figure 6.15.

Solution: Comparing equations (6.3) and (6.14), $\Delta\sigma = \sigma\,m\ln(\dot{\varepsilon}_2/\dot{\varepsilon}_1) = m'\ln(\dot{\varepsilon}_2/\dot{\varepsilon}_1)$ indicates that m′ = mσ. Drawing a smooth curve through the data in Figure 6.15 and picking several values, m′ = 0.05 × 200 MPa = 10 MPa, m′ = 0.025 × 400 MPa = 10 MPa, m′ = 0.0125 × 800 MPa = 10 MPa. It is apparent that m′ = 10 MPa describes the data well.

For steels, the strain hardening exponent, n, decreases with increasing strain rate. This is a result of the additive nature of the rate dependence. For power law strain hardening, the constant, C, in equation (6.13) is $M\varepsilon^{n'}$, so equation (6.13) becomes

$$\sigma = K\varepsilon^{n'} + m'\ln\dot{\varepsilon}. \qquad (6.15)$$

Figure 6.16 is a plot of equation (6.15) for K = 520 MPa, n′ = 0.22, and m′ = 10 MPa for several strain rates. In a plot of the same stress–strain curves on logarithmic scales (Figure 6.17), the average slope decreases as the strain rate increases.

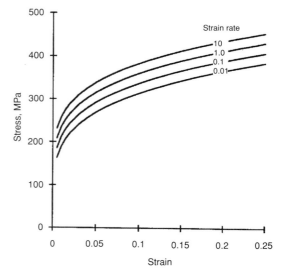

Figure 6.16. True stress–strain curves calculated with $\sigma = 520\varepsilon^{0.22} + 10\ln\dot\varepsilon$ for several strain rates.

Because the conventional strain hardening exponent, n (taken as the average slope), decreases with strain rate (Figure 6.18).

Temperature Dependence

As temperature increases, the whole level of the stress–strain curve generally drops. Figure 6.19 shows the decrease of tensile strengths of copper and aluminum. Usually, the rate of work hardening also decreases at high temperatures, as shown for aluminum in Figure 6.20.

Combined Temperature and Strain Rate Effects

The effects of temperature and strain rate are interrelated. Decreasing temperature has the same effect as increasing strain rate, as shown schematically in Figure 6.21. This effect occurs even in temperature regimes where the rate sensitivity is negative.

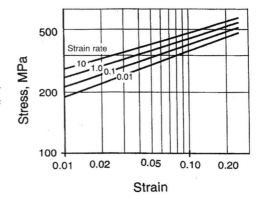

Figure 6.17. The same stress–strain curves as in Figure 6.16, plotted on logarithmic scales. Note that the slope decreases with increasing strain rate.

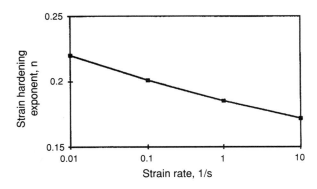

Figure 6.18. The decrease of n with strain rate calculated from Figure 6.17.

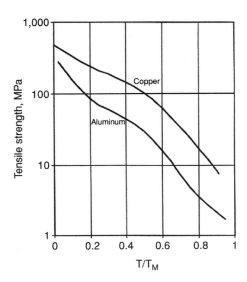

Figure 6.19. Tensile strengths decrease with increasing temperatures. Data from R. P. Carrecker and W. R. Hibbard, Jr., *Trans. TMS-AIME*, v. 209 (1957).

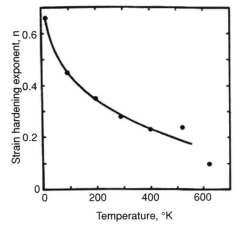

Figure 6.20. Decrease of the strain hardening exponent of aluminum with increasing temperature. From Hosford and Caddell, using data from R. P. Carrecker and W. R. Hibbard, Jr., *Ibid*.

Figure 6.21. Schematic plot showing the temperature dependence of flow stress at two different strain rates. Note that an increased strain rate has the same effects as a decreased temperature. From W. F. Hosford and R. M. Caddell, *Ibid*.

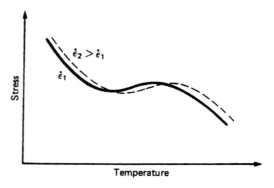

The simplest treatment of temperature dependence of strain rate is that of Zener and Hollomon,[*] who treated plastic straining as a thermally activated rate process. They assumed that strain rate follows an Arrhenius rate law, rate $\propto \exp(-Q/RT)$, that is widely used in analyzing many temperature-dependent rate processes. They proposed

$$\dot{\varepsilon} = A \exp(-Q/RT), \tag{6.16}$$

where Q is the activation energy, T is absolute temperature, and R is the gas constant. At a fixed strain, A is a function only of stress, $A = A(s)$, so equation (6.19) can be written

$$A(\sigma) = \dot{\varepsilon} \exp(Q/RT) \tag{6.17}$$

or

$$A(\sigma) = Z, \tag{6.18}$$

where $Z = \dot{\varepsilon} \exp(Q/RT)$ is called the *Zener-Hollomon parameter*.

Equation (6.20) predicts that a plot of strain at constant stress on a logarithmic scale versus $1/T$ should be linear. Figure 6.22 shows that such a plot for aluminum alloy 2024 is linear for a wide range of strain rates, although the relation fails at high strain rates.

EXAMPLE PROBLEM 6.6: For an aluminum alloy under constant load, the creep rate increased by a factor of 2 when the temperature was increased from 71°C to 77°C. Find the activation energy.

Solution: Because $\dot{\varepsilon} = A \exp(-Q/RT)$, the ratio of the creep rates at two temperatures is $\dot{\varepsilon}_2/\dot{\varepsilon}_1 = \exp[(-Q/R)(1/T_2 - 1/T_1)]$, so $Q = R \ln(\dot{\varepsilon}_2/\dot{\varepsilon}_1)/(1/T_1 - 1/T_2)$. Substituting $T_1 = 273 + 71 = 344 \, \text{K}$, $T_2 = 273 + 77 = 350 \, \text{K}$, $R = 8.314 \, \text{J/mole·K}$, and $\dot{\varepsilon}_2/\dot{\varepsilon}_1 = 2$ and solving $Q = 116 \, \text{kJ/mole}$. Compare this with Figure 6.23.

The Zener-Hollomon development is useful if the temperature and strain rate ranges are not too large. Dorn and coworkers measured the activation energy, Q, for pure aluminum over a large temperature range. They did this by observing the

[*] C. Zener and H. Hollomon, *J. Appl. Phys.*, 15 (1944).

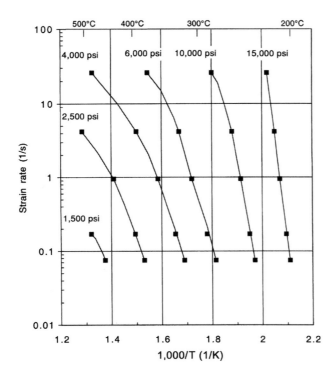

Figure 6.22. Combinations of strain rate and temperature for several stresses in aluminum alloy 2024. The data are taken at an effective strain of 1.0. Data from D. S. Fields and W. A. Backofen, *Ibid.*

change of creep rate under fixed load when the temperature was suddenly changed. Because the stress is constant,

$$\dot{\varepsilon}_2/\dot{\varepsilon}_1 = \exp[(-Q/R)(1/T_2 - 1/T_1)] \tag{6.19}$$

or

$$Q = R \ln(\dot{\varepsilon}_2/\dot{\varepsilon}_1)(1/T_1 - 1/T_2). \tag{6.20}$$

Their results (Figure 6.23) indicated that Q was independent of temperature above 500 K, but very temperature dependent at lower temperatures. This observation was later explained by Z. S. Basinski and others. The basic argument is that the stress helps thermal fluctuations overcome activation barriers. This is illustrated in Figure 6.24. If there is no stress, the activation barrier has a height of Q, with random fluctuations. The rate of overcoming the barrier is proportional to $\exp(-Q/RT)$. An applied stress skews the barrier so that the effective height of the barrier is reduced to $Q - \sigma v$, where v is a parameter with the units of volume. Now the rate of

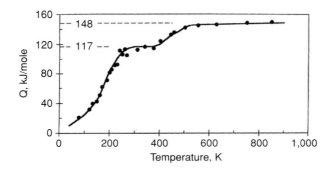

Figure 6.23. Measured activation energies for creep of aluminum as a function of temperature. The change of Q with temperature indicates that equation (6.20) will lead to errors if it is applied over temperature range in which Q changes. Data from O. D. Sherby, J. L. Lytton, and J. E. Dorn, *AIME Trans.*, v. 212 (1958).

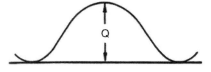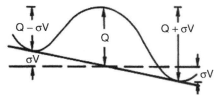

Figure 6.24. Schematic illustration of the skewing of an activation barrier by an applied stress. From W. F. Hosford and R. M. Caddell, *Ibid.*

overcoming the barrier is proportional to $\exp[-Q - \sigma v)/RT]$. The finding of lower activation energies at lower temperatures was explained by the fact that in the experiments, greater stresses were applied at lower temperatures to achieve measurable creep rates.

In Figure 6.24, the rate of the overcoming the barrier from left to right is proportional to $\exp[-(Q - \sigma v)/RT]$, whereas the rate from right to left is proportional to $\exp[-(Q + \sigma v)/RT]$. Thus, the net reaction rate is $C\{\exp[-(Q - \sigma v)/RT] - \exp[-(Q + \sigma v)/RT]\} = C\exp(-Q/RT)\{\exp[(\sigma v)/RT] - \exp[-(\sigma v)/RT]\}$. This simplifies to

$$\dot{\varepsilon} = 2C\exp(-Q/RT)\sinh[(\sigma v)/RT]. \tag{6.21}$$

This expression has been modified, based on some theoretical arguments, to provide a better fit with experimental data,

$$\dot{\varepsilon} = A\exp(-Q/RT)[\sinh(\alpha\sigma)]^{1/m}, \tag{6.22}$$

where A is an empirical constant and the exponent $1/m$ is consistent with equation (6.1). Figure 6.25 shows that equation (6.22) can be used to correlate the combined

Figure 6.25. Plot of the Zener-Hollomon parameter versus flow stress data for aluminum over a wide range of strain rates. Data are from J. J. Jonas, *Trans. Q. ASM*, v. 62 (1968).

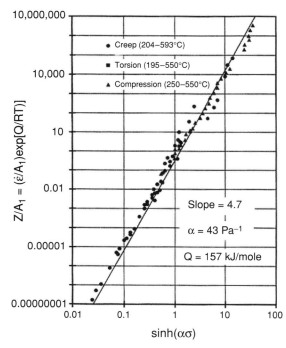

- Creep (204–593°C)
- Torsion (195–550°C)
- Compression (250–550°C)

$Z/A_1 = (\dot{\varepsilon}/A_1)\exp[Q/RT]$

sinh($\alpha\sigma$)

Slope = 4.7
α = 43 Pa^{-1}
Q = 157 kJ/mole

effects of temperature, stress, and strain rate over an extremely large range of strain rates.

If $\alpha\sigma \ll 1$, $\sinh(\alpha\sigma) \approx \alpha\sigma$, equation (6.22) simplifies to

$$\dot{\varepsilon} = A \exp(+Q/RT)(\alpha\sigma)^{1/m} \qquad (6.23)$$

$$\sigma = \dot{\varepsilon}^m A' \exp(-mQ/RT) \qquad (6.24)$$

$$\sigma = A'Z^m \qquad (6.25)$$

where $A' = (\alpha A^m)^{-1}$. Equation (6.25) is consistent with the Zener-Hollomon development. At low temperatures and high stresses, $(\alpha\sigma) \gg 1$, so $\sinh(\alpha\sigma) \to \exp(\alpha\sigma)/2$ and equation (6.22) reduces to

$$\dot{\varepsilon} = C \exp(\alpha'\sigma - Q/RT). \qquad (6.26)$$

Under these conditions, both C and α' depend on strain and temperature. For a constant temperature and strain,

$$\sigma = C + m' \ln \dot{\varepsilon}. \qquad (6.27)$$

Note that this equation is consistent with equation (6.13), but not with the power law rate expression in equation (6.1).

Hot Working

Metallurgists make a distinction between *cold working* and *hot working*. It is often said that hot working is done above the recrystallization temperature and the work material recrystallizes as it is deformed. This is an oversimplification. The strain rates during hot working are often so high that there is not enough time for recrystallization *during* the deformation. It is more meaningful to think of hot working as a process in which recrystallization occurs in the period between repeated operations, as in forging, rolling, and so on, or at least before the work material cools from the working temperature to room temperature.

Tool forces during hot working are lowered not only by the recrystallization itself, but also because of the inherently lower flow stresses at high temperatures. A second advantage of hot working is that the resulting product is in an annealed state. There are, however, disadvantages of hot working. Among these are

1. The tendency of the work metal to oxidize. In the case of steel and copper-base alloys, this results in scale formation and a consequent loss of metal. Titanium alloys must be hot worked under an inert atmosphere to avoid dissolving oxygen.
2. Lubrication is much more difficult and, consequently, friction much higher. This effect somewhat negates the effect of lower flow stress.
3. The scaling and high friction result in tool wear, and shortened tool life.
4. The resulting product surface is rough because of the scale, tool wear, and poor lubrication. There is also loss of precise gauge control.
5. The lack of strain hardening in hot working is, in some cases, undesirable.

Usually, ingot breakdown and subsequent working are done hot until the section size becomes small enough that high friction and the loss of material by oxidation become important. Hot rolling of steel sheet is continued to a thickness of 0.050 to 0.200 in. Then the product is pickled in acid to remove the scale and finished by cold working to obtain a good surface.

Most cold-rolled sheet is sold in an annealed condition. With steels, the term *cold rolled* refers to the surface quality rather than the state of work hardening. The term *warm working* is used to refer to working of metals above room temperature, but at a low enough temperature that recrystallization does not occur.

REFERENCES

Metal Forming: Mechanics and Metallurgy, 2nd ed., W. F. Hosford and R. M. Caddell, Cambridge University Press (2007).
Ferrous Metallurgical Design, John H. Hollomon and Leonard Jaffe, John Wiley & Sons (1947).

Notes

Svante August Arrhenius (1859–1927) was a Swedish physical chemist. He studied at Uppsala where he obtained his doctorate in 1884. It is noteworthy that his thesis on electrolytes was given a fourth-level (lowest) pass because his committee was skeptical of its validity. From 1886 to 1890, he worked with several noted scientists in Germany who did appreciate his work. In 1887, he suggested that a wide range of rate processes follow what is now known as the Arrhenius equation. For years, his work was recognized throughout the world, except in his native Sweden.

Count Rumford (Benjamin Thompson) was the first person to measure the mechanical equivalent of heat. He was born in Woburn, Massachusetts, in 1753, studied at Harvard and taught in Concord, New Hampshire, where he married a wealthy woman. At the outbreak of the American Revolution, having been denied a commission by Washington, he approached the British who did commission him. When the British evacuated Boston in 1776, he left for England, where he made a number of experiments on heat. However, he was forced to depart quickly for Bavaria after being suspected of selling British naval secrets to the French. In the Bavarian army, he rose rapidly to become Minister of War and eventually Prime Minister. While inspecting a canon factory, he observed a large increase in temperature during the machining of bronze canons. He measured the temperature rise, and, with the known heat capacity of the bronze, he calculated the heat generated by machining. By equating this to the mechanical work done in machining, he was able to deduce the mechanical equivalent of heat. His value was a little too low, in part due to the fact that some of the plastic work done in machining is stored as dislocations in the chips.

Thompson was knighted in 1791, choosing the title Count Rumford after the original name of Concord. During his rapid rise to power in Bavaria, he made many enemies so that when the winds of politics changed, he had to leave. He returned to Britain as an ambassador from Bavaria, but the British refused to recognize him. During this period, he established the Royal Institute. At this time, the United

States was establishing the U.S. Military Academy at West Point, and Rumford applied for and was selected to be its first commandant. However, during the negotiations about his title, it was realized that he had deserted the American cause during the American Revolution and the offer was withdrawn. He finally moved to Paris, where he married the widow of the famous French chemist, Lavousier, who had been beheaded during the French Revolution.

Problems

1. During a tension test, the rate of straining was suddenly doubled. This caused the load (force) to rise by 1.2%. Assuming that the strain rate dependence can be described by $\sigma = C\dot{\varepsilon}^m$, what is the value of m?

2. Two tension tests were made on the same alloy, but at different strain rates. Both curves were fitted to a power law strain hardening expression of the form, $\sigma = K\varepsilon^n$. The results are summarized as follows. Assuming that the flow stress at constant strain can be approximated by $\sigma = C\dot{\varepsilon}^m$, determine the value of m.

	Test A	Test B
Strain rate (s^{-1})	2×10^{-3}	10^{-1}
Strain hardening exponent, n	0.22	0.22
Constant, K (MPa)	402	412

3. To achieve weight saving in an automobile, replacement of a low-carbon steel with an HSLA steel is being considered. In laboratory tension tests at a strain rate of 10^{-3}/s, the yield strengths of the HSLA steel and the low-carbon steel were measured to be 400 MPa and 220 MPa, respectively. The strain rate exponents are m = 0.005 for the HSLA steel and m = 0.015 for the low-carbon steel. What percent weight saving could be achieved if the substitution was made so that the forces were the same at the strain rates of 10^{+3}, typical of crash conditions?

4. The thickness of a cold-rolled sheet varies from 0.0322 to 0.0318, depending on where the measurement is made, so strip tensile specimens cut from the sheet show similar variation in cross-section.

    A. For a material with n = 0.20 and m = 0, what will be the thickness of the thickest regions when the thinnest region necks?

    B. Find the strains in the thicker region if m = 0.50 and n = 0 when the strain in the thinnest region reaches

        i. 0.5,

        ii. ∞.

5. Estimate the total elongation of a superplastic material if

    A. n = 0, m = 0.5, and $f = 0.98$,

    B. n = 0, $f = 0.75$, and m = 0.8.

6. In superplastic forming, it is often necessary to control the strain rate. Consider the forming of a hemispherical dome by clamping a sheet over a circular hole and bulging it with gas pressure.

A. Compare gas pressure needed to form a 2.0-in. dome with that needed to form a 20-in. dome if both are formed from sheets of the same thickness and at the same strain rate.

B. Describe (qualitatively) how the gas pressure should be varied during the forming to maintain a constant strain rate.

7. During a constant load creep experiment on a polymer, the temperature was suddenly increased from 100°C to 105°C. It was found that this increase of temperature caused the strain rate to increase by a factor of 1.8. What is the apparent activation energy for creep of the plastic?

8. Figure 6.26 provides some data for the effect of stress and temperature on the strain rate of a nickel-base super-alloy single crystal. The strain rate is independent of strain in the region of this data. Determine as accurately as possible.

A. The activation energy, Q, in the temperature range 700°C to 810°C.

B. The exponent, m, for 780°C.

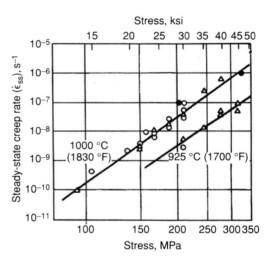

Figure 6.26. Effects of stress and temperature on the strain rate of nickel-base super-alloy single crystals under stress. From *Metals Handbook*, 9th ed., v. 8, ASM (1985).

9. It has been suggested that $\dot{\varepsilon} = A \exp[-(Q - \sigma v)/RT] - A \exp[-(Q + \sigma v)/RT]$ is a better representation of the dependence of strain rate on temperature and stress than the Holloman-Zener approach. Using this equation, derive an expression for the dependence of stress on strain rate if $\sigma v \gg Q$.

10. The stress–strain curve of a steel is represented by $\sigma = (1{,}600\,\text{MPa})\varepsilon^{0.1}$. The steel is deformed to a strain of $\varepsilon = 0.1$ under adiabatic conditions. Estimate the temperature rise in the sample. For iron, $\rho = 7.87$ kg/m^2, $C = 0.46$ J/g°C, $E = 205$ GPa.

11. Evaluate m for copper at room temperature from Figure 6.13.

12. The stress–strain curves for silver at several temperatures and strain rates are shown in Figure 6.27. Determine the strain rate exponent, m for silver at 25°C.

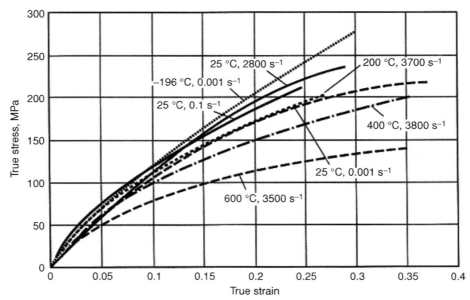

Figure 6.27. Stress–strain curves for silver (fcc). Note that at 25 degrees, the difference, $\Delta\sigma$, between the stress–strain curves at different rates is proportional to the stress level. From G. T. Gray, *ASM Metals Handbook*, v. 8 (2000).

7 Slip and Crystallographic Textures

Introduction

Plastic deformation of crystalline materials usually occurs by *slip*, which is the sliding of planes of atoms over one another (Figure 7.1). The planes on which slip occurs are called *slip planes*, and the directions of the shear are the *slip directions*. These are crystallographic planes and directions that are characteristic of the crystal structure. The magnitude of the shear displacement is an integral number of interatomic distances, so that the lattice is left unaltered. If slip occurs on only part of a plane, there remains a boundary between the slipped and unslipped portions of the plane, which is called a *dislocation*. Slip occurs by movement of dislocations through the lattice. It is the accumulation of the dislocations left by slip that is responsible for work hardening. Dislocations and their movement are treated in Chapters 8 and 9. This chapter is concerned only with the geometry of slip.

Visual examination of the surface of a deformed crystal will reveal slip lines. The fact that we can see these indicates that slip is inhomogeneous on an atomic scale. Displacements of thousands of atomic diameters must occur on discrete or closely spaced planes to create steps on the surface that are large enough to be visible. Furthermore, the planes of active slip are widely separated on an atomic scale. Yet, the scale of the slip displacements and distances between slip lines are small compared to most grain sizes, so slip usually can be considered as homogeneous on a macroscopic scale.

Slip Systems

The slip planes and directions for several common crystals are summarized in Table 7.1. Almost without exception, the slip directions are the crystallographic directions with the shortest the distance between like atoms or ions, and the slip planes are usually densely packed planes.

Schmid's Law

Erich Schmid* discovered that if a crystal is stressed, slip begins when the shear stress on a slip system reaches a critical value, τ_c, often called the *critical resolved*

---

* E. Schmid, *Proc. Internat. Cong. Appl. Mech.*, Delft (1924).

Table 7.1. *Slip directions and planes*

Structure	Slip direction	Slip planes
fcc	<110>	{111}
bcc	<111>	{110}, {112}, {123}, pencil glide*
hcp	<11$\bar{2}$0>	(0001), {1$\bar{1}$00}, {1$\bar{1}$01}
	<11$\bar{2}$3>[†]	{1$\bar{1}$0$\bar{1}$}
dia. cub.	<110>	{111}
NaCl	<110>	{110}
CsCl	<001>	{100}
fluorite	<110>	{001}, {110}[†], {111}[†]

* With pencil glide, slip is possible on all planes containing the slip direction.
[†] <11$\bar{2}$3> slip has been reported on many hcp crystals under special loading
conditions or high temperatures.

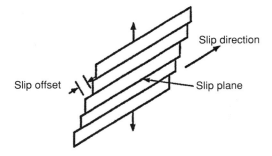

Slip offset

Slip direction

Slip plane

Figure 7.1. Slip by shear between parallel planes of atoms.

shear stress. In most crystals*, slip occurs with equal ease forward or backward, so the condition necessary for slip can be written as

$$\tau_{nd} = \pm\tau_c. \tag{7.1}$$

The subscripts n and d refer to the slip plane normal and the direction of slip, respectively. This simple yield criterion for crystallographic slip is called *Schmid's law*. In a uniaxial tension test along the x-direction, the shear stress can be found from the stress transformation (equation (1.6)),

$$\tau_{nd} = \ell_{nx}\ell_{dx}\sigma_{xx}. \tag{7.2}$$

Schmid's law is usually written as

$$\tau_c = \pm\sigma_x \cos\lambda \cos\varphi, \tag{7.3}$$

where λ is the angle between the slip direction and the tensile axis, and ϕ is the angle between the tensile axis and the slip plane normal (Figure 7.2), so $\ell_{dx} = \cos\lambda$ and $\ell_{nx} = \cos\phi$.

Equation (7.3) can be shortened to

$$\sigma_x = \pm\tau_c/m_x, \tag{7.4}$$

where m_x is the *Schmid factor*, $m_x = \cos\lambda \cos\phi$.

* Bcc crystals are exceptions. The stress required to cause slip in one direction is less than that for the opposite direction.

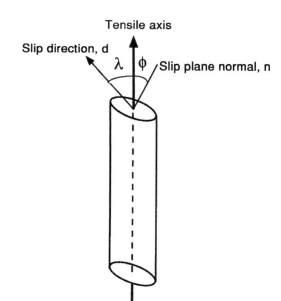

Figure 7.2. Slip elements in uniaxial tension.

The condition for yielding under a general stress state is

$$\pm\tau_c = \ell_{nx}\ell_{dx}\sigma_{xx} + \ell_{ny} + \ell_{dy}\sigma_{yy} + \ldots\ldots\ldots + (\ell_{nx} + \ell_{dy} + \ell_{ny} + \ell_{dx})\sigma_{xy}. \quad (7.5)$$

EXAMPLE PROBLEM 7.1: A tensile stress of $\sigma = 5$ kPa is applied parallel to the [432] direction of a cubic crystal. Find the shear stress, τ, on the $(11\bar{1})$ plane in the [011] direction.

Solution: First find $m = \cos\lambda\cos\phi$ for $(11\bar{1})[011]$ slip. In cubic crystals, the normal to a plane has the same indices as the plane, so the normal to $(11\bar{1})$ is $[11\bar{1}]$. Also, in cubic crystals, the cosine of the angle between two directions is given by the dot product of unit vectors in those directions. Therefore,

$$\cos\phi = (4\cdot1 + 3\cdot1 + 2\cdot\bar{1})/[(4^2 + 3^2 + 2^2)^{1/2}(1^2 + 1^2 + \bar{1}^2)^{1/2}] = 5/(\sqrt{29}\sqrt{3}).$$
$$\cos\lambda = (4\cdot0 + 3\cdot1 + 2\cdot1)/[(4^2 + 3^2 + 2^2)^{1/2}(0^2 + 1^2 + 1^2)^{1/2}] = 5/(\sqrt{29}\sqrt{2}).$$
$$m = \cos\lambda\cos\phi = 25/(29\sqrt{6}) = 0.352.$$
$$\tau = m\sigma = 0.352 \times 5 \text{ kPa} = 1.76 \text{ kPa}.$$

EXAMPLE PROBLEM 7.2: A cubic crystal is subjected to a stress state, $\sigma_x = 15$ kPa, $\sigma_y = 0$, $\sigma_z = 7.5$ kPa, $\tau_{yz} = \tau_{zx} = \tau_{xy} = 0$, where x = [100], y = [010], and z = [001]. What is the shear stress on the $(11\bar{1})[101]$ slip system? See Figure 7.3.

Solution: $\tau_{nd} = \ell_{nx}\ell_{dx}\sigma_x + \ell_{nz} + \ell_{dz}\sigma_z$, where d = [101], and n = $[11\bar{1}]$. Taking dot products,

$$\ell_{nx} = [11\bar{1}]\cdot[100] = (1\cdot1 + 1\cdot0 + \bar{1}\cdot0)/\sqrt{[1^2 + 1^2 + \bar{1}^2)(1^2 + 0^2 + 0^2)]} = 1/\sqrt{3},$$
$$\ell_{dx} = [101]\cdot[100] = (1\cdot1 + 0\cdot0 + 1\cdot0)/\sqrt{[1^2 + 0^2 + 1^2)(1^2 + 0^2 + 0^2)]} = 1/\sqrt{2},$$
$$\ell_{nz} = [11\bar{1}]\cdot[001] = (1\cdot0 + 1\cdot0 + \bar{1}\cdot1)/\sqrt{[1^2 + 1^2 + \bar{1}^2)(0^2 + 0^2 + 1^2)]} = -1/\sqrt{3},$$
$$\ell_{dz} = [101]\cdot[001] = (1\cdot0 + 0\cdot0 + 1\cdot1)/\sqrt{[1^2 + 0^2 + 1^2)(0^2 + 0^2 + 1^2)]} = 1/\sqrt{2}.$$

Substituting, $\tau_{nd} = (1/\sqrt{3})(1/\sqrt{2})15 \text{ kPa} + (-1/\sqrt{3})(1/\sqrt{2})7.5 \text{ kPa} = 7.5/\sqrt{6} = 3.06$ kPa.

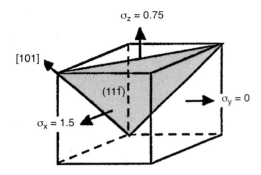

Figure 7.3. Cubic crystal in example problem 7.2.

Strains Produced by Slip

The incremental strain transformation equation (1.26) may be used to find the shape change that results from slip when the strains are small (i.e., when the lattice rotations are negligible). For infinitesimal strains,

$$d\varepsilon_{xx} = \ell_{xn}^2 d\varepsilon_{nn} + \ell_{xd}^2 d\varepsilon_{dd} + \ldots\ldots\ldots \ell_{xn}\ell_{xd}d\gamma_{nd}. \tag{7.6}$$

With slip on a single slip system in the d direction and on the n plane, the only strain term on the right-hand side of equation (7.6) is $d\gamma_{nd}$, so we can write

$$d\varepsilon_{xx} = \ell_{xn}\ell_{xd}d\gamma_{nd}. \tag{7.7}$$

In Schmid's notation this is

$$d\varepsilon_{xx} = \cos\lambda\cos\phi d\gamma = md\gamma \tag{7.8}$$

where $d\gamma$ is the shear strain on the slip system. The other strain components referred to the x, y, and z axes are similarly

$$
\begin{aligned}
d\varepsilon_{yy} &= \ell_{yn}\ell_{yd}d\gamma \\
d\varepsilon_{zz} &= \ell_{zn}\ell_{zd}d\gamma \\
d\gamma_{yz} &= (\ell_{yn}\ell_{zd} + \ell_{yd}\ell_{zn})d\gamma \\
d\gamma_{zx} &= (\ell_{zn}^1\ell_{xd} + \ell_{zd}\ell_{xn})d\gamma \\
d\gamma_{xy} &= (\ell_{xn}\ell_{yd} + \ell_{xd}\ell_{yn})d\gamma.
\end{aligned}
\tag{7.9}
$$

EXAMPLE PROBLEM 7.3: If slip occurs on the $(11\bar{1})[101]$ slip system in example problem 7.2, what will be the ratios of the resulting strains, $d\varepsilon_y/d\varepsilon_x$ and $d\varepsilon_z/d\varepsilon_x$, where x = [100], y = [010], and z = [001]?

Solution: $d\varepsilon_x = \ell_{nx}\ell_{dx}\gamma_{nd} = 1/\sqrt{6}d\gamma_{nd}$ $d\varepsilon_y = \ell_{ny}\ell_{dy}d\gamma_{nd}$. Substituting $\ell_{ny} = [11\bar{1}]\cdot[010] = 1/\sqrt{3}$ and $\ell_{dy} = [101]\cdot[010] = 0$, $d\varepsilon_y = 0$, $d\varepsilon_z = \ell_{nz}\ell_{dz}\gamma_{nd} = -1/\sqrt{6}d\gamma_{nd}$. Therefore, $d\varepsilon_y/d\varepsilon_x = 0$ and $d\varepsilon_z/d\varepsilon_x = -1$.

For a polycrystalline material to have appreciable ductility, each grain must be able to undergo the same shape change as the entire body. That is, each grain in a polycrystal must deform with the same external strains as the whole polycrystal. For an individual grain, this amounts to an imposed set of strains ε_1, ε_2, γ_{23}, γ_{31}, γ_{12} along the crystal axes. Five independent strains, ε_1, ε_2, γ_{23}, γ_{31}, γ_{12} describe an arbitrary strain state or shape change. The third normal strain, ε_3, is not independent because $\varepsilon_3 = -\varepsilon_1 - \varepsilon_2$. If a material has less than five independent slip systems, a polycrystal of that material will have limited ductility unless another deformation mechanism supplies the added freedom necessary. The number of independent slip systems is the same as the number of strain components, $\varepsilon_1, \varepsilon_2, \gamma_{23}, \gamma_{31}, \gamma_{12}$ that can be independently satisfied by slip.

EXAMPLE PROBLEM 7.4: The number of independent slip systems is the same as the number of the strain components, $\varepsilon_1, \varepsilon_2, \gamma_{23}, \gamma_{31}, \gamma_{12}$, that can be accommodated by slip. Find the number of independent slip systems in crystals of the NaCl structure that deform by <110>{1$\bar{1}$0} slip.

Solution: Let the [100], [010], and [001] directions be designated as 1, 2, and 3, respectively, and let the slip systems be designated as follows:

[011]($01\bar{1}$) & [01$\bar{1}$](011) as Aa, [101]($10\bar{1}$) & [101]($10\bar{1}$) as Bb, and [110]($1\bar{1}0$) & [1$\bar{1}$0](110) as Cc.

The strains along the cubic axes can be written as

$$\varepsilon_1 = \ell_{1A}\ell_{1a}\gamma_{Aa} + \ell_{1B}\ell_{1b}\gamma_{Bb} + \ell_{1C}\ell_{1c}\gamma_{Cc} = 0 + (1/2)\gamma_{Bb} + (1/2)\gamma_{Cc}$$
$$\varepsilon_2 = \ell_{2A}\ell_{2a}\gamma_{Aa} = \ell_{2B}\ell_{2b}\gamma_{Bb} + \ell_{2C}\ell_{2c}\gamma_{Cc} = 0 + (1/2)\gamma_{Aa} + (1/2)\gamma_{Cc}$$
$$\gamma_{23} = (\ell_{2A}\ell_{3a} + \ell_{2a}\ell_{3A})\gamma_{Aa} + (\ell_{2B}\ell_{3b} + \ell_{2b}\ell_{3B})\gamma_{Bb} + (\ell_{2C}\ell_{3c} + \ell_{2c}\ell_{3C})\gamma_{Cc}$$
$$\gamma_{31} = (\ell_{3A}\ell_{1a} = \ell_{3a}\ell_{1A})\gamma_{Aa} + (\ell_{3B}\ell_{1b} + \ell_{3b}\ell_{1B})\gamma_{Bb} + (\ell_{3C}1_{1c} + 1_{3c}\ell_{1C})\gamma_{Cc}$$
$$\gamma_{12} = (\ell_{1A}\ell_{2a} = \ell_{1a}\ell_{2A})\gamma_{Aa} + (\ell_{1B}\ell_{2b} + \ell_{1b}\ell_{2B})\gamma_{Bb} + (\ell_{1C}\ell_{2c} + \ell_{1c}\ell_{2C})\gamma_{Cc}.$$

Substituting $\ell_{1A} = [100] \cdot [110] = 1/\sqrt{2}$, $\ell_{1a} = [100] \cdot [1\bar{1}0] = 1/\sqrt{2}$, $\ell_{1B} = [100] \cdot [101] = 1/\sqrt{2}$,

$\ell_{1b} = [100] \cdot [10\bar{1}] = 1/\sqrt{2}$, $\ell_{1C} = [100] \cdot [011] = 0$, $\ell_{1c} = [100] \cdot [01\bar{1}] = 0$,
$\ell_{2A} = [010] \cdot [110] = 1/\sqrt{2}$, $\ell_{2a} = [010] \cdot [1\bar{1}0] = -/\sqrt{2}$, $\ell_{2B} = [010] \cdot [101] = 0$,
$\ell_{2b} = [010] \cdot [10\bar{1}] = 0$, $\ell_{2C} = [010] \cdot [011] = 1/\sqrt{2}$, $\ell_{2c} = [010] \cdot [01\bar{1}] = 1/\sqrt{2}$,
$\ell_{3A} = [001] \cdot [110] = 0$, $\ell_{3a} = [001] \cdot [1\bar{1}0] = 0$, $\ell_{3B} = [001] \cdot [101] = 1/\sqrt{2}$,
$\ell_{3b} = [001] \cdot [10\bar{1}] = -1/\sqrt{2}$, $\ell_{3C} = [001] \cdot [011] = 1/\sqrt{2}$,
$\ell_{3c} = [001] \cdot [01\bar{1}] = -1/\sqrt{2}$.

Substituting $(\ell_{2A}\ell_{3a} + \ell_{2a}\ell_{3A}) = (\ell_{2B}\ell_{3b} + \ell_{2b}\ell_{3B}) = (\ell_{2C}\ell_{3c} + \ell_{2c}\ell_{3C}) = 0$, and so on, it can be concluded that $\gamma_{23} = \gamma_{31} = \gamma_{12} = 0$, so the shear strains γ_{23}, γ_{31}, and γ_{12} cannot be satisfied. Therefore, there are only two independent strains, ε_1 and ε_2, so there are only two slip systems are independent.

Strain Hardening of fcc Single Crystals

Figure 7.4 is a typical stress–strain curve for an fcc single crystal. Initially, slip occurs on a single plane, and the rate of strain hardening is very low. This low strain hardening rate is similar to that in crystals of hcp metals and is called *easy glide* or *stage I*. At some point, slip is observed on other systems. The result is that dislocations on different slip systems intersect, causing much more rapid strain hardening. The

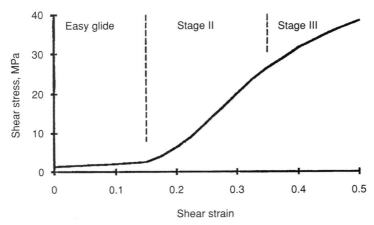

Figure 7.4. Schematic shear stress–shear strain curve for a typical single crystal of a fcc metal, showing an initial stage of easy glide, a second stage with rapid and almost constant hardening rate, and finally a third stage in which the rate of strain hardening decreases.

result is a high and nearly constant rate of strain hardening in *stage II*. Lower temperatures increase the extent of stage II. At higher strains, in *stage III*, the rate of strain hardening decreases.

The extent of easy glide in a crystal depends on its orientation, its perfection, and the temperature. A high degree of crystal perfection and low temperature promote more easy glide. Likewise, the extent of easy glide is greater in orientations for which the resolved shear stress on other potential systems is low. Easy glide does not occur in fcc crystals oriented so that slip occurs simultaneously on many slip systems. Instead, the initial strain hardening rate is rapid and comparable with that in polycrystals (Figure 7.5). No easy glide is observed in bcc single crystals. Extensive easy glide is observed in tension tests of single crystals of the hcp metals that slip primarily on the basal (0001) plane (e.g., Zn, Cd, Mg).

Tensile Deformation of fcc Crystals

It is customary to represent the tensile axis in the basic stereographic triangle with [100], [110], and [111] corners, as shown in Figure 7.6. (Details of the stereographic projection are given in Appendix II.) For all orientations of fcc crystals within this triangle, the Schmid factor for slip in the [101] direction on the (11$\bar{1}$) plane is higher

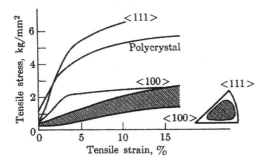

Figure 7.5. Stress–strain curves for polycrystalline aluminum and aluminum single crystals. Note that both the polycrystal and the <111>-oriented crystal strain harden rapidly. Strain hardening in orientations with only a single slip system (*shaded region*) is much lower. From U. F. Kocks, *Acta Met.*, v. 8 (1960).

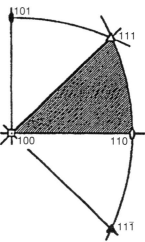

Figure 7.6. Basic orientation triangle. If the tensile axis of an fcc crystal lies in this triangle, the most heavily stressed slip system is $[101](11\bar{1})$. From W. F. Hosford, *The Mechanics of Crystals and Textured Polycrystals*, Oxford University Press (1993).

than for any other slip system. Slip on this *primary* slip system should be expected. If the tensile axis is represented as lying in any other stereographic triangle, the slip elements may be found by examining the remote corners of the three adjacent triangles. The <111> direction in one of the adjacent triangles is the normal to slip plane, and the <110> direction in another adjacent triangle is the slip direction.

Figures 7.7 and 7.8 show the orientation dependence of the Schmid factor, m, within the basic triangle. The highest value, m = 0.5, occurs where the tensile axis lies on the great circle between the slip direction and the slip plane normal with $\lambda = 45°$ and $\varphi = 45°$.

If the tensile axis lies on a boundary of the basic triangle, two slip systems are equally favored. They are the ones most favored in the two triangles that form the boundary. For example, if the tensile axis is on the [100]–[111] boundary, the second system is $[110](1\bar{1}1)$, which is called the *conjugate* system. Two other systems have names: $[10\bar{1}](111)$ is the *critical slip* system, and the system that shares the slip direction with the primary system, $[101](1\bar{1}\,\bar{1})$, is called the *cross slip* system. At the corners there are four, six, or eight equally favored slip systems.

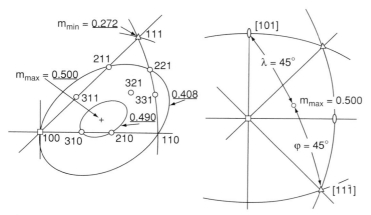

Figure 7.7. Orientation dependence of the Schmid factor for fcc crystals. From W. F. Hosford, *Ibid.*

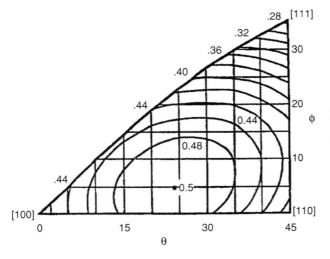

Figure 7.8. Full representation of contours of constant Schmid factor for fcc crystals. From W. F. Hosford, *Ibid.*

Slip in bcc Crystals

The slip direction in bcc metals is always the direction of close packing, <111>. Slip has been reported on various planes: {110}, {123}, and {110}, {112}. These planes contain at least one <111> direction. G. I. Taylor suggested that slip in bcc crystals could be described as *<111>-pencil glide*. This term means that slip can occur on any plane that contains a <111> slip direction. Figure 7.9 shows the orientation dependence of the Schmid factors for uniaxial tension with <111>-pencil glide. Note that the basic orientation triangle is divided into two regions, with a different <111> slip direction in each.

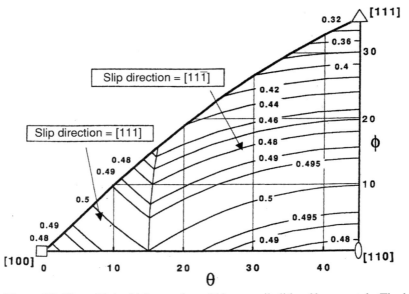

Figure 7.9. Plot of Schmid factors for <111>-pencil glide of bcc crystals. The basic triangle is divided into two regions with different slip directions. The Schmid factor is 0.417 at [100] and [110], and it is 0.314 at [111]. From W. F. Hosford, *Ibid.*

Slip in hcp Crystals

With all hcp metals, the most common slip direction is $<11\bar{2}0>$. This is the direction of close contact between atoms in the basal plane. Several slip planes have been reported as indicated in Table 7.1. Slip in the $<11\bar{2}0>$ direction causes no strains parallel to the c axis, regardless of the plane of slip. Pyramidal slip on $<11\bar{2}3>\{1\bar{1}0\bar{1}\}$ has been reported for a few hcp metals at high temperatures or under special loading conditions. Figure 7.10 shows shear stresses required for several slip systems in beryllium. The stresses required for pyramidal slip are much higher than those required to cause basal and prism slip and are often high enough to cause fracture. Twinning, unlike slip, does produce c-direction strains.

Lattice Rotation in Tension

Slip normally causes a gradual lattice rotation or orientation change. Figure 7.11 shows the elongation of a long single crystal by slip on a single slip system. As the crystal is extended, the orientation of the tensile axis changes relative to the crystallographic elements. In Figure 7.11, the slip plane and slip direction are represented as being fixed in space with the tensile axis rotating relative to these, whereas in a real tension test, the tensile axis would remain vertical and the crystal elements would rotate. The deformation by slip is like shearing a deck of cards or a stack of poker chips on the top of a table. The following geometric treatment of the orientation change during tension is essentially that of Schmid and coworkers.

Slip causes translation of point P parallel to the slip direction to a new position, P′. Points C and C′ are constructed by extending the slip direction through O and constructing perpendiculars from P and P′ to the extension of the slip direction.

The normal distance between the OCC′ and PP′ is not changed by slip so, $\overline{OC'} = \overline{P'C'}$. Substituting $\overline{PC} = \ell_o \sin\lambda_o$ and $\overline{P'C'} = \ell \sin\lambda$, $\ell_o \sin\lambda_o = \ell \sin\lambda$, $\sin\lambda_o / \sin\lambda = \ell/\ell_o = 1 + e$, or

$$\sin\lambda = \sin\lambda_o/(1+e), \qquad (7.10)$$

where $e = (\ell - \ell)_o/\ell_o = \Delta\ell/\ell_o$ is the engineering strain.

Figure 7.10. Stresses necessary to operate several slip systems in beryllium. Compression parallel to the c-axis can be accommodated only by pyramidal slip, which requires a very high stress. From a talk by W. R. Blumenthal, C. N. Tomé, C. R. Necker, G. T. Gray, and M. C. Mataya.

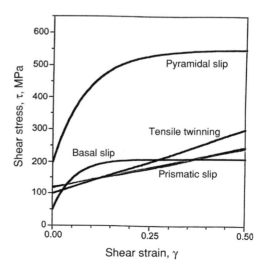

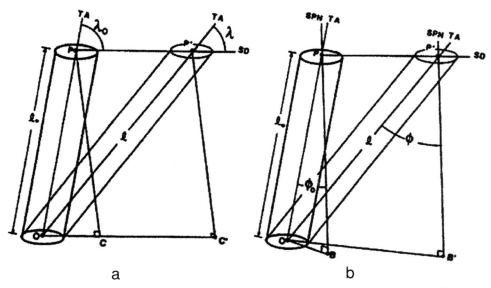

Figure 7.11. Extension of a long thin crystal. Slip causes point P to move to P'. (a) Points C and C' are constructed by extending the slip direction through point O and dropping normals from P and P'. (b) Points B and B' are constructed by dropping normals from P and P' to the slip plane through O.

Similarly, slip does not change the distance between two parallel slip planes. By an analogous construction, it can be shown that

$$\cos\phi = \cos\phi_o/(1 + e). \tag{7.11}$$

The shear strain, γ, associated with the slip is

$$\gamma = \overline{PP'}/\overline{PB} = \overline{OC}/\overline{PB} - \overline{OC}/\overline{PB} \tag{7.12}$$

Substituting $\overline{OC} = \cos\lambda$, $\overline{OC} = \ell_o \cos\lambda_o$ and $\overline{PB} = \ell_o \cos\phi_o = \ell \cos\phi$, equation (7.12) becomes $\gamma = \ell\cos\lambda/\ell\cos\phi - \ell_o\cos\lambda_o/\ell_o\cos\phi_o$, or

$$\gamma = \cos\lambda/\cos\phi - \cos\lambda_o/\cos\phi_o. \tag{7.13}$$

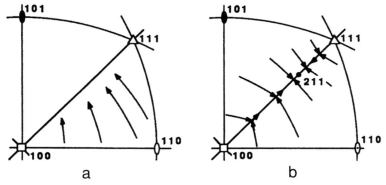

Figure 7.12. (a) Primary slip in the [101] direction causes the tensile axis to rotate toward [101]. (b) Once the tensile axis reaches the [100]–[111] symmetry line. Simultaneous slip in the [101] and [110] directions causes a rotation toward [211]. From W. F. Hosford, *Ibid.*

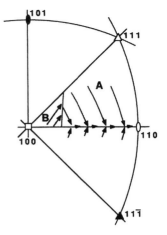

Figure 7.13. Rotation of the tensile axis for a bcc crystal deforming by <111>-pencil glide. If the tensile axis is in region B, rotation toward [111] will carry it into region A. In region A, slip in the [11$\bar{1}$] direction will rotate the tensile axis to the [100]–[110] symmetry line, where duplex slip will cause a net rotation toward [110]. From W. F. Hosford, *Ibid.*

It is clear from Figure 7.11 and equations (7.11) and (7.12) that ϕ increases and λ decreases during tensile extension. The orientation change in tension is a simple rotation of the slip direction toward the tensile axis. For an fcc crystal oriented in the basic triangle, the tensile axis will rotate toward [101] until it reaches the [100]–[111] boundary (Figure 7.12a). At this point, slip starts on [110]($1\bar{1}1$), which is the system most favored in the conjugate triangle. With equal slip on the two systems, the net rotation of the tensile axis is toward [211] (Figure 7.12b). The rotation along the [100]–[111] boundary becomes slow as the tensile axis approaches [211], which would be the stable end orientation if the crystal could be extended infinitely.

The tensile axis of a bcc crystal deforming by pencil glide rotates toward the active <111> slip direction. For orientations near [110] and [111] (region A), the tensile axis will rotate toward [11$\bar{1}$], as indicated in Figure 7.13. For crystal orientations in the basic triangle near [100] (region B), the rotation will be toward [111]. Once the tensile axis enters the region B, the rotation will be toward [11$\bar{1}$]. When the [100]–[110] boundary is reached, combined slip in the [111] and [11$\bar{1}$] directions will cause rotation toward [110], which is the stable end orientation.

Lattice Rotation in Compression

For the compression of thin flat single crystals, the compression axis rotates toward the slip plane normal, as sketched in Figure 7.14. For an fcc crystal, the compression

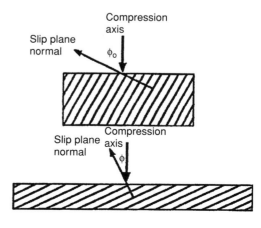

Figure 7.14. Compression causes a rotation of the slip plane normal toward the compression axis, which is equivalent to a rotation of the compression axis toward the slip plane normal.

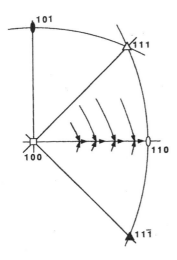

Figure 7.15. In an fcc crystal, the compression axis rotates toward [11$\bar{1}$] until it reaches the [100]–[110] symmetry line, where duplex slip causes a net rotation toward [110]. From W. F. Hosford, *Ibid.*

axis rotates toward [11$\bar{1}$] until it reaches the [100]–[110] boundary, as shown in Figure 7.15. Then simultaneous slip on the (11$\bar{1}$) and (111) planes will cause a net rotation toward [110], which is the stable end orientation.

For pencil glide, rotation toward the compression axis is equivalent to rotation away from the active slip direction. Figure 7.16 shows that for bcc crystals, orientations initially in region A will end up rotating to [111], whereas those initially in region A will rotate toward [100].

Texture Formation in Polycrystals

In polycrystalline metals, the grains undergo similar rotations, and these lead to *crystallographic textures* or *preferred orientations*. Table 7.2 shows that the experimentally observed textures developed in tensile extension and compression of polycrystals are in most cases the same as the end orientations predicted for single crystals.

Clearly, the preferred orientations (crystallographic textures) are the result of the lattice rotation that accompanies slip. Such preferred orientations cause

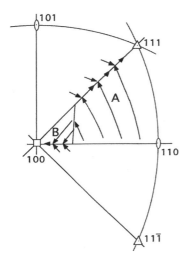

Figure 7.16. For a bcc crystal deforming by <111>-pencil glide, compression causes a rotation of orientations in region B away from [111] toward [100]. Orientations in region A rotate away from [11$\bar{1}$]. When they reach the [100]–[111] symmetry line, simultaneous slip in the [111] and [11$\bar{1}$] directions causes a net rotation toward [111].

Table 7.2. *Comparison of theoretical end orientations with preferred orientations*

Crystal structure and slip systems	Single crystals: End orientations		Polycrystals: Preferred orientation	
	Tension	Compression	Tension	Compression
fcc {111}<110>	<112>	<110>	<100>&<111>	<110>
bcc <111>-pencil glide	<110>	<100>&<111>	<110>	<100>&<111>
hcp (0001)<11$\bar{2}$0>	<11$\bar{2}$0>	[0001]	<11$\bar{2}$0>	[0001]
hcp {1$\bar{1}$00}<11$\bar{2}$0>&{1$\bar{1}$01}<11$\bar{2}$0>	<10$\bar{1}$0>	[0001]	<10$\bar{1}$0>	[0001]

anisotropy of both elastic and plastic properties. A rigorous analysis of slip in poly-crystals and texture formation is much more complicated. However, the agreement between the end orientations predicted for single crystals and the textures observed in polycrystals suggests that slip on the most highly stressed slip system plays the dominant role in texture formation.

Approximate Calculations of *R* Values

For textured sheets, it is possible to predict R values (ratio of width-to-thickness strains, $\varepsilon_w/\varepsilon_t$, in tension tests). To do this, attention is focused on the most highly stressed slip system. Once this system has been identified, the strains in the x, y, z coordinate system can be calculated in terms of the shear strain, γ, on the slip system. The ratio, $\varepsilon_w/\varepsilon_t$, is the R value.

EXAMPLE PROBLEM 7.5: Consider a copper sheet with a {01$\bar{1}$}<211> texture. Predict the R values in a tension test parallel to the [211] prior rolling direction.

Solution: Let x = [211] and z = [01$\bar{1}$]. The transverse direction must be y = [211] × [01$\bar{1}$] = [1$\bar{1}$$\bar{1}$]. Copper is fcc. The most favored slip systems would be (11$\bar{1}$)[101] and (1$\bar{1}$1)[110]. Consider tension applied along [211]. For slip on (11$\bar{1}$)[101], the resulting strains can be found as $\varepsilon_x = \gamma \cos \lambda_x \cos \phi_x$,

$$\cos \lambda_x = [211] \cdot [101] = \left\{ 2 \cdot 1 + 1 \cdot 0 + 1 \cdot 1 \right) / [(2^2 + 1^2 + 1^2)(1^2 + 0^2 + 1^2)]^{1/2} \right\}$$
$$= 3/\sqrt{(6 \cdot 2)}$$

$$\cos \phi_x = [211] \cdot [11\bar{1}] = \left\{ 2 \cdot 1 + 1 \cdot 1 + 1 \cdot \bar{1} \right) / [2^2 + 1^2 + 1^2)(1^2 + 1^2 + \bar{1}^2)]^{1/2}$$
$$= 2 3/\sqrt{(6 \cdot 3)}$$

so $\varepsilon_x = \gamma(1/\sqrt{6})$,

$$\varepsilon_z = \gamma \cos \lambda_z \cos \phi_z;$$

$$\cos \lambda_z = [01\bar{1}] \cdot [101] = \left\{ (0 \cdot 1 + 1 \cdot 0 + \bar{1} \cdot 1) / [(0^2 + 1^2 + \bar{1}^2)(1^2 + 0^2 + 1^2)]^{1/2} \right\}$$
$$= \bar{1}\sqrt{(2 \cdot 2)}$$

$$\cos \phi_z = [01\bar{1}] \cdot [11\bar{1}] = \left\{ (0 \cdot 1 + 1 \cdot 1 + \bar{1} \cdot \bar{1}) / [(0^2 + 1^2 + \bar{1}^2)(1^2 + 1^2 + \bar{1}^2)]^{1/2} \right\}$$
$$= 2\sqrt{(2 \cdot 3)}$$

so $\varepsilon_z = \gamma(-1/\sqrt{6})$,

$$\varepsilon_y = \gamma \cos \lambda_y \cos \phi_y;$$

$$\cos \lambda_y = [1\bar{1}\bar{1}] \cdot [101] = \{(1 \cdot 1 + \bar{1} \cdot 0 + \bar{1} \cdot 1)/[(1^2 + 1^2 + \bar{1}^2)(1^2 + 0^2 + 1^2)]^{1/2}\}$$
$$= 0/\sqrt{(3 \cdot 2)}$$
$$\cos \phi_y = [1\bar{1}\bar{1}] \cdot [11\bar{1}] = \{(1 \cdot 1 + \bar{1} \cdot 1 + \bar{1} \cdot \bar{1})/[(1^2 + 1^2 + \bar{1}^2)(1^2 + 1^2 + \bar{1}^2)]^{1/2}\}$$
$$= 1/\sqrt{(3 \cdot 3)}$$

so $\varepsilon_y = 0$.

Dividing, $R = \varepsilon_y/\varepsilon_z = 0$.

Analysis of slip on the $(1\bar{1}1) \cdot [110]$ system would have resulted in exactly the same conclusion.

Deformation of Polycrystals

Each grain in a polycrystal is surrounded by other grains, so it must deform in such a way that its change of shape is compatible with its neighbors. To satisfy this compatibility, there must be slip on many systems within a grain. G. I. Taylor* assumed that each grain of a polycrystal undergoes the same shape change (set of strains) as the whole polycrystal, $\varepsilon_x, \varepsilon_y, \ldots \gamma_{xy}$, expressed relative to the external axes. For each orientation of grain, these strains may be written in terms of the shear strains on the twelve slip systems, $\gamma_a, \gamma_b, \ldots, \gamma_j$,

$$\begin{aligned}
\varepsilon_x &= m_{xa}\gamma_a + m_{xb}\gamma_b + \cdots m_{xj}\gamma_j, \\
\varepsilon_y &= m_{ya}\gamma_a + m_{yb}\gamma_b + \cdots m_{yj}\gamma_j, \\
\lambda_{yz} &= m_{yza}\gamma_a + m_{yzb}\gamma_b + \cdots m_{yzj}\gamma_j, \\
\lambda_{zx} &= m_{zxa}\gamma_a + m_{zxb}\gamma_b + \cdots m_{zxj}\gamma_j, \\
\gamma_{xy} &= m_{xya}\gamma_a + m_{xyb}\gamma_b + \cdots m_{xyj}\gamma_j.
\end{aligned} \qquad (7.14)$$

The terms m_{xa}, m_{xb}, m_{xya}, and so on, are the resolving factors (products of cosines of the angles between slip elements and the external axes). There are five independent equations in this set. The corresponding equation for ε_z is not independent because $\varepsilon_z = -\varepsilon_x - \varepsilon_y$. With $\varepsilon_x, \varepsilon_y, \ldots, \gamma_{xy}$ fixed, a solution to equation (7.14) requires that at least five of the terms, $\gamma_a, \gamma_b, \ldots, \gamma_j$ must be finite. If a crystal is to deform with an arbitrary shape change, at least five slip system (a, b, ..., j) must be active. Taylor concluded that an imposed shape change would be achieved with the least work when only five were active.

For a uniaxial tensile test, $\varepsilon_y = \varepsilon_z = -(1/2)\varepsilon_x$ and $\gamma_{yz} = \gamma_{zx} = \gamma_{xy} = 0$. Taylor solved equation (7.15) for fcc crystals that deform by {111}<110> slip, assuming various combinations of 5 of the 12 shear strain terms, $\gamma_a, \gamma_b, \ldots, \gamma_j$ were finite. For each orientation of crystal, he selected the combination with the lowest sum of shear strains, $\Sigma|\gamma_i|$. Taylor assumed that the shear stress to cause slip would be the same on all active systems. The incremental work/volume expended by slip in a given grain is given by $dw = \Sigma\tau_i d\gamma_i$, where τ_i is the shear stress required for slip

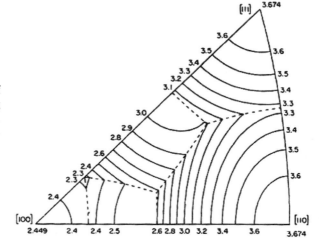

Figure 7.17. Orientation dependence of $M = \sigma_x/\tau$ for fcc crystals undergoing axisymmetric deformation. From G. Y. Chin and W. L. Mammel, *Trans TMS-AIME*, v. 239, 1967.

on the ith slip system and γ_i is the shear strain on that system. Removing τ from the summation, $dw = \tau \Sigma d|\gamma_i|$ or simply $\tau d\gamma$, where now $d\gamma$ means $\Sigma d|\gamma_i|$. For uniaxial tension, the work/volume can also be expressed in terms of the external stress and strain, $dw = \sigma_x d\varepsilon_x$. Equating internal and external work,

$$\sigma_x d\varepsilon_x = \tau d\gamma. \tag{7.15}$$

The values of

$$M = d\gamma/d\varepsilon_x = \sigma_x/\tau \tag{7.16}$$

for different orientations of fcc crystals undergoing axisymmetric deformation (as in a tension test of a randomly oriented polycrystal) are plotted in Figure 7.17.

The value of M averaged over all orientations is $\overline{M} = 3.067$. The results are similar for bcc metals. They are identical for bcc metals that slip on only <111>{110} systems. For <111>-pencil glide, $\overline{M} = 2.733$.

Using the value of $\overline{M} = 3.067$, the stress–strain curve for a randomly oriented polycrystal can be used to predict the τ-γ curve for a single crystal. Points on the σ-ε curve, $\varepsilon_x = \gamma/\overline{M}$ and $\sigma = \overline{M}\tau$, are calculated from corresponding points on the τ-γ curve. The results are in good agreement with experiments on single crystals that exhibit no easy glide (Figure 7.18).

Texture Strengthening

Most polycrystals have crystallographic textures. The grains are not randomly oriented. Some grains are more favorably orientations for slip than others. The Schmid factors for the most favorably oriented and least favorably oriented single crystals in fcc and bcc metals differ by a factor of a little less than 2 (0.5–0.272 at [111] for fcc 0.5–0.314 at [111] for bcc). For polycrystals in axially symmetric flow, the variation in Taylor factors is somewhat less, 3.674 to about 2.228 for fcc and 3.182 to about 2.08 for bcc.

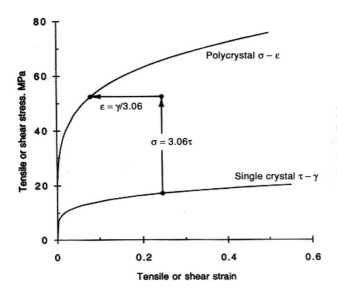

Figure 7.18. Comparison of stress–strain curves for single and polycrystalline aluminum. The solid curve for the polycrystalline aluminum was constructed from the single crystal curve by taking $\sigma = 3.06\tau$ and $\varepsilon = \gamma/3.06$.

Figure 7.18 can be used to predict the strength of textured fcc polycrystals undergoing axisymmetric deformation. For example, $M = 1.5\sqrt{6} = 3.67$ for wires with a <111>-fiber texture, and $M = \sqrt{6} = 2.45$ for wires with a <100>-fiber texture. Similar calculations for other shape changes have been used to predict the yield loci of textured fcc and bcc* metals.

For hcp metals, crystallographic texture has a large effect on yielding. The Schmid factor for basal slip varies with orientation from 0.5 to 0. Grains oriented unfavorably for $<11\bar{2}0>$ slip, can deform by twinning or slip on other systems that require much higher resolved shear stresses. Analytic treatment of the problem is difficult because several types of slip and twinning systems contribute to the deformation. For sheets of many hcp metals, the crystallographic textures can be approximately described as having the basal plane, (0001), parallel to the plane of the sheet, as schematically shown in Figure 7.19. If this texture were perfect, the $<11\bar{2}0>$ would lie in the plane of the sheet and slip could not cause any thinning of a sheet with this ideal texture. Therefore, the yield strengths under biaxial tension are much higher than those under uniaxial tension. Figure 7.20 shows the stress–strain curves for a plate of zirconium rolled in such a way as to produce a texture with a strong alignment of the basal plane with the rolling plane. The yield locus of a textured sheet of titanium is shown in Figure 7.21.

Effects of Texture on Microstructure

Peck and Thomas[†] observed that the microstructures of drawn wires of tungsten metals contain grains that are elongated in along the wire axis but seem to curl about one another, as shown in Figure 7.22.

A similar microstructure was observed in draw iron wires (Figure 7.23).

* R. W. Logan and W. F. Hosford, *Int. J. Mech. Sci.*, v. 22 (1980).
† J. F. Peck and D. A. Thomas, *Trans. TMS-AIME*, v. 221 (1961).

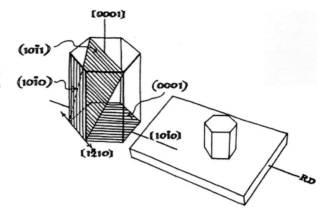

Figure 7.19. Schematic drawing of the idealized texture of an hcp sheet and slip elements. From W. F. Hosford and W. A. Backofen, *Fundamentals of Deformation Processing*, Syracuse University Press (1964).

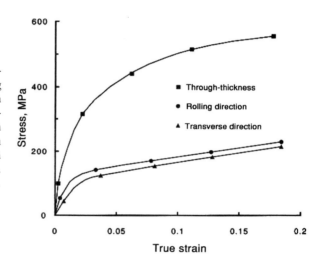

Figure 7.20. Stress–strain curves for a cold-rolled plate of zirconium. The rolling direction was rotated 90 degrees between passes. The basal plane was oriented parallel to the plate causing the strength in the through-thickness direction to be much higher than in directions in the plate. Data from P. J. Maudlin, G. T. Gray, C. M. Cody, and G. C. Kaschnev, *Phil. Trans. R. Soc. London*, v. 357A (1999).

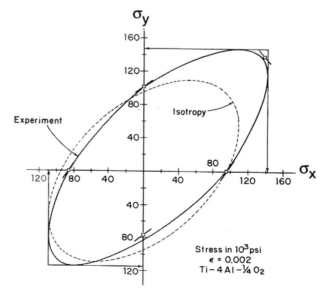

Figure 7.21. Yield locus of a textured sheet of titanium alloy, Ti-4Al-1/4O. The strength in plane–strain tension is about 40% greater than in uniaxial tension. From D. Lee and W. A. Backofen, *Trans. TMS-AIME*, v. 242 (1968).

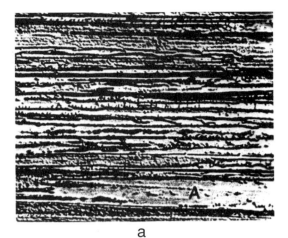

a

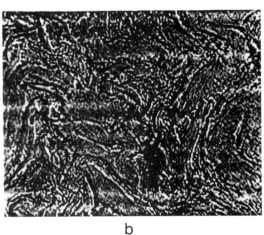

b

Figure 7.22. Microstructure of a tungsten wire. Longitudinal section (a) and cross-section (b). The normal to the plane of view for grain A is a <110> direction. From J. F. Peck and D. A. Thomas, *Trans. TMS-AIME*, v. 221 (1961).

Figure 7.23. Curly grain structure of an iron wire drawn to a strain of 2.7. Courtesy of J. F. Peck and D. A. Thomas.

Figure 7.24. Schematic drawing of the shape of grains in drawn wires of bcc metals. From W. F. Hosford, *Mechanics of Crystals and Textured Polycrystals*, Oxford Science Publ. (1993).

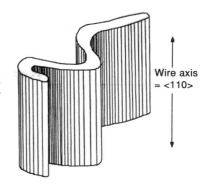

Wire axis = <110>

These microstructures suggest that the grain in the drawn wires are shaped like long ribbons, folded about the wire axis as illustrated schematically in Figure 7.24.

The reason for this microstructure can understood in terms of the <110> crystallographic texture in wires of bcc metals (Figure 7.25). With a <110> direction parallel to the wire axis, two of the <111> slip directions allow thinning parallel to the <001> lateral direction, but not to the <1$\bar{1}$0> lateral direction. The other two <111> slip directions, are perpendicular to the wire axis and should not operate.

This explains the appearance of a few wide grains in the longitudinal section, such as grain A in Figure 7.23. A grain appears to be wide when the plane of the section is parallel to [0$\bar{1}$1].

Swaging of wires by hammering with rotating dies produces the microstructure in which the grains are oriented so that their <100> directions are oriented in the radial direction, producing a <011>{001} cylindrical texture.

The microstructure of pearlite after wire drawing is similar to that of pure bcc metals. During wire drawing, the ferrite develops a <110> wire texture.

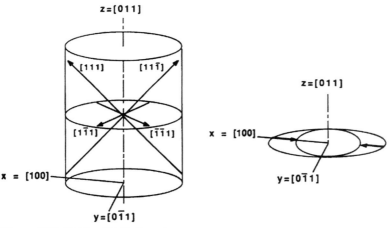

Figure 7.25. With a [011] direction parallel to the wire axis, the [111] and [11$\bar{1}$] slip directions cause thinning parallel to the lateral [100] direction without any thinning parallel to [0$\bar{1}$1]. The other two <111> slip directions are unstressed. From W. F. Hosford, *Ibid.*

A similar relation between texture and microstructure during heavy compression fcc metals.* With the <110> compression texture, slip should occur on only two of the four {111} planes, with the result that lateral spreading occurs only in the <001> direction and grains must curl about each other. Hexagonal close-packed metals develop similar microstructures during wire drawing as a result of their wire textures (B. C. Wonsiewitcz, personal communication, circa 1963).

REFERENCES

J. F. W. Bishop and R. Hill, *Phil. Mag.*, ser. 7, v. 42 (1951), pp. 414–427, 1298–1307.
E. Schmid and W. Boas, *Kristallplastizität mit besonderer Berücksichtigung der Metalle*, Springer-Verlag, Berlin (1935).
G. I. Taylor, *Proc. Roy. Soc.*, V. A116 (1927).
W. F. Hosford, *The Mechanics of Crystals and Textured Polycrystals*, Oxford University Press (1993).
C. S. Barrett, *Structure of Metals*, McGraw-Hill, New York and London (1943).
B. D. Cullity, *Elements of X-Ray Diffraction*, Addison-Wesley (1956).
W. A. Backofen, *Deformation Processing*, Addison-Wesley (1972).
W. F. Hosford, *The Mechanics of Crystals and Textured Polycrystals*, Oxford University Press (1993).
W. C. Leslie, *The Physical Metallurgy of Steels*, McGraw-Hill (1981).

Notes

Crystallographic slip was reported as early as 1867 by Reusch, who observed it in NaCl crystals. The first report of slip lines in metals was by Ewing and Rosenhain in 1899. Slip was extensively studied in the early part of the twentieth century. Two classic books were written in 1935 and 1936. One was *Kristallplastizität mit besonderer Berücksichtigung der Metalle,* Springer-Verlag (1935) by Erich Schmid and W. Boas. This was written in German and under Hitler it was forbidden to have a German book translated into another language. After World War II, the Allies took German copyrights, and the book was translated and published by F. A. Hughes and Co. Ltd. (1950) under the title *Plasticity of Crystals with Special Reference to Metals.* The second book was *Distortion of Crystals* (1936) by Constance F. Elam, who later publisher under the name Mrs. G. H. Tipper. Later, W. Boas published *An Introduction to the Physics of Metals and Alloys*, Melbourne University Press (1947).

Erich Schmid (May 4, 1896–October 22, 1983) was born in Bruck-on-the-Mur in Austria and studied in Vienna. His studies were interrupted by service in World War I. He obtained his doctorate in 1920 and remained at the Technische Hochschule in Vienna for 2 years. In 1922, he left Austria for 30 years to work in various places in Germany and Switzerland. In Berlin, he did research on metal crystals and formulated what is known today as "Schmid's law." During this period, he introduced physical concepts into the study of metals to supplement the earlier chemical concepts. His research with Dehlinger and Sachs forms the basis for much of the present understanding of mechanical behavior of metals. During World War II, Schmid worked on developing and improving substitute materials. In 1951, he returned to the University of Vienna in Austria and remained there until 1967.

* W. F. Hosford, Jr., *Trans. TMS-AIME*, v. 230 (1964).

 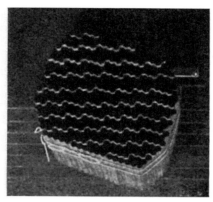

Figure 7.26. Taylor's illustration of shearing a pack of pencils. The direction of shearing is fixed (parallel to the pencils), but shearing can occur on any plane. From G. I. Taylor, *Proc. Roy. Soc.*, v. 62 (1926).

He served as president of the Austrian Academy of Science for 10 years and was responsible for the establishment of several research institutes. He was awarded honorary doctorates by several technical universities in Germany, and in 1979, was recognized by Austria's highest achievement award. Scientific organizations and institutes in Germany, Austria, and Japan have given him numerous honors.

In the 1930s, Boas emigrated from Germany to Australia to escape the Nazi persecution of Jews. He spent the remainder of his career at the Melbourne University.

Geoffrey Ingram Taylor (1886–1975) was born in London. He contributed a steady stream of important papers, principally on fluid mechanics and plasticity. His grandfather was George Boole, who is renowned for developing the foundations of what is now called *Boolean algebra*. Taylor did both his undergraduate studies and graduate research at Cambridge. Taylor never received a doctorate, simply because Cambridge did not grant them until many years later. During World War I, he did research at Farnsborough on aircraft. After the war, he was appointed lecturer at Cambridge and remained there until his death. Taylor must be regarded as one of the most outstanding scientists of the first half of the twentieth century. His work with turbulent flow is fundamental. Taylor's studies of plasticity ranged from testing continuum models, to experiments and analysis of slip in single crystals, to pioneering work that bridged the gap between crystal and continuum mechanics. He is regarded as one of the founders of dislocation theory, along with Orowan and Polanyi. Taylor coined the term *pencil glide* to describe slip in bcc metals on any plane containing a <111>-slip direction. He chose the term because slip in one direction on any plane is similar to the behavior of a pile of pencils. Figure 7.26 is a copy of his illustration.

Problems

1. If a single crystal of aluminum were stressed in uniaxial tension applied along the [0$\bar{2}$1] direction, which slip system (or systems) would be most highly stressed?

2. A single crystal of copper is loaded under a stress state such that $\sigma_2 = -\sigma_1, \sigma_3 = \tau_{23} = \tau_{31} = \tau_{12} = 0$. Here, $1 = [100]$, $2 = [010]$, and $3 = [001]$.

    A. When the stress $\sigma_1 = 6$ kPa, what is the shear stress, τ, on the
        i. $(111)[10\bar{1}]$ slip system?
        ii. $(111)[1\bar{1}0]$ slip system?
        iii. $(111)[01\bar{1}]$ slip system?
    B. On which of these systems would you expect slip to first occur as the applied stresses increased?
    C. Assuming slip on that system, determine the ratios of the strains, $\varepsilon_2/\varepsilon_1$ and $\varepsilon_3/\varepsilon_1$.

3. A single crystal of aluminum was grown in the shape of a tensile bar with the $[32\bar{1}]$ direction aligned with the tensile axis.

    A. Sketch a standard cubic projection with $[100]$ at the center and $[001]$ at the North Pole, and locate the $[32\bar{1}]$ direction on this projection.
    B. Which of the $\{111\}<110>$ slip system(s) would be most highly stressed when tension is applied along $[32\bar{1}]$? Show the slip plane normal(s) and the slip direction(s) of the system(s) on your plot.
    C. What will be the ratio of the shear stress on the slip system to the tensile stress applied along $[32\bar{1}]$?

4. An aluminum single crystal is subjected to a tensile stress of $\sigma_x = 250$ kPa parallel to x $= [100]$ and a compressive stress, $\sigma_y = -50$ kPa parallel to y $= [010]$ with $\sigma_z = \tau_{yx} = \tau_{zx} = \tau_{xy} = 0$. What is the shear stress on the (111) plane in the $[1\bar{1}0]$ direction?

5. Consider an aluminum single crystal that has been stretched in tension applied parallel to x $= [100]$, $(\sigma_x = 250$ kPa$)$ and compressed parallel to y $= [010]$, $(\sigma_y = -50$ kPa$)$ with $\sigma_z = 0$, where z $= [001]$. Assume that slip occurred on the (111) in the $[1\bar{1}0]$ direction and only on that slip system. Also assume that the strains are small.

    A. Calculate the ratios of resulting strain, $\varepsilon_y/\varepsilon_x$ and $\varepsilon_z/\varepsilon_x$.
    B. If the crystal were strained until $\varepsilon_x = 0.0100$, what will the angle be between the tensile axis and $[100]$?

6. NaCl crystals slip on $\{110\}<1\bar{1}0>$ slip systems. There are six systems of this type. Consider a crystal subjected to uniaxial compression parallel to z $= [110]$.

    A. On which of the $\{110\}<1\bar{1}0>$ slip systems would the shear stress be the highest? That is, on which of the systems would slip be expected?
    B. Let the lateral directions be x $= [1\bar{1}0]$ and y $= [001]$. Determine the shape change that occurs as the crystal deforms on one of these systems by finding the ratios, $\varepsilon_y/\varepsilon_z$ and $\varepsilon_x/\varepsilon_z$. Describe the shape change in words. (*Hint:* Analyze one of the slip systems in your answer to A. Be careful about signs.)

7. Predict the R value for a sheet of an hcp metal with (0001) parallel to the rolling plane and $[10\bar{1}0]$ parallel to the rolling direction.

8. Determine the number of independent slip systems for crystals with each combination of slip systems listed here. (The simplest way to do this is to determine how many of the strains ε_1, ε_2, γ_{23}, γ_{31}, and γ_{12} can be independently imposed on the crystal.)

    A. Cubic crystal that deforms by {100}<011> slip.
    B. Cubic crystal that deforms by slip on {100}<011> and {100}<001> systems.
    C. Cubic crystal that deforms by {110}<$\overline{1}$10> slip.
    D. Hcp crystal that deforms by slip on (0001)<11$\overline{2}$0> and {10$\overline{1}$0}<11$\overline{2}$0> systems.

9. A tetragonal crystal slips on {011}<111> systems. How many of the strains, ε_x, ε_y, ε_z, γ_{yz}, γ_{zx}, γ_{xy}, can be accommodated? (Note that in a tetragonal crystal the {011} family does not include (110) or ($\overline{1}$10).) See Figure 7.27.

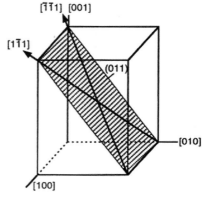

Figure 7.27. A tetragonal crystal with {011}<111> slip systems.

10. A. Consider an fcc single crystal extended in uniaxial tension parallel to [321]. Will the Schmid factor, $m = \cos \lambda \cos \phi$, for the most highly stressed slip system increase, decrease, or remain constant as the crystal is extended?
    B. Consider an hcp single crystal that slips easily only on (0001)<11$\overline{2}$0> slip systems. If an hcp crystal is extended in uniaxial tension in a direction oriented 45 degrees from the c axis and 45 degrees from the most favored <11$\overline{2}$0> slip direction, will the Schmid factor, $m = \cos \lambda \cos \phi$, for the most highly stressed slip system increase, decrease, or remain constant?

11. Consider a sheet of an fcc metal that has a {110}<001> texture. That is, a {110} plane is parallel to the plane of the sheet, and a <001> direction is parallel to the prior rolling direction.

    A. Predict the value of R_o (the strain ratio measured in a rolling direction tension test).
    (*Hint:* Let x, y, and z be the rolling, transverse, and sheet normal directions.) Assign specific indices [hkℓ] to the rolling and sheet normal directions. Find the specific indices of the transverse direction, y. Then sketch a standard cubic projection showing these directions. (It is convenient to choose x, y, and z so that they lie in the hemisphere of the projection.) For uniaxial tension along x, determine which slip systems will be active, and assume an equal shear strain, γ_i, on each. For each system, calculate the resulting strains, ε_x, ε_y, and ε_z, in terms of γ_i and sum these over all slip systems. Assume equal amounts of slip on all equally favored slip systems. Now predict the strain ratio R_o.
    B. Predict R_{90}.

12. For a unit elongation along a <111> direction in a bcc metal, determine the ratio of the amount of slip required for axially symmetric flow to that required for plane strain.

13. Predict the ratio of the flow stresses for copper wire with a <111> texture to that with a <100> texture. Assume power law hardening with $n = 0.3$.

8 Dislocation Geometry and Energy

Introduction

It was well know in the late nineteenth century that crystals deformed by slip. In the early twentieth century, the stresses required to cause slip were measured by tension tests of single crystals. Dislocations were not considered until after it was realized that the measured stresses were far lower than those calculated from a simple model of slip. In the mid-1930s, G. I. Taylor,[*] M. Polanyi,[**] and E. Orowan[†] independently postulated that preexisting crystal defects (dislocations) were responsible for the discrepancy between measured and calculated strengths. It took another two decades and the development of the electron microscope before dislocations were observed directly.

Slip occurs by the motion of dislocations. Many aspects of the plastic behavior of crystalline materials can be explained by dislocations. Among these are how crystals can undergo slip, why visible slip lines appear on the surfaces deformed crystals, why crystalline materials become harder after deformation, and how solute elements affect slip.

Theoretical Strength of Crystals

Once it was established that crystals deformed by slip on specific crystallographic systems, physicists tried to calculate the strength of crystals. However, the agreement between their calculated strengths and experimental measurements was poor. The predicted strengths were orders of magnitude too high, as indicated in Table 8.1.

The basis for the theoretical calculations is illustrated in Figure 8.1. Each plane of atoms nestles in pockets formed by the plane below (Figure 8.1a). As a shear stress is applied to a crystal, the atoms above the plane of shear must rise out of the stable pockets in the plane below and slide over it until they reach the unstable position shown in Figure 8.1b. From this point, they will spontaneously continue to shear to the right until they reach a new stable position (Figure 8.1c).

---

[*] G. I. Taylor, *Proc. Roy. Soc. (London),* v. A145 (1934).
[**] M. Polanyi, *Z. Physik,* v. 89 (1934).
[†] E. Orowan, *Z. Physik,* v. 89 (1934).

Table 8.1. *Critical shear stress for slip in several materials*

| | | Critical Shear Stress (MPa) | |
Metal	Purity %	Experiment*	Theory
Copper	>99.900	1.00	414
Silver	99.990	0.60	285
Cadmium	99.996	0.58	207
Iron	99.900	~7.00	740

* There is considerable scatter caused by experimental variables,
  particularly purity.

For simplicity, consider a simple cubic crystal. An applied shear stress, τ, will displace one plane relative to the next plane, as shown in Figure 8.2.

When the shear displacement, x, is 0, d, $2d$, or nd (i.e., $\gamma = 0, 1, 2, \ldots n$), the lattice is restored, so it should be zero. The shear stress, t, is also zero when the displacement is $x = (1/2)d, (3/2)d$, etc. ($\gamma = 1/2, 3/2, \ldots$). A sinusoidal variation of τ with γ, as shown in Figure 8.3, seems reasonable, so

$$\tau = \tau_{max} \sin(2\pi\gamma). \tag{8.1}$$

Here, τ_{max} is the theoretical shear stress required for slip. If the stress is less than τ_{max}, the shear strain is elastic and will disappear when the stress is released. For very low values of γ (Figure 8.4), Hooke's law should apply,

$$\tau = G\gamma. \tag{8.2}$$

This can be expressed as

$$(d\tau/d\gamma)_{\gamma\to0} = G. \tag{8.3}$$

Differentiating equation (8.1), $(d\tau/d\gamma)_{\gamma\to0} = 2\pi\tau_{max}\cos(2\pi\gamma)_{\gamma\to0} = 2\pi\tau_{max}$. Substituting into equation (8.3),

$$\tau_{max} = G/2\pi \tag{8.4}$$

A somewhat more sophisticated analysis for real crystal structures predicts something close to

$$\tau_{max} = G/10. \tag{8.5}$$

In Table 8.1, the theoretical strength values predicted by equation (8.5) are orders of magnitude higher than experimental measurements. The poor agreement

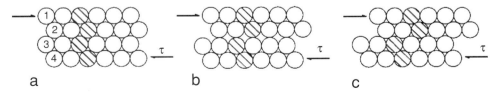

Figure 8.1. Model of slip occurring by sliding of planes 1 and 2 over planes 3 and 4. At the unstable condition (b), the planes are at attracted equally to the stable configurations in (a) and (c).

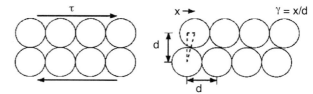

Figure 8.2. Model used to calculate the theoretical shear stress for slip.

between experimental and theoretical strengths indicated that there was something wrong with the theory. The problem with the theoretical calculations is that it was assumed that slip occurs by one entire plane of atoms sliding over another at the same time. Taylor, Orowan, and Polanyi realized that it is not necessary for a whole plane to slip at the same time. They postulated that crystals have preexisting defects that are boundaries between regions that are already displaced relative to one another by a unit of slip. These boundaries are called *dislocations*. Movement of a dislocation allows slip to occur. The critical stress for slip is the stress required to move a dislocation. At any instant, slip need only occur at the dislocation rather than over the entire slip plane.

The Nature of Dislocations

One special form of a dislocation is an *edge* dislocation sketched in Figure 8.5. The geometry of an edge dislocation can be visualized as having cut part way into a perfect crystal and then inserted an extra half plane of atoms. The dislocation is the bottom edge of this extra half plane. The *screw dislocation* (Figure 8.6) is another special form. It can be visualized as a spiral-ramp parking structure. One circuit around the axis leads one plane up or down. Planes are connected in a manner similar to the levels of a spiral parking ramp.

An alternate way of visualizing dislocations is illustrated in Figure 8.7. An edge dislocation is created by shearing the top half of the crystal by one atomic distance perpendicular to the end of the cut (Figure 8.7b). This produces an extra half plane of atoms, the edge of which is the center of the dislocation. The other extreme form of a dislocation is the *screw dislocation*. A screw dislocation is generated by cutting into a perfect crystal and then shearing half of it by one atomic distance in a direction parallel to the end of the cut (Figure 8.7c). The end of the cut is the dislocation.

In both cases, the dislocation is a boundary between regions that have and have not slipped. When an edge dislocation moves, the direction of slip is perpendicular to the dislocation. In contrast, movement of a screw dislocation causes slip in the direction parallel to itself. The edge and screw are extreme cases. A dislocation may be neither parallel nor perpendicular to the slip direction.

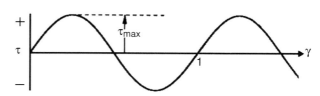

Figure 8.3. Theoretical variation of shear stress with displacement.

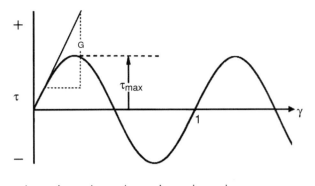

Figure 8.4. The shear modulus, G, is the initial slope of the theoretical τ versus γ curve.

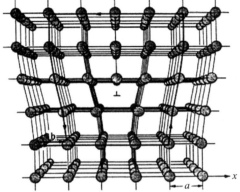

Figure 8.5. An edge dislocation is the edge of an extra half plane of atoms. From A. G. Guy, *Elements of Physical Metallurgy*, Addison-Wesley (1951).

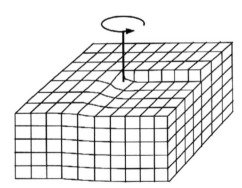

Figure 8.6. A screw dislocation. Traveling on the lattice around the screw dislocation is like following the thread of a screw.

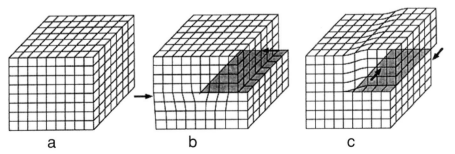

a   b   c

Figure 8.7. Consider a cut made in a perfect crystal (a). If one half is sheared by one atom distance parallel to the direction of the cut, an edge dislocation results (b). If one half is sheared by one atom distance perpendicular to the direction of the cut, a screw dislocation results (c).

Figure 8.8. The Burgers vector of a dislocation can be determined by drawing a clockwise circuit that would close if it were drawn in a perfect crystal. If the circuit is drawn around a dislocation, the closure failure is the Burgers vector.

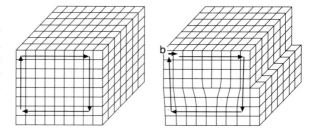

Burgers Vectors

Dislocations are characterized by *Burgers vectors*. Consider an atom-to-atom circuit in Figure 8.8 that would close on itself if made in a perfect crystal. If this same circuit is constructed so that it goes around a dislocation, it will not close. The closure failure is the Burgers vector, denoted by b. The Burgers vector can be considered a *slip vector* because its direction is the slip direction and its magnitude is the magnitude of the slip displacement caused by its movement of the dislocation.

A dislocation may wander through a crystal with its orientation changing from place to place, but its Burgers vector is the same everywhere. See Figure 8.9. If the dislocation branches into two dislocations, the sum of the Burgers vectors of the branches equals its Burgers vector.

There is a simple notation system for describing the magnitude and direction of the Burgers vector of a dislocation. The direction is indicated by direction indices, and the magnitude by a scalar preceding the direction. For example, $b = (a/3)[2\bar{1}1]$ in a cubic crystal means that the Burgers has components of $2a/3$, $-a/3$, and $a/3$ along the [100], [010], and [001] directions, respectively, where a is the lattice parameter. Its magnitude is $|b| = [(2a/3)^2 + (-a/3)^2 + (a/3)^2]^{1/2} = a\sqrt{6}/3$.

A dislocation in an fcc crystal corresponding to a full slip displacement that restores the lattice would have a Burgers vector (a/2)<110>. In this case, the magnitude is $a\sqrt{2}/2$.

EXAMPLE PROBLEM 8.1: Using proper notation, write the Burgers vector for a dislocation corresponding to the shortest repeat distance in a bcc crystal.

Solution: The shortest repeat distance is half of the body diagonal of the unit cell. $b = (a/2)<111>$.

Figure 8.9. A dislocation may wander through a crystal, but everywhere it has the same Burgers vector. Where it is parallel to its Burgers vector, it is a screw (a). Where it is perpendicular to its Burgers vector, it is an edge (b). Elsewhere it has mixed character.

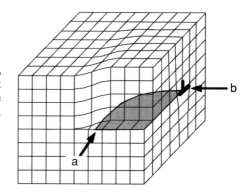

Energy of a Screw Dislocation

The energy associated with a screw dislocation is the energy required to distort the lattice surrounding the dislocation elastically. The distortion is severe near the dislocation, but decreases with distance from it. Consider an element of length, L, and thickness, dr, at a distance, r, from the center of a screw dislocation (Figure 8.10a). The volume of the element is $2\pi r L dr$. Imagine unwrapping this element so that it lies flat (Figure 8.10b). Unwrapping causes no strain because the element is differentially thin.

The shear strain associated with this element is

$$\gamma = b/(2\pi r), \tag{8.6}$$

where b is the Burgers vector of the screw dislocation. The energy/volume associated with an elastic distortion is

$$U_v = (1/2)\tau\gamma, \tag{8.7}$$

where τ is the shear stress necessary to cause the shear strain, γ. According to Hooke's law,

$$\tau = G\gamma, \tag{8.8}$$

where G is the shear modulus. Combining equations (8.6), (8.7), and (8.8), the energy/volume is

$$U_v = G\gamma^2/2 = (1/2)Gb^2/(2\pi r)^2. \tag{8.9}$$

The elastic energy, dU, associated with the element is its energy/volume times its volume,

$$dU = [(1/2)Gb^2/(2\pi r)^2](2\pi r L dr) = Gb^2 L/(2\pi)dr/r. \tag{8.10}$$

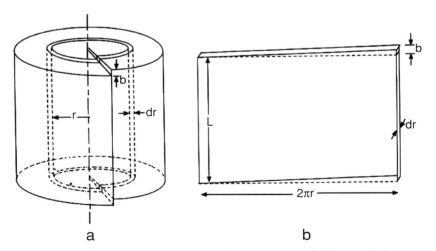

Figure 8.10. (a) Screw dislocation in a cylindrical crystal; (b) flattened element.

The total energy of the dislocation per length, L, is obtained by integrating equation (8.10):

$$U/L = [Gb^2/(4\pi)] \int_{r_0}^{r_1} dr/r = [Gb^2/(4\pi)] \ln(r_1/r_0). \qquad (8.11)$$

EXAMPLE PROBLEM 8.2: Equation (8.9) gives the energy/volume as a function of r. Calculate the distance from the core of a screw dislocation at which the energy per volume equals to the heat of vaporization, H_v, and express this distance in terms of r/b. Evaluate this critical value of r/b for copper. $r = 8.93$ Mg/m^3, $G = 77$ GPa, $b = 0.255$ nm, $H_v = 4.73$ MJ/kg.

Solution: Solving equation (8.9) for r/b,

$$(r/b)^2 = G/(8\pi^2 U_v) = (77 \times 10^9 \text{ Pa})/(8\pi^2 8.93 \times 10^3 \text{ kg/m}^3 \times 4.73 \times 10^6 \text{ J/kg})$$
$$= 0.024, \quad r/b = 0.16.$$

For copper, $r = 0.16 \times 0.255 = 0.04$ nm.

Clearly, equation (9.9) is not valid for values of $r < 0.04$ nm because it predicts that the material within this radius would vaporize.

The problem, now, is to decide the appropriate values for the lower limit, r_0, and the upper limit, r_1, of the integral in equation (8.11). We might be tempted to let r_0 be 0 and r_1 be ∞, but both would cause the value of U in equation (8.11) to be infinite. First, consider the lower limit, r_0. Equation (8.6) predicts an infinite strain at $r = 0$, which corresponds to an infinite energy/volume at the core of the dislocation, $(r_0 = 0)$. This is clearly unreasonable. The energy/volume cannot possibly be higher than the heat of vaporization. This discrepancy is a result of the breakdown of Hooke's law at the high strains that exist near the core of the dislocation. The stress, t, is proportional only for small strains. A lower limit, r_0, can be chosen so that the neglected energy of the core in the region, $0 < r < r_0$, is equal to the over-estimation of the integral for $r > r_0$. See Figure 8.11. A value of $r_0 = b/4$ has been suggested as reasonable.

The upper limit, r_1, cannot be any larger that the radius of the crystal. A reasonable value for r_1 is half of the distance between dislocations. At greater distances, the stress field is probably neutralized by stress fields of other dislocations. The value of r_1 is often approximated by $10^5 b$, which corresponds to a dislocation density of 10^{10} dislocations per m^2. This is convenient because

Figure 8.11. Energy density near core of a screw dislocation.

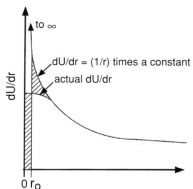

$\ln(10^5 b/0.25b) = \ln(4 \times 10^5) = 12.9 \approx 4\pi$. With this approximation, the energy per length, U_L, from equation (8.11) simplifies to

$$U_L \approx Gb^2. \tag{8.12}$$

The process of choosing the limits of integration so that the expression for the energy/length simplifies to equation (8.12) may seem arbitrary. Yet the results, although approximate, are reasonable. The derivation of the energy of an edge dislocation is more complicated because the stress field around an edge dislocation is more complex. For edge dislocations,

$$U_L \approx Gb^2/(1 - u), \tag{8.13}$$

where u is Poisson's ratio. Thus, the energy of an edge dislocation is greater than that of a screw by a factor of $1/(1 - v) \approx 1.5$.

There are two important features of equations (8.12) and (8.13). One is that the energy of a dislocation is proportional to its length. Energy per length is equivalent to line tension, or a contractile force. The units of energy/length are J/m, which is the same as the units of force, N.

The second important feature is that the energy of a dislocation is proportional to b^2. This controls the energetics of reactions between parallel dislocations.

Reactions between Parallel Dislocations and Frank's Rule

Two parallel dislocations may combine and form a third dislocation. If they do, the Burgers vector of the third dislocation will be the vector sum of the Burgers vectors of the two reacting dislocations. That is, if $b_1 + b_1 \rightarrow b_3$ then $b_1 + b_1 \rightarrow b_3$. The reaction is energetically favorable if it lowers the energy of the system. Frank's rule states that because the energy of a dislocation is proportional to b^2, the reaction is favorable if $b_1^2 + b_2^2 > b_3^2$. Similarly, a dislocation, b_1, may spontaneously dissociate into two parallel dislocations, b_2 and b_3, if $b_1^2 > b_2^2 + b_3^2$. It is energetically favorable for dislocations with large Burgers vectors to react with one another to form dislocations with smaller Burgers vectors. As a consequence, dislocations in crystals tend to have small Burgers vectors.

> **EXAMPLE PROBLEM 8.3:** Consider the reaction of two parallel dislocations of Burgers vectors $b_1 = (a/2)[110]$ and $b_2 = a[0\bar{1}1]$. What would be the Burgers vector of the product dislocation if they react with each other? Would the reaction be energetically favorable?
>
> **Solution:**
> $b_3 = b_1 + b_2 = (a/2)[110] + a[0\bar{1}1] = [a/2 + 0, \ a/2 - a, \ 0 + a] = (a/2)[1\bar{1}2]$
> $b_1^2 + b_2^2 = (a/2)^2(1^2 + 1^2 + 0^2) + (a/2)^2[0^2 + (-1)^2 + 1^2] = a^2$.
> $b_3^2 = (a/2)^2[1^2 + (-1)^2 + 2^2] = 1.5a^2$. Because $b_1^2 + b_2^2 < b_3^2$, the reaction would be energetically unfavorable.

Stress Fields around Dislocations

Around a dislocation, the atoms are displaced from their normal positions. These displacements are equivalent to the displacements caused by elastic strains arising from external stresses. In fact, the lattice displacements or strains can be completely described by these stresses, and Hooke's laws can be used to find them.

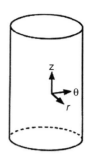

Figure 8.12. Coordinate system for describing a location near a screw dislocation. The dislocation is parallel to z and at a distance, $r = 0$.

It should be noted that the equations given here are based on the assumption of isotropic elasticity. The coordinate system is shown in Figure 8.12. For a screw dislocation with a Burgers vector, b, parallel to the z-axis,

$$\tau_{z\theta} = -Gb/(2\pi r) \quad \text{and} \quad \tau_{r\theta} = \tau_{rz} = \sigma_r = \sigma_\theta = \sigma_z = 0, \tag{8.14}$$

where G is the shear modulus and r is the radial distance from the dislocation. The minus sign indicates that the repulsive force is inversely proportional to the distance between the dislocations.

Equation (8.14) indicates that a screw dislocation creates no hydrostatic tension or compression because $\sigma_H = (\sigma_r + \sigma_\theta + \sigma_z)/3$. Therefore, there should be no dilatation (volume strain) associated with a screw dislocation. (However, real crystals are elastically anisotropic, so there may be a small dilatations associated with screw dislocations.)

For an edge dislocation that lies parallel to z and that has its Burgers vector parallel to x.

$$\tau_{xy} = Dx(x^2 - y^2)/(x^2 + y^2)^2, \tag{8.15a}$$

$$\sigma_x = -Dy(3x^2 - y^2)/(x^2 + y^2)^2, \tag{8.15b}$$

$$\sigma_y = Dy(x^2 - y^2)/(x^2 + y^2)^2, \tag{8.15c}$$

$$\sigma_z = \upsilon(\sigma_x + \sigma_y) = -2D\upsilon y/(x^2 + y^2), \quad \tau_{yz} = \tau_{zx} = 0. \tag{8.15d}$$

where x and y are the coordinates of the dislocation as shown in Figure 8.13 and $D = Gb/[2\pi(1 - \upsilon)]$.

One of the important features of these equations is that there is a hydrostatic stress, $\sigma_H = (\sigma_x + \sigma_y + \sigma_z)/3$, around an edge dislocation. Combining equations (8.15b), (18.15c), and (18.15d),

$$\sigma_H = -(2/3)Dy(1 + \upsilon)/(x^2 + y^2) \quad \text{or}$$

$$\sigma_H = -A(y)/(x^2 + y^2), \tag{8.16}$$

where $A = Gb(1 + \upsilon)/[3\pi(1 - \upsilon)]$. Figure 8.14 shows how the hydrostatic stress varies near an edge dislocation. There is hydrostatic compression (negative σ_H) above the edge dislocation (positive y) and hydrostatic tension below it.

Figure 8.13. Coordinates around an edge dislocation.

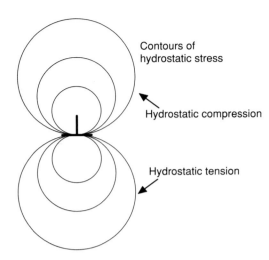

Contours of hydrostatic stress

Hydrostatic compression

Hydrostatic tension

Figure 8.14. Contours of hydrostatic stress around an edge dislocation. Note that the level of hydrostatic stress increases near the dislocation.

The dilatation causes interactions between edge dislocations. Given sufficient mobility, edge dislocations of like signs (same Burgers vectors) tend to form walls with one dislocation directly over another, as shown in Figure 8.15. The hydrostatic tension caused by one dislocation is partially annihilated by the hydrostatic compression of its neighbor. This relatively low energy and therefore stable configuration forms a low-angle grain boundary.

In substitutional solutions, solute atoms that are larger than the solvent atoms are attracted to the region just below the edge dislocation where their larger size helps relieve the hydrostatic tension. Similarly, substitutional solute atoms that are smaller than the solvent atoms are attracted to the region just above the edge. In either case, edge dislocation will attract solute atoms (Figure 8.16). In interstitial solid solutions, all solute atoms are attracted to the region just below the edge dislocation where they help relieve the tension. It is this attraction of edge dislocations in iron for carbon and nitrogen that is responsible for the yield point effect and strain aging phenomenon in low-carbon steel. (See Chapter 12.)

Forces on Dislocations

A stress in crystals causes a force on dislocations. Consider a dislocation of length, L, and Burgers vector, b, on a plane, as shown in Figure 8.17. A shear stress, τ, acting on that plane will cause a force on the dislocation, per unit length, F_L, of

$$F_L = -\tau \cdot b \tag{8.17}$$

Note that a dot product is possible here because once the plane of the stress is fixed, the stress can be treated as a vector (force). The stress, τ, may result from the stress

Figure 8.15. A low-angle grain boundary formed by edge dislocations. The misorientation angle, $\theta = b/d$.

Figure 8.16. If small substitutional solute atoms migrate to the extra half-plane, they help relieve the crowding there. Large substitutional solute atoms will help relieve the tension below.

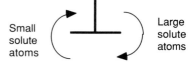

field of another dislocation. Thus, two screw dislocations exert an attractive force on each other of

$$F_L = -Gb_1 \cdot b_2/(2\pi r), \tag{8.18}$$

where b is the Burgers vector of the dislocation of concern. The minus sign means that they repel one another if the dot product is positive. An equivalent statement is *two dislocations repel each other if Frank's rule predicts that their combination would result in an energy increase.* If the angle between b_1 and b_2 is greater than 90 degrees, $|b_1 + b_2| > |b_1| + |b_2|$.

The interaction of two parallel edge dislocations is somewhat more complex. The shear stress field for one edge dislocation that lies parallel to z with a Burgers vector parallel to x is given by equation (8.15a), $\tau_{xy} = Dx(x^2 - y^2)/(x^2 + y^2)^2$, where $D = Gb/[2\pi(1 - \upsilon)]$. The mutual force on that plane is

$$F_L = -G\{b_1 b_2/[2\pi(1 - \upsilon)]\}x(x^2 - y^2)/(x^2 + y^2)^2. \tag{8.19}$$

For dislocations with like sign ($b_1 \cdot b_2 > 0$), there is mutual repulsion in the region $x > y$ and attraction in the region $x < y$. This is equivalent to saying that there is mutual repulsion if Frank's rule predicts that a reaction would cause an increase of energy and mutual attraction if it would cause a decrease of energy. Figure 8.18 shows the regions of attraction and repulsion.

The stress, τ_{xy}, is zero at $x = 0$, $x = y$, and $x = \infty$.

Between $x = 0$ and $x = y$, τ_{xy} is negative, indicating that the stress field would cause attraction of edge dislocations of the same sign to each other. Therefore, edge dislocations of the same sign tend to line up one above the other, as shown in Figure 8.15. For x greater than y, the stress, τ_{xy}, is positive indicating that the stress field would tend to repel another edge dislocation of the same sign.

Partial Dislocations in fcc Crystals

In fcc crystals, slip occurs on {111} planes and in <110> directions (Figure 8.19). Consider the specific case of (111)[$\bar{1}$10] slip. The Burgers vector corresponding to displacements of one atom diameter is (a/2)[$\bar{1}$10]. A dislocation with this Burgers vector can dissociate into two partial dislocations,

$$(a/2)[\bar{1}10] \rightarrow (a/6)[\bar{2}11]\,(a/6)[\bar{1}2\bar{1}]. \tag{8.20}$$

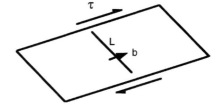

Figure 8.17. The force per length on a dislocation, $F_L = -\tau b$.

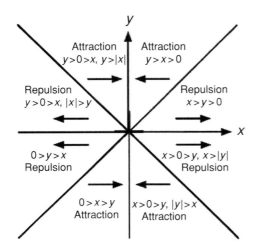

Figure 8.18. Stresses around an edge dislocation either attract or repel another parallel dislocation having the same Burgers vector, depending on how the two are positioned relative to one another.

We can check that this reaction is vectorially correct by noting that $b_1 = (-a/2, a/2, 0)$ does equal $b_2 = (-a/3, a/6, a/6) + b_3 = (-a/6, a/3, -a/6)$. Figure 8.20 is a geometrical representation of these two partials.

EXAMPLE PROBLEM 8.4: When an $(a/2)[\bar{1}10]$ dissociates into two partial dislocations, $(a/6)[\bar{2}11]$ and $(a/6)[\bar{1}\,2\,\bar{1}]$, on what $\{111\}$ plane must the partial dislocations lie?

Solution: The normal to the plane must be at 90 degrees to both Burgers vectors, so it can be found by taking their cross-product, $[\bar{2}11] \times [\bar{1}\,2\,\bar{1}] = -3, -3, -3$ or (111).

An alternative way is to try all of the $\{111\}$ possibilities and see which has a zero dot product with both $[\bar{2}11]$ and $[\bar{1}\,2\,\bar{1}]$.

EXAMPLE PROBLEM 8.5: Into what other pair of $(a/6) <211>$ partials could the $(a/2)[\bar{1}10]$ dislocation in example problem 9.4 dissociate? On what plane would they lie?

Solution: By inspection, $(a/2)[\bar{1}10] \rightarrow (a/6)[\bar{2}1\bar{1}] + (a/6)[\bar{1}21]$. The $(a/6)[\bar{2}1\bar{1}]$ and $(a/6)[\bar{1}21]$ partials must lie in a $(11\bar{1})$ plane. Note that $[\bar{2}1\bar{1}] \cdot [11\bar{1}] = 0$ and $[\bar{1}21] \cdot [11\bar{1}] = 0$

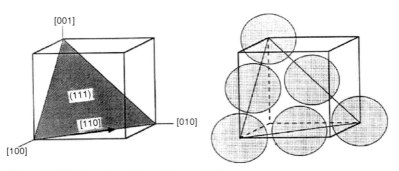

Figure 8.19. Slip systems in an fcc crystal.

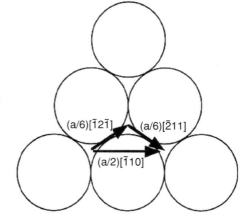

Figure 8.20. Burgers vectors for partial slip dislocations in an fcc crystal.

Stacking Faults

If a single $(a/6)<112>$ partial dislocation passes through an fcc crystal, it leaves behind a region in which the sequence of stacking of the close-packed {111} planes does not correspond to the normal fcc lattice. The correct stacking order is not restored until the second partial dislocation passes. The normal stacking order in fcc and hcp lattices is shown in Figure 8.21. In hcp crystals, the third close-packed plane (A) lies directly over the first (A). In fcc crystals, the third close-packed plan (C) is over neither the first (A) nor the second (B){111} plane. Figure 8.22 shows that a $(a/6)<112>$ partial dislocation changes the position of the third plane so that it is directly over the first and therefore produces a local region of hcp packing.

In Figure 8.23, the stacking order near a stacking fault in an fcc crystal is compared with the stacking in fcc and hcp lattices and near a twin boundary in an fcc crystal.

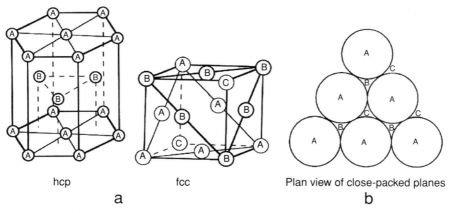

Figure 8.21. (a) Stacking of close-packed planes in hcp and fcc crystals. (b) Plan view of stacking of close-packed planes.

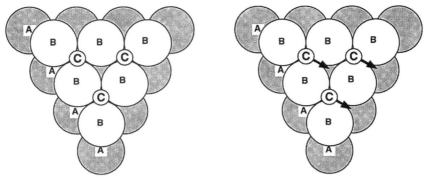

Figure 8.22. Stacking of close-paced planes in fcc and hcp crystals. When a $(a/6)<112>$ partial dislocation moves plane C so that it is directly over plane A, it creates a region where the packing sequence is hcp rather than fcc.

The stacking sequence near a stacking fault in an fcc crystal is similar to the packing sequence in the hcp lattice. Because this is not the equilibrium structure of fcc, the stacking fault raises the energy, and the increase of energy depends directly on the area of the fault. The stacking fault energy, γ_{SF}, is the energy per area of fault and can be regarded as a surface tension pulling the partial dislocations together. A stacking fault has twice as many incorrect second-nearest neighbors as a twin boundary. The similarity of the packing sequences at a twin boundary and at a stacking fault is clear in Figure 8.23. For most metals, the stacking fault energy is about twice the twin boundary energy, as shown in Figure 8.24. The frequency of annealing twins is much higher in fcc metals of low stacking fault energy (Ag, brass, Al-bronze, γ-stainless) than in those with higher stacking fault energy (Cu, Au). Annealing twins in the microstructures of aluminum alloys are rare. Table 8.2 lists the values of the stacking fault energy for a few fcc metals. The values of γ_{SF} for brass (Cu-Zn), aluminum bronze (Cu-Al), and austenitic stainless steel are still lower than the value for Ag.

The stacking fault between two $(a/6)<112>$ partial dislocations is illustrated in Figure 8.25. The mutual repulsion of two partials tends to drive them away from each other. However, balancing this mutual repulsion is the attraction that results from the surface tension of the stacking fault between them (Figure 8.26).

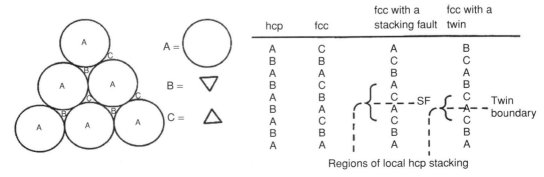

Figure 8.23. Stacking of close-packed planes.

Table 8.2. *Stacking fault energies of several fcc metals*[*]

Ag	Al	Au	Cu	Ni	Pd	Pt	Rh	Ir
16	166	32	45	125	180	322	750	300 mJ/m^2

[*] From listing in J. P. Hirth and J. Lothe, *Theory of Dislocations*, Wiley, New York (1982).

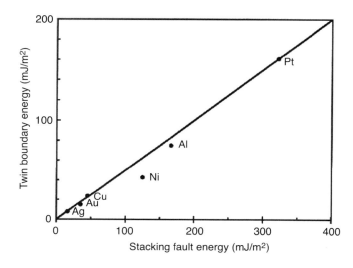

Figure 8.24. The relationship of stacking fault energy and twin boundary energy. The line represents $\gamma_{\text{stacking fault}} = 2\gamma_{\text{twin}}$.

Figure 8.25. Two partial dislocations separated by a stacking fault.

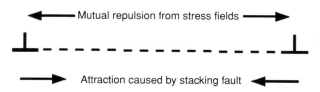

Figure 8.26. Equilibrium spacing between two partial dislocations corresponds to the separation at which the mutual repulsion and attraction balance.

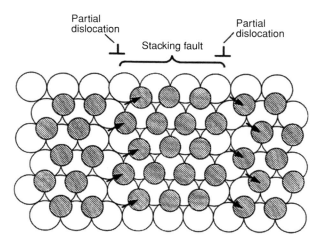

EXAMPLE PROBLEM 8.6: Use equation (8.18) to make an approximate calculation of equilibrium separation of the partial dislocations. Neglect the fact that both partials cannot be pure screws.

Solution: Taking $b_1 = (a/6)[\bar{2}11]$ and $b_2 = (a/6)[\bar{1}2\bar{1}]$, $b_1 \cdot b_2 = (a^2/36)(\bar{2} \cdot \bar{1} + 1 \cdot 2 + \bar{1} \cdot 1) = (a^2/12)$. Substituting into equation (8.18), $f_L = -Ga^2/(24\pi r)$. The attractive force per length tending to pull the partials together is the stacking fault energy, γ. At equilibrium, $\gamma + f_L = 0$.

$$r = Ga^2/(24 \cdot \pi \gamma). \tag{8.21}$$

Hirth and Loethe give a more exact relation for the force between two parallel dislocations

$$F_L = [G/(2\pi r)][(b_1 \cdot \xi)(b_2 \cdot \xi) + (b_1 \times \xi) \cdot (b_2 \times \xi)/(1 - v)], \tag{8.22}$$

where ξ is a unit vector parallel to the two dislocations. The two terms inside the brackets are the effects of the screw and edge components of the dislocations. Thus, the equilibrium separation of two partial dislocations depends inversely on the stacking fault energy, which varies from one fcc metal to another.

REFERENCES

A. H. Cottrell, *Dislocations and Plastic Flow in Crystals*, Oxford (1953).
W. T. Read, *Dislocations in Crystals*, McGraw-Hill (1953).
J. Weertman and J. R. Weertman, *Elementary Dislocation Theory*, Oxford (1992).
D. Hull and D. J. Bacon, *Introduction to Dislocations*, 3rd ed., Butterworth Heinemann (1997).
J. P. Hirth and J. Loethe, *Theory of Dislocations*, Wiley, New York (1982).

Notes

Postulation of dislocations: Calculations of the strength of crystals by J. Frenkel, *Z. Phys.*, 37 (1926), p. 572, showed the great discrepancy between theory and experimental observation. It was not until 1934 that G. I. Taylor, E. Orowan, and M. Polanyi more or less independently proposed edge dislocations to explain the difference between s_{theor} and s_{exper}. Screw dislocations were postulated later by J. M. Burgers, *Proc. Kon. Ned. Akad. Wetenschap.*, 42 (1939), p. 378. Orowan's interest in why crystals are as weak as they are was stimulated by an accident during research for his bachelor degree at the Technische Universität Berlin. He dropped his only zinc single crystal on the floor. Rather than quit, he straightened the crystal and then tested it in creep. He later commented that reflection on the nature of the results contributed to his postulation of dislocations.

Discovery of the strength of whiskers: As part of an investigation of the failure of some electrical condensers at the Bell Telephone Laboratories, whiskers of tin about 2×10^{-6} m diameter were found growing from the condenser walls. What distinguished this investigation from earlier findings of metal whiskers is that these whiskers were tested in bending (C. Herring and J. K. Galt, *Phys. Rev.*, v. 85, 1060 (1952)). The surprising finding was that these whiskers could be bent to a strain of 2% to 3% without plastic deformation. This meant that the yield strength was 2% to 3% of Young's modulus. For tin, $E = 16.8 \times 10^6$ psi, so $Y > 330,000$ psi. Thus, overnight, tin became the strongest material known to man. Once the Herring-Galt observation was reported, many others began testing whiskers of other metals such as copper and iron with similar results and tin lost this distinction.

Problems

1. A crystal of aluminum contains 10^{12} m of dislocation per m^3.

    A. Calculate the total amount of energy per m^3 associated with dislocations. Assume that half of the dislocations are edges and half are screws.

    B. What would be the temperature rise if this energy could be released as heat?

Data for aluminum: atomic diameter = 0.286 nm, crystal structure = fcc, density = 2.70 Mg/m^3, atomic mass = 27 g/mole, $C = 0.215$ cal/g °C, $G = 70$ GPa, $\upsilon = 0.3$.

2. Calculate the average spacing between dislocations in a 1/2-degree tilt boundary in aluminum. See Figure 8.15 and look up any required data.

3. On which {110} planes of bcc iron can a dislocation with a Burgers vector $(a/2)[11\bar{1}]$ move?

4. A single crystal of aluminum was stretched in tension. Early in the test, the specimen was removed from the testing machine and examined at high magnification (Figure 8.27). The distance between slip lines was found to be 100 μm, and the average offset at each slip line was approximately 500 nm. Assume for simplicity that both the slip direction and the slip plane normal are oriented at 45 degrees to the tensile axis.

    A. How many dislocations, on average, must have emerged from the crystal at each observable slip line?

    B. Find the shear strain on the slip system calculated over the whole crystal.

    C. Find the tensile strain (measured along the tensile axis) that must have occurred. What is the percent elongation when the test was stopped?

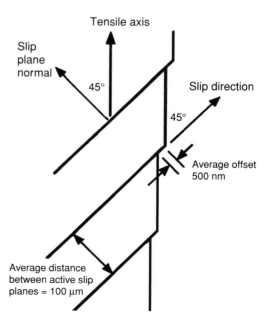

Figure 8.27. Profile of a crystal surface showing offsets caused by slip.

5. Consider the reactions between parallel dislocations given here. In each case, write the Burgers vector of the product dislocation, and determine whether the reaction is energetically favorable.

   A. $(a/2)[1\bar{1}0] + (a/2)[110] \rightarrow$
   B. $(a/2)[101] + (a/2)[01\bar{1}] \rightarrow$
   C. $(a/2)[1\bar{1}0] + (a/2)[101] \rightarrow$

6. Consider the dislocation dissociation reaction $(a/2)[110] \rightarrow (a/6)[21\bar{1}] + (a/6)[121]$ in an fcc crystal. Assume that the energy/length of a dislocation is given by $E_L = Gb^2$ and neglect any dependence of the energy on the edge versus screw nature of the dislocation. Assume that this reaction occurs and the partial dislocations move very far apart. Neglect the energy associated with the stacking fault between the partial dislocations.

   A. Express the total decrease in energy/length of the original $(a/2)[110]$ dislocation in terms of a and G.
   B. On which $\{111\}$ must these dislocations lie?

7. A dislocation in an fcc crystal with a Burgers vector, $b = (a/2)[011]$, dissociates on the $(1\bar{1}1)$ plane into two partial dislocations of the $(a/6)<211>$ type.

   A. Give the specific indices of the two $(a/6)<211>$ partial dislocations.
   B. Onto what other plane of the $\{111\}$ family could the $b = (a/2)[011]$ dislocation have dissociated?
   C. Give the specific indices of the two $(a/6)<211>$ partials that would be formed if the $(a/2)[011]$ dislocation had dissociated on the plane in B.

8. Consider a circular dislocation loop of diameter, d, in a crystal under a shear stress, τ. The region inside the circle has slipped relative to the material outside the circle. The presence of this dislocation increases the energy of the crystal by the dislocation energy/length times the length of the dislocation. The slip that occurs because of the formation of the dislocation under the stress, τ, lowers the energy of the system by τAb. (τA is the shear force, and b is the distance the force works through.) If the diameter of the loop is small, the energy will be reduced if the loop shrinks. If the loop is large enough, the loop will spontaneously expand. Find the diameter of critical size loop in terms of b, G, and τ. For simplicity, take $E_L = Gb^2$.

9. Referring to Figure 8.23, find the ratio of the wrong second-nearest neighbors across a stacking fault to the number across a twin boundary. If the surface energies are proportional to the number of wrong second-nearest neighbors, what is γ_{SF}/γ_{TB}?

10. Using the values of γ_{SF} from Table 8.2 and equation (8.21), make an approximate calculation of the equilibrium separation, r, of two partial dislocations resulting from dissociation of screw dislocation in aluminum and in silver. Express the separation in terms of atom diameters. The shear modulus of aluminum is 70 GPa and that of silver is 75 GPa. The atom diameters of Al and Ag are 0.286 and 0.289 nm.

11. Plot the variation of the force, F_L, with x on an edge dislocation caused by another edge dislocation at a fixed level of y according to equation (8.19). Let the units of F_L be arbitrary.

9 Dislocation Mechanics

Introduction

Once the concept of dislocations was accepted, there were three important questions to be answered. First, when a single crystal is deformed, slip occurs with shear offsets of thousands of atom distances on relatively widely spaced planes, rather than uniformly throughout the crystal. See Figure 9.1. Why does slip not occur uniformly at an atomic scale?

Second, cold working increases the dislocation content of crystals even though dislocations must run out of the crystals. See Figures 9.2 and 9.3. Where do the additional dislocations come from?

Third, the yield stress increases as the number of dislocations increases even though without any dislocations, the strength would be even higher. See Figure 9.4. Why does the yield stress increase with dislocation density?

Frank-Read Sources

The first two questions can be answered in terms of the *Frank-Read source,* which generates dislocations. Suppose that there is a finite length of a dislocation, AB, in a slip plane (Figure 9.5). The dislocation leaves the plane at A and B, but the end points are pinned at A and B. A shear stress, τ, acting on the plane, will create a force that causes dislocation to bow. This bowing is resisted by the line tension of the dislocation. As the shear stress is increased, the dislocation will bow out until it spirals back on itself. The sections that touch annihilate each other, leaving a dislocation loop that can expand under the stress and a restored dislocation segment between the pinning points. The process can repeat itself, producing many loops. Frank-Read sources explain how the number of dislocations increases with deformation. Their existence on some planes and not others explains why slip occurs with large offsets (corresponding to the passing of many dislocations) on widely spaced planes.

A quantitative treatment of the operation of a Frank-Read source can be made by considering a balance of forces acting on the bowed segment of the dislocation (Figure 9.6.) The applied shear stress, τ, causes a force, $\tau \cdot bd$, where d is the distance between the pinned ends. The line tension of the dislocation (energy/length),

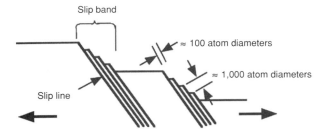

Figure 9.1. Sketch showing slip on widely spaced planes. The offsets of slip are thousands of atom diameters.

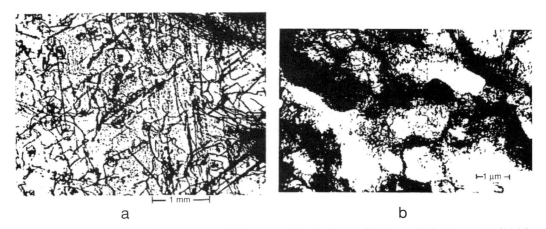

a   b

Figure 9.2. Dislocation structure (a) in annealed aluminum (B. Nost, *Phil. Mag.,* v. 11 (1965), p. 183) and (b) in aluminum after 10% reduction of area at 77K (P. R. Swann, in *Electron Microscopy and Strength of Crystals*, Thomas and Washburn, eds. Wiley (1963)). Note the great difference in magnification. The dislocation density in (a) is about 10^2/cm^2, whereas in (b) it is in the order of 10^{10}.

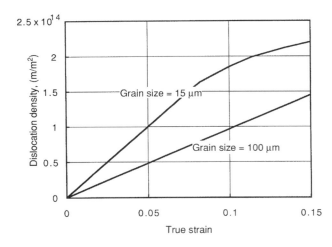

Figure 9.3. Increase of dislocation density with strain in iron. Data from A. Keh and S. Weissmann, in *Electron Microscopy and Strength of Crystals*, G. Thomas and J. Washburn, eds., Wiley (1963).

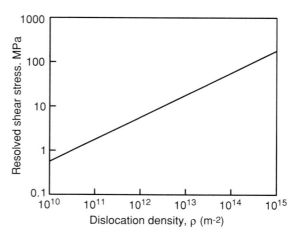

Figure 9.4. Dependence of the critical shear stress for slip in copper as a function of dislocation density. Data are for copper and include experiments on polycrystals as well as single crystals oriented for single slip and multiple slip. Data from H. Weidersich, *J. Inst. Metals*, v. 16 (1964).

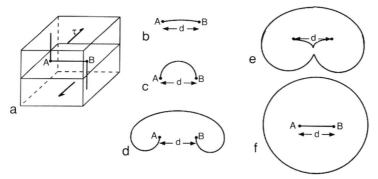

Figure 9.5. Sketch illustrating the operation of a Frank-Read source (adapted from T. H. Courtney, *Mechanical Behavior of Materials,* 2nd ed, McGraw-Hill (2000)). A segment of a dislocation of length, d, is pinned at points A and B (a). The segment bows out under a shear stress (b). The shear stress reaches a maximum when the segment becomes a semicircle (c). As the dislocation segment continues to expand under decreasing stress (d), it eventually recombines with itself (e), forming a dislocation loop. The loop continues to expand, and a new dislocation segment, AB, is formed. The process can repeat itself sending out many loops.

Figure 9.6. Force balance on a pinned dislocation that is bowed by a shear stress, τ.

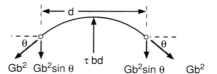

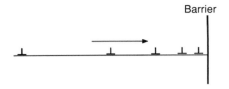

Figure 9.7. Pile-up of dislocations at an obstacle.

$U_L \approx Gb^2$ (equation 8.12), acts parallel to the dislocation line and tends to keep it from moving. Considering both ends, the vertical component of this force is $2Gb^2 \sin \theta$. This force reaches a maximum, $2Gb^2$, when the dislocation is bowed into a semicircle ($\theta = 90°$). Equating these two forces and assuming the shear stress is parallel to b, $\tau bd = 2Gb^2$ or

$$\tau = 2Gb/d. \tag{9.1}$$

Thus, the stress necessary to operate a Frank-Read source is inversely proportional to the size of the source, d.

> **EXAMPLE PROBLEM 9.1:** Slip was found to first occur in some aluminum crystals when the resolved shear stress reached 1.0 MPa. Assuming that this is the stress necessary to operate the largest Frank-Read sources, calculate the size, d, of these sources.
>
> **Solution:** Substituting $\tau = 10^6$ Pa, $G = 25 \times 10^9$ Pa, and $b = 0.286 \times 10^{-9}$ m into equation (9.1), $d \approx 2bG/\tau \approx 14\,\mu$m.

Dislocation Pile-Ups

When dislocations from a Frank-Read source come to an obstacle such as a grain boundary or hard particle, they tend to form a pile-up (Figure 9.7). Because they are of like sign, they repel one another. The total repulsion of a dislocation by a pile-up is the sum of the repulsions of each dislocation in the pile-up. With n dislocations in the pile-up, the stress on the leading dislocation, τ_n, will be

$$\tau_n = n\tau. \tag{9.2}$$

It takes a relatively small number in a pile-up to effectively stop further dislocation movement. A pile-up creates a back stress on the source so that the stress to continue to operate the source must rise. This higher stress allows other sources of slightly smaller spacing, d, to operate.

Cross-Slip

A dislocation cannot move on a plane unless the plane contains both the dislocation and its Burgers vector. This requirement uniquely determines the slip plane, except in the case of screw dislocations. Screw dislocations are parallel to their Burgers vector so they can slip on any plane in which they lie. When a screw dislocation gliding on one plane changes to another, it is said to undergo *cross-slip*. In principle, screw dislocations should be able to cross-slip with ease to avoid obstacles (Figure 9.8).

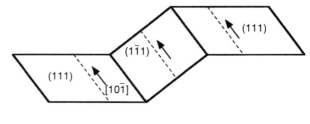

Figure 9.8. A screw dislocation on one slip plane can move onto another plane containing the Burgers vector. This is called *cross-slip*.

- - - - - - - - - - - - - - - Successive positions of dislocation

If, however, a screw dislocations is separated into partials connected by a stacking fault, both partials cannot be screws. They must recombine to form a screw dislocation before they can dissociate on a second plane (Figure 9.9).

Such recombination increases the total energy, and this energy must be supplied by the applied stresses aided by thermal activation. How much energy is required depends on the degree of separation of the partials and therefore on the stacking fault energy. If the stacking fault energy is high, the separation of partial dislocations is small, so the force required to cause recombination is low. In metals of high stacking fault energy (e.g., aluminum), cross-slip occurs frequently. In crystals of low stacking fault energy (e.g., brass), the separation of partials is large. A high force is required to bring them together so cross-slip is rare. This is in accord with observation of slip traces on polished surfaces. Slip traces are very wavy in aluminum and very straight in brass (Figure 9.10).

Double cross-slip has been proposed as a variant of the original Frank-Read source. In this case, a segment of dislocation, which has undergone double cross-slip, can act as a Frank-Read source on its new plane, as shown in Figure 9.11.

Dislocation Intersections

The number of dislocations increases as deformation proceeds, and the increased number makes their movement become more difficult. This increased difficulty of movement is caused by intersection of dislocations moving on different planes. During easy glide, the rate of work hardening is low because slip occurs on parallel planes, and there are few intersections. As soon as slip occurs on more than one set

Figure 9.9. Cross-slip of a dissociated screw dislocation (a). It must first recombine (b) and then dissociate onto the cross-slip plane (c) before it finally can glide on the cross-slip plane (d).

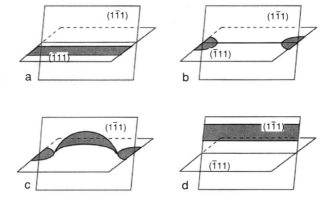

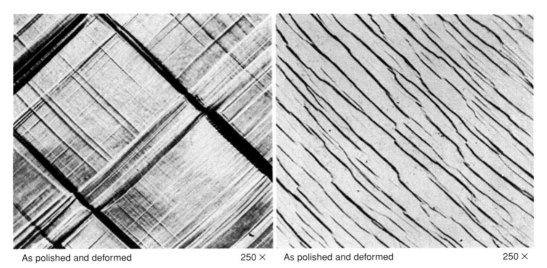

As polished and deformed   250 ×   As polished and deformed   250 ×

Figure 9.10. Straight parallel slip lines in an alloy of low stacking fault energy (*left*). Wavy slip lines in aluminum resulting from frequent cross slip (*right*). From G. Y. Chin, *Metals Handbook,* v. 8, 8th ed. (1973).

of slip planes, dislocations on different planes will intersect, and these intersections impede further motion, causing rapid work hardening.

The nature of dislocations intersections can be understood by considering several types of intersections in simple cubic crystals as illustrated in Figure 9.12. When two dislocations intersect, a jog is created in each dislocation. The direction of the jog is parallel to the Burgers vector of the intersecting dislocation, and the length of the jog equals the magnitude of the Burgers vector of the intersecting dislocation.

For dislocations a and b in Figure 9.12, the jogs create no problem. If the upper part of dislocation a and the right side of dislocation b move slightly faster, the jogs will disappear. The same is true for dislocation c if its left side moves faster. The jog in dislocation d simply represents a ledge in the extra half plane, and it can move with the rest of the dislocation. However, the jogs in dislocations e and f cannot move conservatively. The jogs have an edge character, and the direction of motion is not in their slip plane. Figure 9.13 is an enlarged view of dislocations e and f in Figure 9.12. Figure 9.14a shows that continued motion of these jogged dislocations would force atoms into interstitial positions. Figure 9.14b shows that

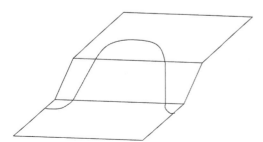

Figure 9.11. Double cross-slip. The segment of dislocation on the top plane can act as a Frank-Read source on that plane.

Before intersection:

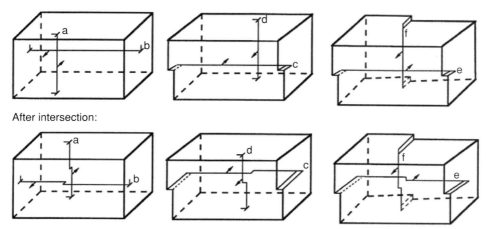

After intersection:

Figure 9.12. Intersection of dislocations. The arrows indicate the direction of motion.

Figure 9.13. Jogs produced in dislocations e and f by their intersection. The
arrows indicate the shear caused by continued movement of the dislocations.
The jogs are such that they must create
interstitial defects as they move.

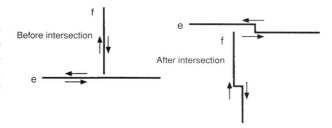

Figure 9.14. (a) Continued motion of jogged
dislocations e and f will create interstitial
atoms. This is like trying to slide two blocks
with ledges over one another. It requires that
the same volume be occupied twice. (b) With
jogs of the opposite sign, motion of the dislocations create vacancies. This is like opening
holes by sliding blocks with ledges over one
another.

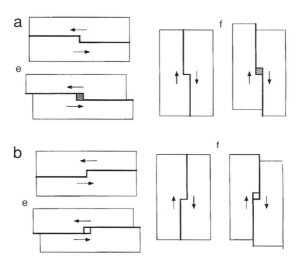

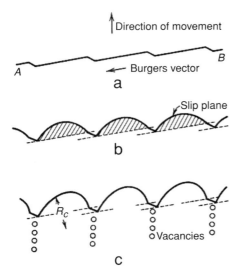

Figure 9.15. Rows of vacancies produced by movement of jogged screw dislocations. From D. Hull, *Introduction to Dislocations Pergamon Press* (1965).

with jogs of the opposite sense, vacancies would be produced. If the intersection had been in the opposite direction, movement of the jogs would create a row of vacancies.

Figure 9.15 shows rows of vacancies being created by moving jogs.

EXAMPLE PROBLEM 9.2: Consider the energy expended as a dislocation with vacancy-producing jogs moves through a crystal. Assume that the distance between jogs is d, and the energy to produce a vacancy is U_v. Find the drag force per length, f_L, caused by the jogs.

Solution: Every time a segment of dislocation of length d moves a distance, b, a jog must be produced. Making an energy balance, the work done by the dislocation, $f_L bd$, must equal U_v, so $f_L = U_v/(bd)$.

A high energy is required to produce a row of interstitials. Therefore, an interstitial-producing jog will likely be pinned rather than produce a row of interstitials. This causes a drag on any motion of the dislocation. Intersections causing interstitial-producing jogs are geometrically much more likely than intersections causing vacancy-producing jogs. The reason for this can be understood in terms of the slip systems likely to be simultaneously activated. Figure 9.16 illustrates this for a

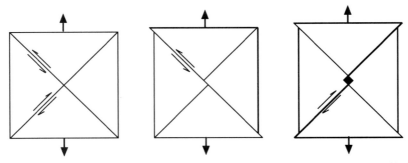

Figure 9.16. Schematic showing why interstitial-producing jogs are more likely that vacancy-producing jogs.

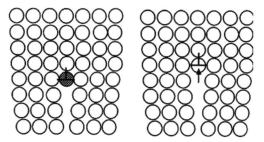

Figure 9.17. Climb of an edge dislocation by diffusion of vacancies.

crystal with two slip systems oriented at 45 degrees to the axis of tension or compression. Intersection of these slip systems will create only interstitial-producing jogs. In real crystals, it is possible to find stress states that can activate slip systems that will create vacancy-producing jogs, but they are fewer than those that create interstitials.

With multiple intersections, the number of pinning points increases, and the stress to continue moving the dislocation increases. These intersections act as anchors, greatly decreasing the mobility of the dislocations. As the number of dislocations increases, the frequency of intersections also increases, making further slip more difficult.

Climb

The movement of an edge dislocation out of its glide plane can occur only if there is a net diffusional flux of atoms away from or to the dislocation. Such motion is called *climb*. Removal of atoms from a dislocation causes upward climb, and addition of atoms to the dislocation causes downward climb (Figure 9.17). Because climb requires diffusion, it is important only at elevated temperatures during creep. Diffusion-controlled climb allows dislocations to avoid obstacles that impede their glide and may become a controlling mechanism during creep (Chapter 16).

REFERENCES

A. H. Cottrell, *Dislocations and Plastic Flow in Crystals*, Oxford (1953).
W. T. Read, *Dislocations in Crystals*, McGraw-Hill (1953).
J. Weertman and J. R. Weertman, *Elementary Dislocation Theory*, Oxford (1992).
J. P. Hirth and J. Lothe, *Theory of Dislocations, 2nd Ed.* Wiley (1982).
D. Hull and D. J. Bacon, *Introduction to Dislocations*, 3rd ed. Butterworth Heinemann (1997).
F. McClintock and A. Argon, *Mechanical Behavior of Materials*, Addison-Wesley (1966).

Note

The Frank-Read source was postulated to explain why the number of dislocations in a crystal increased during deformation (rather than decrease as dislocations leave the crystal). F. C. Frank of the University of Bristol and W. T. Read of the Bell Telephone Laboratories each conceived the idea independently as they traveled to the Symposium on Plastic Deformation of Crystalline Solids at the Carnegie Institute of Technology in Pittsburgh. Each became aware of the other's ideas during informal discussions before the formal conference. In a Pittsburgh pub, they worked out the theory together and decided to make a joint presentation (F. C. Frank and W. T. Read, *Symposium on Plastic Deformation of Crystalline Solids*, Carnegie Inst. Tech. v. 44 (1950) and F. C. Frank and W. T. Read, *Phys. Rev.*, v. 79 (1950)).

Problems

1. Typical values for the dislocation density in annealed and heavily deformed copper are 10^7 and 10^{11} cm^{-2}.

    A. Calculate the average distance between dislocations for both cases. For simplicity, assume the dislocations are parallel and in a square pattern.

    B. Calculate with equation (9.11) the energy/length of dislocation line for both cases, assuming the dislocations are screws.

For copper: lattice parameter $a = 0.361$ nm, $E = 110$ GPa, $v = 0.30$, $\rho = 8.96$Mg/m^3.

2. For a typical annealed metal, the yield stress in shear is $10^{-4}G$. Deduce the typical spacing of a Frank-Read source using a typical value for b.

3. Several theoretical models predict that the dislocation density, ρ, should increase parabolically with strain, $\rho = C\varepsilon^{1/2}$. Assuming this and the dependence of τ on ρ shown in Figure 9.4, predict the exponent n in power law approximation of the true stress–true strain curve, $\sigma = K\varepsilon^n$.

4. Orowan showed that the shear strain rate is given by $\dot{\gamma} = d\gamma/dt = \rho b \bar{v}$, where ρ is the dislocation density, and $\bar{v}$ is the average velocity of the dislocation. In a tension test, the tensile strain rate, ($\dot{\varepsilon}$, is half of the shear strain rate, $\dot{\gamma}$, $\dot{\varepsilon} = (1/2)\dot{\gamma}$, if the Schmid factor $= 1/2$. Otherwise, $\dot{\varepsilon} < (1/2)\dot{\gamma}$). In a typical tension test, the cross-head rate is 0.2 in./min, and the gauge section is 2 in. in length. Estimate $\bar{v}$ in a typical tension test for a typical metal with an initial dislocation density of 10^{10}/cm^2.

5. Consider the possible intersections of dislocations sketched in Figure 9.18. In the sketches, dislocations B and D are moving to the right, and A and C are moving to the left.

    A. Sketch the nature of the jogs or kinks in each dislocation after they intersect one another.

    B. Which, if any, would leave a trail of vacancies or interstitials?

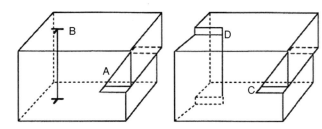

Figure 9.18. Sketch for problem 5.

6. The sketches in Figure 9.19 show pairs of dislocations that are about to intersect. The arrows show the directions of motion. (Dislocations a, c, and e are moving into the paper, dislocations b and f are moving out of the paper, and dislocation d is moving to the left.)

A. Which of these dislocations (a, b, c, d, e, or f) after intersection, would have jogs that would produce point defects if the dislocations continued to move in the direction shown?

B. Indicate for each whether the point defects would be vacancies or interstitials.

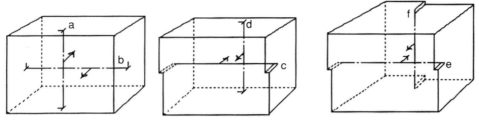

Figure 9.19. Sketch for problem 6.

7. Consider how dislocation intersections affect the stress necessary to continue slip. Let the density of screw dislocations be ρ (meters of dislocations/m^3), and for simplicity, assume that these dislocations are equally divided into three sets of dislocations, each set being parallel to one of the three orthogonal axes.

A. Assume each set of dislocations is arranged in a square pattern. Express the distance between the dislocations in a set in terms of ρ.

B. Now, consider the jogs formed as a screw dislocation moves. What will be the distance between jogs?

C. Assume the energy of forming a row of interstitials is so high that the jogs essentially pin the dislocation. Express the shear stress necessary to continue the motion in terms of G, b, and ρ.

D. Now assume that the dislocation density is proportional to $\sqrt{\varepsilon}$. What would this model predict as the value of n in $\sigma = K\varepsilon^n$?

8. A. Derive an equation for estimating the shear stress, τ, necessary to move a dislocation with vacancy-producing jogs in terms of the Burgers vector, b; the distance between jogs, d; and the energy to create a vacancy, E_v. Note that there must be a vacancy produced each time a segment of dislocation of length, d, advances a distance, b. The work done by the shear stress would be bdf_L, where the force per length of dislocation is $f_L = \tau b$.

B. Evaluate τ for $d = 1$ mm, $E_v = 0.7$ eV, and $b = 0.3$ nm.

9. Assume that a resolved shear stress of 1.4 MPa is applied to a crystal and this causes dislocations to pile up at a precipitate particle. Assume that the shear strength of the particle is 7.2 MPa. What is the largest number of dislocations that can pile up at the precipitate before it yields?

10 Mechanical Twinning and Martenitic Shear

Introduction

Most crystals can deform by twinning. Twinning is particularly important in hcp metals because hcp metals do not have enough easily activated slip systems to produce an arbitrary shape change.

Mechanical twinning, like slip, occurs by shear. A twin is a region of a crystal in which the orientation of the lattice is a mirror image of that in the rest of the crystal. Normally, the boundary between the twin and the matrix lies in or near the mirror plane. Twins may form during recrystallization (*annealing twins*), but the concern here is formation of twins by uniform shearing (*mechanical twinning*), as illustrated in Figure 10.1. In this figure, plane 1 undergoes a shear displacement relative to plane 0 (the mirror plane). Then plane 2 undergoes the same shear relative to plane 1, and plane 3 relative to plane 2, and so on. The net effect of the shear between each successive plane is to reproduce the lattice, but with the new (mirror image) orientation.

Both slip and twinning are deformation mechanisms that involve shear displacements on specific crystallographic planes and in specific crystallographic directions. However, there are important differences.

1. With slip, the magnitude of the shear displacement on a plane is variable, but it is always an integral number of interatomic repeat distances, nb, where b is the Burgers vector. Slip occurs on only a few of the parallel planes separated by relatively large distances. With twinning, on the other hand, the shear displacement is a fraction of an interatomic repeat distance, and every atomic plane shears relative to its neighboring plane.
2. The twinning shear is always directional in the sense that shear in one direction is not equivalent to shear in the opposite direction. Twinning in fcc crystals occurs by shear on the (111) plane in the $[11\bar{2}]$ direction, but not by shear in the $[\bar{1}\bar{1}2]$ direction. In fcc metals, slip can occur on a (111) plane in either the $[\bar{1}10]$ or the $[1\bar{1}0]$ direction.
3. With slip, the lattice rotation is gradual. Twinning reorients the lattice abruptly.

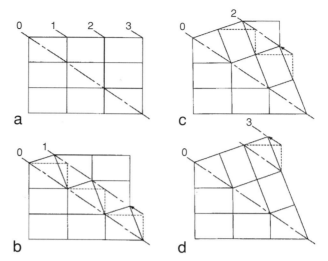

Figure 10.1. Progressive formation of twins by shearing between each parallel plane of atoms. From W. F. Hosford, *The Mechanics of Crystals and Textured Polycrystals* (1993). Used by permission of Oxford University Press.

Formal Notation

There is a formal notation system that describes twinning. Figure 10.2 illustrates a sphere of material, the top half of which has undergone a twinning shear. There are two material planes that remain undistorted. The first undistorted plane, K_1, is the *twinning plane* (i.e., the mirror plane). The second undistorted plane is denoted by K_2. The direction of shear is the first characteristic direction, η_1. The second characteristic direction, η_2, lies in K_2 and is perpendicular to the intersection of K_1 and K_2. The plane containing η_1, η_2, and the normals to K_1 and K_2 is the plane of shear. Table 10.1 provides the twinning elements reported for twins in several metals.

Twinning Shear

For the twins listed in Table 10.1, K_2 and η_2 have the same forms as K_1 and η_1. In these cases, twinning can be regarded as a shear creating a mirror image in an

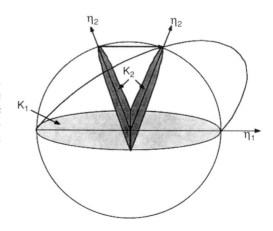

Figure 10.2. Homogeneously sheared hemisphere. The two undistorted planes are K_1 and K_2. The shear direction is h_1. The second characteristic direction, h_2, is the direction in K_2 that is perpendicular to the intersection of K_1 and K_2. From W. F. Hosford, *Ibid.*

Table 10.1. *Twinning elements for simple twins*

Structure	K_1	K_2	η_1	η_2
fcc & dia. cubic	{111}	{11$\bar{1}$}	<11$\bar{2}$>	<112>
bcc	{112}	{11$\bar{2}$}	<11$\bar{1}$>	<111>
hcp	{10$\bar{1}$2}	{$\bar{1}$012}	<10$\bar{1}$1>	<$\bar{1}$011>

orthorhombic cell, as shown in Figure 10.3. This simple form of twinning can be analyzed as follows: let the twinning plane, K_1, be (01$\bar{1}$) and the twinning direction, η_1, be [011]. In Figure 10.3, the shears are projected onto the (100) plane. When the upper left-hand portion of the crystal undergoes the twinning shear, point A moves to a new position, E. The shear strain, γ, is defined as $\overline{A\,E}/\overline{E\,F}$. The length, $\overline{E\,F}$, can be related to w and to h by noting that the triangle EFB is similar to triangle DCB, so that $\overline{E\,F}/\overline{D\,C} = \overline{E\,B}/\overline{D\,B}$. Substituting, $w = \overline{D\,C}, h = \overline{E\,B} = \overline{B\,C}$, and $\overline{D\,B} = \sqrt{(w^2 + h^2)}, \overline{E\,F} = wh/\sqrt{(w^2 + h^2)}$. Because triangle EFB is similar to triangle DCB, $\overline{D\,C}/\overline{E\,B} = \overline{B\,C}/\overline{D\,C}$, so the length $\overline{D\,C} = h^2/\sqrt{(w^2 + h^2)}$. Therefore, $\overline{A\,E} = \overline{D\,B} - 2\overline{F\,B} = \sqrt{(w^2 + h^2)} - 2h^2/\sqrt{(w^2 + h^2)}$. Substituting, $\gamma = \overline{A\,E}/\overline{E\,F} = [\sqrt{(w^2 + h^2)} - 2(h^2/\sqrt{(w^2 + w^2)})]/[wh/\sqrt{(w^2 + h^2)}]$. Simplifying,

$$\gamma = w/h - h/w. \tag{10.1}$$

If the center of the cell (point H in Figure 10.4) is connected to points A and E as shown in Figure 10.4, it is clear that line $\overline{H\,A}$ in the parent is equivalent to line $\overline{H\,E}$ in the twin. Therefore, $K_2 = (011)$ and $\eta_2 = [0\bar{1}1]$. The angle of reorientation, θ, of the [010] direction and (001) plane is given by

$$\tan(\theta/2) = h/w. \tag{10.2}$$

Table 10.2 lists the values of h and w in equations (10.1) and (10.2) for several crystal structures.

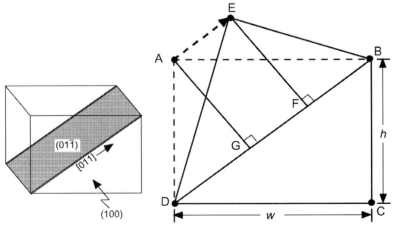

Figure 10.3. Sketch of a simple orthorhombic cell undergoing twinning. The shear strain is $\overline{A\,E}/\overline{E\,F}$.

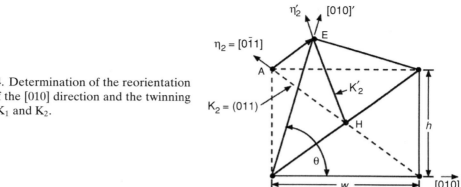

Figure 10.4. Determination of the reorientation angle, θ, of the [010] direction and the twinning elements, K_1 and K_2.

Table 10.2. *Values of h, w, and, γ*

Crystal Structure	h	w	γ
fcc and dia. cubic	a	$a\sqrt{2}$	$\sqrt{2}/2$
bcc	a	$a/\sqrt{2}$	$\sqrt{2}/2$
hcp	c	$a\sqrt{3}$	$c/(\sqrt{3}a) - \sqrt{3}a/c$

Twinning in fcc Metals

Twinning in fcc metals occurs on $K_1 = \{111\}$ planes in $\eta_1 = <11\bar{2}>$ directions. The atomic movements are shown in Figures 10.5 and 10.6. It is clear that $K_2 = \{11\bar{1}\}$ and $\eta_2 = <112>$. The shear strain can be calculated from equation (10.1) by substituting $w = a\sqrt{2}$ and $h = a$, $\gamma = a\sqrt{2}/a - a/(a\sqrt{2}) = \sqrt{2}/2 = 0.707$. The angle of reorientation of the [001] direction is, from equation (10.2), $\theta = 2\arctan(h/w) = 2\arctan(1/\sqrt{2}) = 70.5°$.

Twinning in bcc Metals

Figure 10.7 is a plan view of the (110) plane in a bcc metal. The upper right half has undergone the twinning shear on the $(\bar{1}12)[1\bar{1}1]$ system. The shear strain is $\gamma = \sqrt{2}/2$ and produces a tensile elongation parallel to [001]. Thus, the deformation is equal in magnitude to that in fcc twinning but of the opposite sign. There is another important difference, however. In fcc crystals, all nearest-neighbor distances near a twin boundary are correct so that every atom near the boundary has 12 near neighbors at the same distance. In bcc, some near-neighbor distances between atoms near the boundary are not correct.

EXAMPLE PROBLEM 10.1: The distance $\overline{A\,G}$ in Figure 11.7c is a little smaller than the normal interatomic distance (e.g., $\overline{A\,F}$) as shown in Figure 10.7c. By what percentage is this distance too small?

Solution: Consider the atom at A. The nearest neighbors at B and C (two each) and at D, E, and F (one each) are at the correct distance from A (i.e., $\sqrt{(3/2)}$). However, the atom at G is not at exactly the correct distance. The distance $\overline{A\,G}$

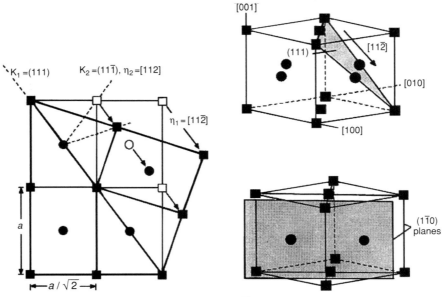

Figure 10.5. Atomic displacement in {111}<11$\bar{2}$> twinning of fcc metals.

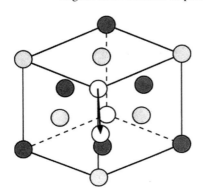

Figure 10.6. Plan view of {111}<11$\bar{2}$> twinning in fcc crystals. The atom positions are projected onto the shear plane, (1$\bar{1}$0). From W. F. Hosford, *Ibid.*

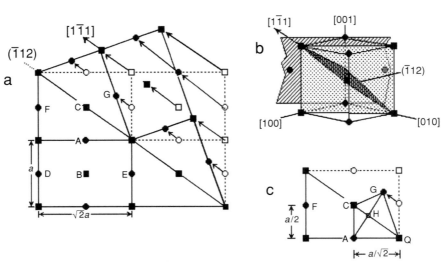

Figure 10.7. (a) Plan view of the (110) plane in a bcc crystal showing twinning on the ($\bar{1}$12)[1$\bar{1}$1] system in the upper right half of the figure. The atoms in several unit cells are indicated. Those in one (110) plane are shown as squares, while those one plane forward and one plane back are indicated by circles. (b) Perspective sketch showing the twinning elements. (c) Nearest-neighbor distances. The distance between A and G is too small. (See example problem 10.1.) From W. F. Hosford, *Ibid.*

Table 10.3. *The c/a ratios of various hcp metals*

Metal	c (nm)	a (nm)	c/a
Be	0.3584	0.2286	1.568
Cd	0.5617	0.2979	1.886
Hf	0.5042	0.3188	1.582
Mg	0.5210	0.3209	1.624
Ti	0.4683	0.2590	1.587
Zn	0.4947	0.2665	1.856
Zr	0.5148	0.3231	1.593
Spherical atoms			$\sqrt{(8/3)} = 1.633$

can be found by realizing that $\overline{AG} = 2\overline{AH}$ and that triangle AHQ is similar to triangle CAQ, so $\overline{AH}/\overline{AC} = \overline{AQ}/\overline{QC}$. Therefore, $\overline{AH} = \overline{AC} \times \overline{AQ}/\overline{QC} = (a/2)(a/\sqrt{2})/[(a/2)^2 + (a/\sqrt{2})^2]^{1/2} = [a^2/(2\sqrt{2})]/[\sqrt{(1/4 + 1/2)}] = a/\sqrt{6}$. The distance, $\overline{AQ} = 2\overline{AH} = 2a/\sqrt{6} = 0.8165a$. Comparing this with the correct "near-neighbor" distance of $a\sqrt{3}/2 = 0.8662a$, there is a difference of $(0.8165 - 0.8662)/0.8662 = 0.0574$, or 5.7%.

Twinning in hcp Metals

This is in contrast to the cubic metals (both fcc and bcc) in which slip can produce any shape change. Therefore, twinning is much more important in the hcp metals than in the cubic metals. The number of slip systems in the hcp metals is limited. The easy slip $<11\overline{2}0>$ directions are perpendicular to the c axis, and therefore, slip does not produce any elongation or shortening parallel to the c axis.

The most common twinning system in hcp metals is $\{10\overline{1}2\}<\overline{1}011>$. This is a simple mode of the form in Figure 10.3 and equations (10.1) and (10.2). Figure 10.8 shows that the direction of shear associated with this twinning system depends on the c/a ratio. These are given in Table 10.3 for a number of hcp

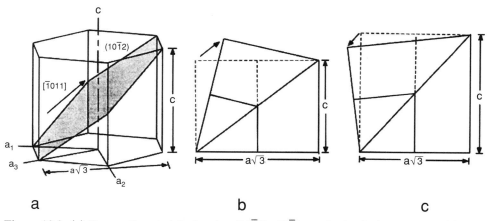

Figure 10.8. (a) Perspective sketch showing $\{10\overline{1}2\}<10\overline{1}1>$ twinning in hcp crystals. (b) For $c/a < \sqrt{3}$, the direction of shear is $[\overline{1}011]$, and twinning will occur under tension parallel to the c-axis. (c) For $c/a > \sqrt{3}$, the direction of shear is $[10\overline{1}\,\overline{1}]$, and twinning will occur under compression parallel to the c axis. From W. F. Hosford, *Ibid.*

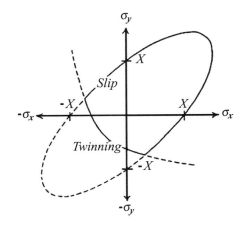

Figure 10.9. Sketch of the yield locus for a textured hcp metal with $c/a < \sqrt{3}$. From W. F. Hosford and W. A. Backofen, *Fundamentals of Deformation Processing*, Syracuse University Press (1964).

metals. Only zinc and cadmium have c/a ratios greater than $\sqrt{3}$. Therefore, only these metals twin when compression is applied parallel to the c axis (or tensile stresses perpendicular to it). For all other hcp metals, $c/a < \sqrt{3}$, so twins are formed by tensile stresses parallel to the c axis (or compressive stresses perpendicular to it). It has been found that the frequency of twins in Cd-Mg solid solution alloys decreases as the composition approaches that for which $c/a = \sqrt{3}$.

> **EXAMPLE PROBLEM 10.2:** Consider the anisotropic yield locus of titanium sheet with a texture such that the c axis is nearly but not completely aligned with the sheet normal. Will $\{10\bar{1}2\}<10\bar{1}1>$ twinning occur in tension or in compression in the plane of the sheet? Considering your answer to A, sketch the shape of the yield locus, taking into account the possibility of yielding by either slip on the prism and pyramidal planes as well as on twinning.
>
> **Solution:** The c/a ratio for titanium is less than $\sqrt{3}$, so titanium will twin in compression but not in tension perpendicular to the c axis. Slip will cause little thinning of the sheet because the $<11\bar{2}0>$ slip directions are nearly parallel to the plane of the sheet. If only slip occurred, the yield locus would be approximated by an ellipse, elongated into the first and third quadrants. The $\{10\bar{1}2\}<10\bar{1}1>$ twinning mode under compression will foreshorten the elongation into the third quadrant, as shown in Figure 10.9.

The atomic motions in twinning of the hcp lattice are complex because the atoms do not all move in the direction of shear. Figure 10.10 shows the atom positions on adjacent $(\bar{1}2\bar{1}0)$ planes of a hypothetical hcp metal with a c/a ratio of $(11/12)\sqrt{3} = 1.588$. The atom motions in are shown in Figure 10.11. Different atoms move different amounts, and there are components of motion toward and away from the mirror plane. These atomic movements have been described as a *shear* plus a *shuffle*.

There are other twinning modes for hcp metals that are not the simple type described by equations (10.1) and (10.2) and illustrated in Figure 10.3. Some of

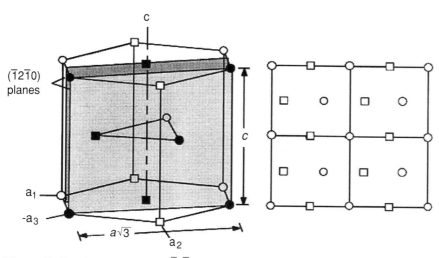

Figure 10.10. Plan view of the $(\bar{1}2\bar{1}0)$ plane of showing the atom positions on two adjacent $(\bar{1}2\bar{1}0)$ planes. From W. F. Hosford, *The Mechanics of Crystals and Textured Polycrystals*, Oxford University Press (1993).

these are listed in Table 10.4. The twin planes are shown in Figure 10.12. The twinning shear strains for the various modes depend on the c/a ratio as shown in Figure 10.13, but only changes sign for $\{10\bar{1}2\}<10\bar{1}1>$.

Shapes of Twins

Deformation twins are generally lenticular (lens shaped), as shown in Figure 10.14a. This is in contrast to twins that are formed during recrystallization. Annealing twins are bounded by almost perfectly flat planes parallel to K_1, as shown in Figure 10.14b. Because deformation twins are lenticular, the boundaries between the twinned and untwinned regions do not coincide exactly with the twinning plane (K_1). The

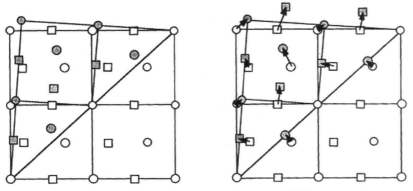

Figure 10.11. The atom movements on two adjacent $(\bar{1}2\bar{1}0)$ planes during $(10\bar{1}2)$ twinning. Note that the shear is not homogeneous. Some atoms move toward the twinning plane and some away from it. From W. F. Hosford, *Ibid*.

Table 10.4. *Other twinning elements in hcp metals*

Metal	K_1	K_2	η_1	η_2	γ
Mg	$\{10\bar{1}1\}$	$\{\bar{1}013\}$	$<10\bar{1}2>$	$<30\bar{3}2>$	
Mg	$\{10\bar{1}3\}$	$\{\bar{1}011\}$	$<\bar{3}032>$	$<\bar{1}012>$	$c/(\sqrt{3}a)-(3/4)\sqrt{3}a/c$
Zr, Ti	$\{11\bar{2}1\}$	(0001)	$<11\bar{2}\bar{6}>$	$<11\bar{2}0>$	a/c
Zr, Ti	$\{11\bar{2}2\}$	$\{11\bar{2}4\}$	$<\bar{1}\,123>$	$<\bar{2}\,243)$	$[c/a-2a/c]=0.224$ for Zr,
and	$\{11\bar{2}4\}$	$\{11\bar{2}2\}$			0.218 for Ti

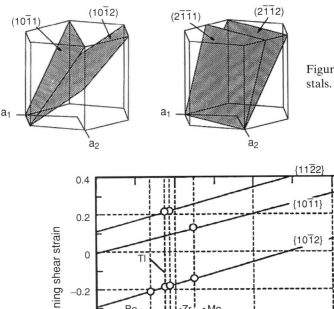

Figure 10.12. Twinning planes in hcp crystals. From W. F. Hosford, *Ibid.*

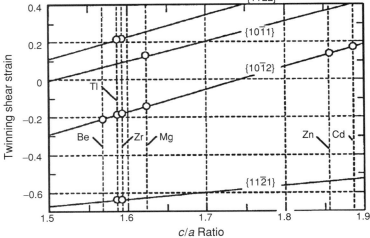

Figure 10.13. Variation of the shear strain, γ, for several twinning modes in hcp metals with c/a ratio. By convention, a positive shear strain causes elongation parallel to the c-axis. For $\{11\bar{2}1\}$ twinning, γ is negative, whereas for $\{11\bar{2}2\}$ and $\{10\bar{1}1\}$ twinning, γ is positive. For $\{10\bar{1}2\}$ twinning, the sign of γ depends on c/a. From W. F. Hosford, *Ibid.*

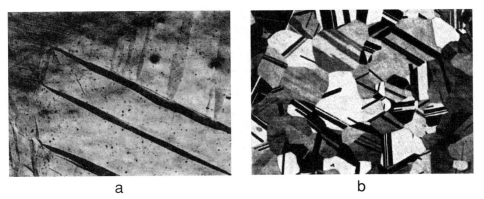

a   b

Figure 10.14. Different appearance of deformation twins and annealing twins. (a) Deformation twins in magnesium. (b) Annealing twins in brass. From W. F. Hosford, *Ibid.*

central plane is however approximately parallel to K_1. In fcc and bcc metals, twins are very narrow and usually look like parallel lines. In contrast, $\{10\bar{1}2\}<10\bar{1}\bar{1}>$ twins in hcp metals are relatively fat. The shape of the deformation twins is probably related to the overall energy change as a twin forms. There are two main contributions to this overall energy change. One is from the introduction of new surface (the twin boundary), which has a surface energy. The other is the strain energy that results when one portion of material (the twinned region) undergoes a plastic shear, while the surrounding material does not. There is mismatch between the twins and the matrix that must be accommodated by elastic distortion of the parent and/or slip in the parent. This accommodation is minimized when the twins are long and narrow. In other words, the accommodation energy is low when the aspect ratio (ratio of length to thickness) of the twin is very large. On the other hand, the surface energy is minimized for a given volume of twin when the aspect ratio approaches unity (a spherical shape). For twins with a high shear strain, the strain energy term should predominate so that high aspect ratio would be expected. In contrast, a high interfacial energy promotes fat twins.

Mechanism of Twinning

Deformation twins form with such extreme rapidity that their formation is often accompanied by audible sounds. The sound emitted by tin as it twins is called "tin cry." In a tension test, load drops occur as a twin is formed and cause the stress–strain curve to be serrated. Figure 10.15 is the stress–strain curve for a cadmium single crystal.

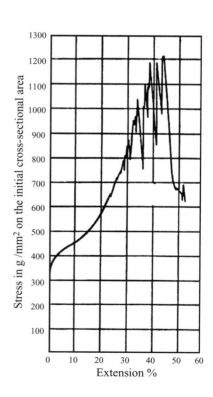

Figure 10.15. Stress–strain curve of a cadmium single crystal. Each serration corresponds to the formation of a twin. From E. Schmid and W. Boas, *Kristallplastizität*, Springer-Verlag (1935).

Twinning, like slip, occurs by the passage of dislocations. The Burgers vector of these dislocations must have the magnitude and direction of the shear between successive planes. In the case of fcc metals, the Burgers vector for twinning is $(a/6)<211>$, which is the same as the Burgers vectors of the partial dislocations formed by dissociation of $(a/2)<110>$ slip dislocations. For twinning, unlike slip, a single $(a/6)<211>$ dislocation must pass on every successive $\{111\}$ plane. Cottrell and Bilby[*] proposed a mechanism to explain how this might occur. In their *pole mechanism* (Figure 10.16), there is a screw dislocation with a $(a/3)<111>$ Burgers vector normal to the $\{111\}$ plane containing the $(a/6)<211>$ partial. When the moving partial encounters this pole, one-half of it winds its way upward around the pole while the other half winds its way downward. In this way, the same partial dislocation passes on every successive $\{111\}$ plane and, at the same time, expands outward.

It is not clear what yield criterion is appropriate for twinning. It is tempting to assume a critical shear stress criterion similar to Schmid's law for slip. Indeed, this has been assumed in most analytic treatments. However, several investigations have questioned the validity of critical shear stress criterion. Deviations from a critical shear stress criterion may be due to the necessity for some slip to nucleate twinning, or to a sensitivity of twinning to the stress normal to the twin plane. In discussing this problem, it is important to recognize that the stress necessary to nucleate a twin may be quite a bit higher that for its propagation. The jerkiness of the stress–strain curves when deformation is by twinning suggests this. The reason for a high nucleation stress is not hard to rationalize. As a twin is initiated, the ratio of its surface area to volume is very high, and the work to produce the surface must come from the mechanical work expended. On a per-volume basis, this can be expressed as

$$dw = \sigma\,d\varepsilon = \tau\gamma\,df, \qquad (10.3)$$

where dw is the incremental work/volume expended in causing an incremental volume fraction, df, to undergo twinning, γ is the shear strain associated with twinning,

[*] A. H. Cottrell and B. A. Bilby, *Phil. Mag.* v. 42 (1951), p. 573.

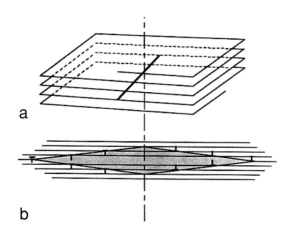

a

b

Figure 10.16. Pole mechanism for twin formation. (a) A partial dislocation approaches a screw dislocation normal to the plane on which it is moving. Part of the partial dislocation runs up the spiral ramp and part runs down it, both spiraling outward. In this way, the same partial dislocation will move on each successive plane. (b) A cross-section showing the lens-shaped twin that is formed. From W. F. Hosford, *Ibid.*

and τ is the required shear stress. For small twins, dw/df will be high, so $\tau = (1/\gamma)dw/df$ must also be high. As a twin thickens, the ratio of surface area to volume decreases, so the stress for growth may be lower than for nucleation.

It is possible that the nucleation of a twin is sensitive to the normal stress across the twin plane. According to the principle of normality, if the shear stress required for twinning depends on the normal stress, there must be a plastic strain normal to the twin plane. The atomic misfit at a twin boundary may produce a small dilatation. However, as the twin thickens, the dilated region simply moves with the boundary without increasing the macroscopic dilatational strain. This argument implies that the stress for nucleation of a twin may depend on the normal stress, but the stress for propagation of the twin should not. If dilatation does occur at a twin boundary, the size of the affected region would be so small that the dilatation would not be detectable by macroscopic measurements.

Grain size has two effects on twinning in polycrystals. The size of the first twins formed in a grain is limited by the grain size. With smaller grain sizes, the average twin size must also be smaller and have a larger ratio of surface area to volume; therefore, more energy (higher average stress) is required to produce the same volume fraction twin in fine-grain polycrystals. Decreasing grain size also tends to reduce the jerkiness of the macroscopic stress–strain curves. Whether slip or twinning occurs as a load is applied depends on which requires the lower stress. Mechanical twinning is rare in most fcc and bcc metals except at low temperatures or high strain rates. This implies that decreasing temperature and increasing strain rates tend to increase the stresses required for slip more than the stresses required for twinning.

Composition is also important. Twinning is more common in fcc metals and alloys of low stacking fault energy (SFE). Figure 10.17 shows how the twinning stress increases with SFE in copper-base alloys. Silver and gold, which have very low SFEs twin at room temperature, whereas mechanical twinning is not observed in metals like aluminum.

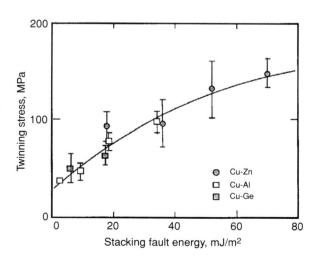

Figure 10.17. Dependence of twinning stress on stacking fault energy in copper-base alloys. From W. F. Hosford, *Ibid*, p. 184. Data from J. A. Venables, *Deformation Twinning*, TMS-AIME (1964).

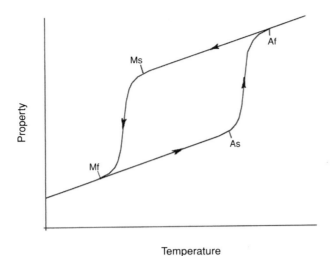

Figure 10.18. On cooling, the high temperature (austenite) phase starts to transform to the low temperature phase (martensite) at Mf, and the transformation is complete at Mf. On heating, the reverse reaction starts at As and finishes at Af.

Martensite Transformation

Martensitic transformations are very similar to mechanical twinning. It occurs by sudden shearing of the lattice, with atoms moving only a fraction of a normal interatomic distance. Originally, the term *martensitic* transformation was reserved for steel, but now it is generally applied to all phase transformations that occur by shear without any composition change. Likewise, the terms *austenite* and *martensite* are used to identify the high and low temperature phases. The extent of a martensitic transformation can be monitored by measuring changes of property, as shown in Figure 10.18. On cooling, the transformation starts at the M_s temperature and finishes at the M_f temperature. On heating, the martensite starts to revert to austenite at A_s and finishes at A_f. There is considerable hysteresis. The A_s and A_f temperatures are higher than the M_s and M_f temperatures. Between the A_s and M_s temperatures, martensite can be formed by deformation.

Shape Memory and Superelasticity

Shape memory: This is an effect in which plastic deformation (or at least what appears to be plastic deformation) is reversed on heating. The alloys that exhibit this effect are invariably ordered solid solutions that undergo a martensitic transformation on cooling. The alloy TiNi at 200°C has an ordered bcc structure. On cooling, it transforms to a monoclinic structure by a martensitic shear. The shear strain associated with this transformation is about 12%. If only one variant of the martensite was formed, the strain in the neighboring untransformed lattice would be far too high to accommodate. To decrease this compatibility strain, two mirror image variants form in such a way that there is no macroscopic strain. The macroscopic shape is the same as before the transformation. Figure 10.19 illustrates this. The boundaries between the two variants are highly mobile so if the resulting structure is deformed, the deformation is easily accommodated by movement of the boundaries in what

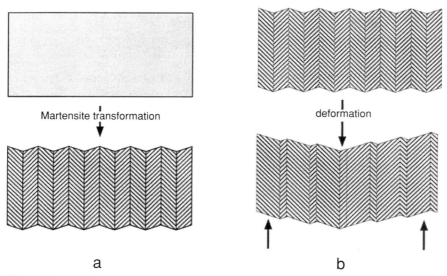

Figure 10.19. (a) As the material is cooled, it undergoes a martensitic transformation. By transforming to equal amounts of two variants, the macroscopic shape is retained. (b) Deformation occurs by movement of variant boundaries so the more favorably oriented variant grows at the expense of the other. Adapted from a sketch by D. Grummon.

appears to be plastic deformation. Figure 10.20 shows a stress–strain curve. Heating the deformed material above the A_f temperature causes it to transform back to the ordered cubic structure by martensitic shear. For both variants, the martensite shears must be of the correct sign to restore the correct order. The overall effect is that the deformation imposed on the low-temperature martensitic form is reversed. The critical temperatures for reversal in TiNi are typically in the range of 80°C to 100°C, but are sensitive to very minor changes in composition so material can be produced with specific reversal temperatures.

Superelasticity: This phenomenon is closely related to the shape memory effect except that the deformation temperature is above the normal A_f temperature. However, the A_f temperature is raised by applied stress. According to the Clausius-Clapyron equation,

$$d(A_f)/d\sigma = T\varepsilon_0/\Delta H, \tag{10.4}$$

where ε_0 is the normal strain associated with the transformation, and ΔH is the latent heat of transformation (about 20 J/g for TiNi). The terms $d(A_s)/d\sigma$, $d(M_s)/d\sigma$, and $d(M_f)/d\sigma$ could be substituted for $d(A_f)/d\sigma$ in equation (10.4). Figure 10.21 shows this effect. If a stress is applied at a temperature slightly above the A_f, the A_f, A_s, M_s, and M_f temperatures are all crossed as the stress increases. The material transforms to its low-temperature structure as it deforms by a martensitic shear. However, when the stress is released, the material again reverts to the high-temperature form. A stress–strain curve for Fe_3Be is shown in Figure 10.22.

For both the memory effect and superelasticity, the alloy must be ordered, there must be a martensitic transformation, and the variant boundaries must be mobile. The difference between the shape memory effect and superelasticity is

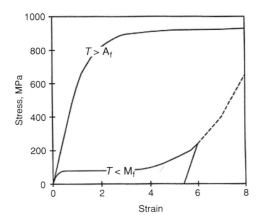

Figure 10.20. Stress–strain curve for a shape memory material. The lower curve is for deformation when the material is entirely martensitic. The deformation occurs by movement of variant boundaries. After the material is of one variant, the stress rises rapidly. The upper curve is for the material above its Af temperature. Adapted from a sketch by D. Grummon.

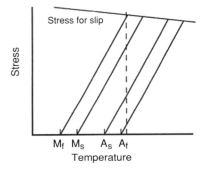

Figure 10.21. As stress is applied to a super-elastic material, the A_f, A_s, M_s, and M_f temperatures for the material undergoes martensitic shear strains. When the stress is removed, the material reverts to its high temperature form reversing all of the martensitic deformation. Adapted from a sketch by D. Grummon.

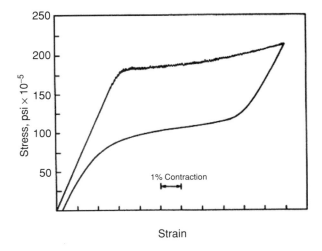

Figure 10.22. The stress–strain curve for superelastic Fe_3Be. After the initial Hookian strain, the material deforms by martensitic transformation. On unloading the reverse martensitic transformation occurs at a lower stress. From R. H. Richman, in *Deformation Twinning*, TMS-AIME (1963).

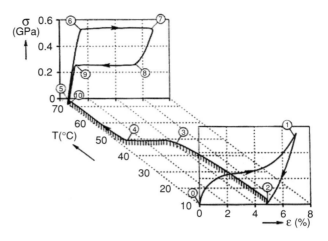

Figure 10.23. Schematic illustration of the difference between shape memory and superelastic effects. For shape memory, the deformation occurs at a temperature for which the material is martensitic. A superelastic effect occurs when the deformation occurs just above the A_f temperature. From J. A. Shaw, *Int. J. of Plasticity*, v. 16 (2000).

shown schematically in Figure 10.23. For the superelastic effect, deformation starts when the material is austenitic, whereas for the shape memory effect, the martensite phase must be deformed.

REFERENCES

E. O. Hall, *Twinning and Diffusionless Transformations*, Butterworths (1954).

R. E. Reed-Hill, J. P. Hirth, and H. C. Rogers, eds., *Deformation Twinning, Deformation Twinning*, TMS-AIME (1964).

W. F. Hosford, *The Mechanics of Crystals and Textured Polycrystals*, Oxford (1993).

K. Otuka and C. M. Wayman, eds., *Shape Memory Materials*, Cambridge University Press (1998).

E. Schmid and W. Boas, *Plasticity of Crystals with Special Reference to Metals*, Chapman and Hall (1950). (This is an English translation of *Kristallplastizität*, Springer-Verlag (1935).)

Note

For many years, it was widely believed that annealing twins were common in fcc metals, but not in bcc or hcp metals, and that mechanical twinning may occur in bcc and hcp metals, but never in fcc metals. The reasons for this inverse relationship were not explained. However, it was so strongly believed, that when in 1957, Blewitt, Coltman, and Redman [*J. Applied Physics*, v. 28 (1957), p. 651] first reported mechanical twinning in copper at very low temperatures, other workers were very skeptical until these findings were confirmed by other researchers.

Swedish researcher Arne Olader first observed the shape-memory effect in 1932 in a gold-cadmium alloy. Other shape-memory alloys include CuSn, InTi, TiNi, and MnCu. The superplastic effect in TiNi was first found by William Buehler and Frederick Wang in 1962 at the Naval Ordnance Laboratory. They called the alloy *Nitinol* after *Ni*ckel *Ti*tanium *N*aval *O*rdnance *L*aboratory.

Problems

1. The rolling texture of most hcp metals can be roughly described by an alignment of the c-axis with the rolling plane normal. Zinc is an exception. The c axis tends to be rotated as much as 80 degrees from the rolling plane normal toward the rolling direction. Explain this observation in terms of $\{10\bar{1}2\}<10\bar{1}1>$ twinning and the fact that easy slip occurs only on the basal plane.

2. For extruded bars of the magnesium alloy AZ61A, the *Metals Handbook* (ASM, v. 1, 8th ed. p. 1106) reports the tensile yield strength as 35,000 psi and the compressive yield strength as 19,000 psi. Deduce how the c axis must be oriented relative to the rod axis in these extrusions, assuming the difference is due to the directionality of twinning.

3. When magnesium twins on the $\{10\bar{1}2\}$ planes in $<\bar{1}01\bar{1}>$ directions, what is the angle between the c axes (i.e., the [0001] directions) of the twin and the untwinned material? For Mg, the lattice parameters are: a = 0.32088 nm and c = 0.52095 nm.

4. Suppose an investigator has reported $<1\bar{1}0>\{110\}$ in a cubic crystal. Comment on this claim. (Think about the resulting atomic arrangement.)

5. How many different $\{111\}<11\bar{2}>$ twinning systems are there in an fcc crystal?

6. Figure 10.24 represents a crystal, which is partially sheared by twinning. The shear direction and the normal to the mirror plane are in the plane of the paper.

    A. Indicate the shear strain, γ, associated by twinning in terms of dimensions on the sketch. Put appropriate dimensions on the sketch and indicate, γ, in terms of these dimensions.

    B. On the drawing, clearly mark the first and second undistorted directions, η_1 and η_2.

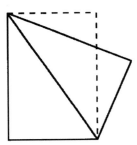

Figure 10.24. Partially twinned crystal.

7. Figure 10.25 illustrates twinning in a bcc crystal. Indicate on the sketch the undistorted direction, η_2, and give its indices.

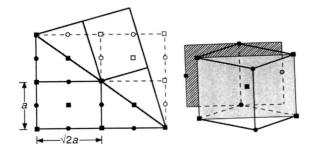

Figure 10.25. Twinning shear in a bcc crystal.

8. Consider an ordered bcc alloy, ordered into a B1 structure. (One specie of atoms occupies the body-centering positions and the other the corner positions, so each atom is surrounded by eight atoms of the other specie.) Suppose this crystal is subjected to exactly the same $<11\bar{2}>\{111\}$ shear as would produce a twin in a disordered bcc alloy. By drawing a $\{1\bar{1}0\}$ plan view, deduce the atomic arrangement after the shear. How many near-neighbors of an atom are of the opposite species? Is this a true twin, or is the crystal structure changed?

9. The shear strain associated with the martensitic transformation of a shape-memory material is 0.18. What is the maximum tensile strain associated with the shape-memory effect?

11 Hardening Mechanisms in Metals

Introduction

The strengths of metals are sensitive to microstructure. Most hardening mechanisms involve making dislocation motion more difficult. These include decreased grain size, strain hardening, solid solution hardening, and dispersion of fine particles. With finer grain sizes, there are more grain boundaries to impede dislocation motion. Metals strain harden because deformation increases the number of dislocations, and each interferes with the movement of others. In solid solutions, solute atoms disrupt the periodicity of the lattice. Fine dispersions of hard particles create obstacles to dislocation motion. Martensite formation and strain aging in steels are sometimes considered separate mechanisms, but these are related to the effects of interstitial solutes on dislocations.

Crystal Structure

Crystal structure strongly affects hardness and yield strength. Fcc metals tend to have higher strengths than bcc metals and lower strengths than hcp metals. The effect of crystal structure is most apparent in metals that transform from one crystal structure to another. Figure 11.1 shows how the hardness of nearly pure iron changes with temperature. The general decrease of hardness with increasing temperature is interrupted by the $\alpha \rightarrow \gamma$ transformation at 910°C. At 910°C, the fcc structure is harder than the bcc structure. The reason the bcc structure is softer is at least partially explained by its greater number of slip systems. Similarly, the hardness of titanium would be expected to undergo a sharp drop as it transforms from hcp because there are fewer slip systems in hcp crystals than bcc ones.

Grain Size

In the early stages of deformation, grain boundaries are important obstacles to slip, so fine-grain materials are stronger than coarse-grain materials. The dependence of

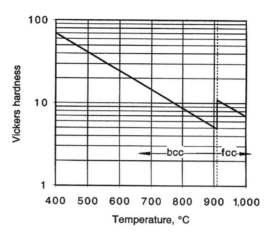

Figure 11.1. Hardness of iron as a function of temperature. Note the discontinuity at 910°C, which indicates that at that temperature the fcc structure is harder than the bcc structure. Adapted from H. C. Chao, PhD thesis, University of Michigan (1964).

initial yield strength on grain size is often expressed by the Hall-Petch* relation

$$\sigma = \sigma_o + K_Y d^{-1/2}. \tag{11.1}$$

This dependence has been explained in terms of dislocation pile-ups at grain boundaries. The number of dislocations in these pile-ups is proportional to the grain size, d. The stress concentration in the grain across the boundary thus increases the number of dislocations in the pile-ups and hence with the grain size. Therefore, with a larger grain size, a lower applied stress is required for slip. Figure 11.2 shows this relation for iron alloys.

A similar equation can be written for the dependence of hardness, H, on grain size.

$$H = H_o + K_H d^{-1/2}. \tag{11.2}$$

where H_o and K_H are constants. Equation (11.2) is equivalent to equation (11.1) if the hardness is expressed in terms of normal stresses.

EXAMPLE PROBLEM 11.1: Evaluate the constants σ_o and K_Y in equation (11.1) for ferrite using the data in Figure 11.2.

Solution: If the yield strengths at two grain sizes are compared, $\sigma_1 - \sigma_2 = K_Y(d_1^{-1/2} - d_2^{-1/2})$. $K_Y = (\sigma_1 - \sigma_2)/(d_1^{-1/2} - d_2^{-1/2})$. Taking $\sigma_1 = 410\,\text{MPa}$ at $d_1^{-1/2} = 11\,\mu\text{m}^{-1/2}$ and $\sigma_2 = 260\,\text{MPa}$ at $d_2^{-1/2} = 4\,\mu\text{m}^{-1/2}$, $K_Y = 21.4\,\text{MPa}(\mu\text{m})^{1/2}$. $\sigma_o = \sigma_1 - K_Y d_1^{-1/2} = 410 - 21.4(11) = 175\,\text{MPa}$.

For coarse-grain metals, there is an additional effect that depends on the ratio of grain size to specimen size.[†] In grains at the free surface, the deformation is less constrained by neighboring grains than in interior grains. Surface grains do not require five active slip systems, so a lower stress is necessary to deform surface grains.

---

[*] E. O. Hall, *Proc. Phys. Soc. London*, v. B64 (1951), p. 747, and N. J. Petch, *Iron and Steel Inst.*, v. 174 (1953), p. 25.
[†] R. L. Fleischer and W. F. Hosford, *Trans. AIME*, v. 221 (1961), pp. 244–247.

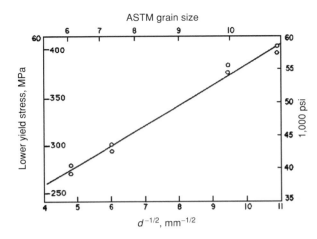

Figure 11.2. The dependence of yield strength of ferrite on strength on grain size. Plotting against $d^{-1/2}$ results in a straight line. This verifies equation (11.4). Data from W. B. Morrison, *J. Iron Steel Inst.* v. 201 (1963). Adapted from R. A. Grange, *ASM Trans.*, v. 59 (1966).

The net effect is that the overall strength depends on the volume fraction surface grains,

$$\sigma = V_f \sigma_s + (1 - V_f)\sigma_i, \tag{11.3}$$

where σ_s and σ_i are the stresses necessary to deform surface and interior grains, respectively, and V_f is the volume fraction of surface grains. For tensile specimen with a circular cross-section, the volume fraction surface grains, f_s, can be approximated as $f_s \approx (\pi/2)(d/D)$, where D is the specimen diameter and d is the average grain diameter. This approximation is reasonable for $d < D/5$. Note that if $d/D = 10$, $f_s = 16\%$ and $\sigma = 0.16\sigma_s + 0.84\sigma_i$. If it is further assumed that $\sigma_s = 0.5\sigma_i$, $\sigma = 0.92\sigma_i$. The same effect applies to flat sheet specimens where $f_s \approx \pi d/t$, t being the thickness.

Strain Hardening

For fcc metals, there is a strong correlation between their stacking fault energy and their strain hardening exponent, as shown in Figure 11.3. In metals with low stacking

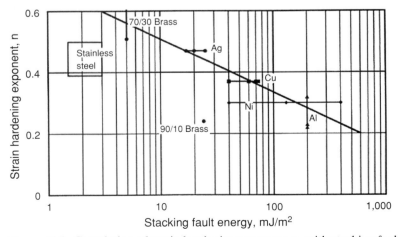

Figure 11.3. Correlation of strain hardening exponent, n, with stacking fault energy. Ag, Cu, Ni, and Al designate points for silver, copper, nickel, and aluminum. Also shown are data for austenitic stainless steels and 70/30 brass.

fault energy, work hardening is more persistent because cross-slip is more difficult. The roles of cross-slip and dislocation intersections in strain hardening are discussed in Chapter 9.

Solid Solution Strengthening

Elements in solid solution usually strengthen crystals. For substitutional solutions in copper (Figure 11.4), the yield strength increases proportional to their concentration of solutes. The rate of this increase is proportional to the 4/3 power of a misfit parameter defined as $\varepsilon = (\mathrm{d}a/a)/\mathrm{d}c$, where $\mathrm{d}a/a$ is the fractional change in lattice parameter with concentration, c, expressed as atomic fraction (Figure 11.5),

$$\Delta\tau/\Delta c = CG\varepsilon^{4/3}, \tag{11.4}$$

where G is the shear modulus of copper and C is a constant.

The effect of substitutional solutes is mainly attributable to interaction of the solutes with the dilatational stress field around edge dislocations. Substitutional solutes have little interaction with screw dislocations. At low temperatures, many solutes in iron actually cause a solid solution softening instead of hardening (Figure 11.6). The explanation for solid solution softening is beyond the scope of this text.

Solid solution hardening by interstitial solutes (C, N, etc.) is very high. Martensite formation in steels is often considered a separate strengthening mechanism. However, martensite can be regarded as a supersaturated solution of carbon in ferrite, and its hardness attributed to solid solution hardening. The crystal structure of martensite is a body-centered tetragonal, with the degree of tetragonality increasing linearly with the amount of carbon. As amount of carbon approaches zero, the lattice and yield strength approach those of bcc ferrite (Figure 11.7). The slope of this plot corresponds to about 575 MPa per atomic % carbon. For comparison, the strengthening of copper by tin (Figure 11.4) is only about 20 MPa/atomic % tin.

Figure 11.4. Effect of solute concentration on shear yield strength of dilute copper-base alloys. Note that the strength is proportional to concentration. Adapted from F. McClintock and A. Argon, *Mechanical Behavior of Materials*, Addison-Wesley (1966).

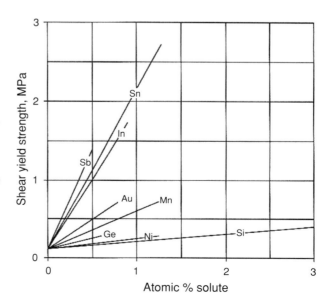

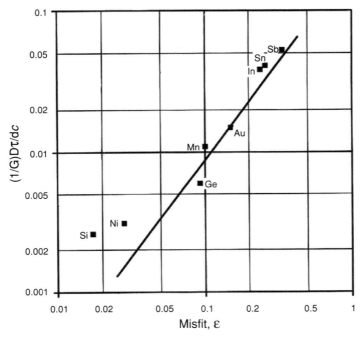

Figure 11.5. The effect of the solute misfit parameter, $\varepsilon = (1/a)da/dc$, on the solute hardening of copper. The solid line is equation (11.4), with C = 0.215. Data from F. McClintock and A. Argon, *Ibid*.

Dispersion Strengthening

Dispersions of fine particles are very effective in strengthening. There are a number of ways of producing such dispersions. These include precipitation from supersaturated solid solution (age hardening), pressing and sintering of metal powder mixed

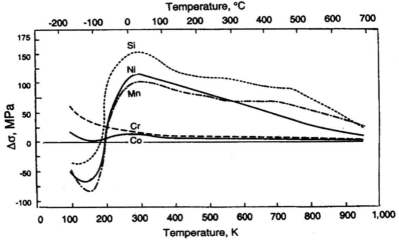

Figure 11.6. The effect of substitutional solutes on the yield strength of iron. Note the solid solution softening at low temperatures. From Leslie, *The Physical Metallurgy of Steels*, McGraw-Hill (1981).

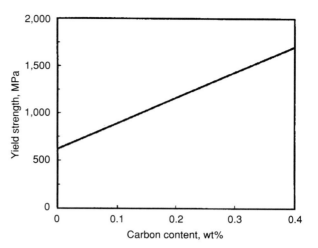

Figure 11.7. The yield strength of martensite as a function of carbon content. Data from W. C. Leslie and E. Hornbogen, in *Physical Metallurgy*, v. 4, Cahn and Haasen, eds., Elsevier (1996).

with fine particles, and internal oxidation (e.g., precipitation of MgO in a silver-base Mg alloy). Hard precipitate particles on slip planes form barriers to dislocation motion. For slip to continue, there are two possibilities: either dislocations bow out between the particles, or they cut through the particles.

Figure 11.8 illustrates the first possibility, which was suggested by Orowan. The stress required to bow a dislocation between particles is greatest when the dislocation is semicircular. At that point, twice the line tension ($\approx Gb^2$) equals the force on the dislocation, $\tau \cdot bd$. Then, $2Gb^2 = \tau \cdot bd$, or

$$\tau = 2Gb/d, \tag{11.5}$$

which is identical to the equation for operation of a Frank-Read source. The distance between particles,* d, replaces the distance between pinning points. After the dislocation bows between the particles, it rejoins other bowed segments and passes on, leaving a loop around each particle (Figure 11.9). These loops in turn repel subsequent dislocations, effectively decreasing the value of d in equation (11.5). The result is that a higher stress is required for them to pass. This explains the very high initial rate of strain hardening in age-hardened alloys.

For a given volume fraction, V_f, the distance, d, decreases with finer particles. An estimate of the effect of particle size can be made by assuming the particles intersecting the slip plane are arranged in a square pattern and that r is the radius of the particle as it is cut on the slip plane. Then the area fraction equals the volume fraction, $V_f = \pi r^2/d^2$. Solving for $d = \bar{r}(\pi/V_f)^{1/2}$, and substituting into equation (11.5),

$$\tau = 2Gb(V_f/\pi)^{1/2}/r. \tag{11.6}$$

Because the increase of tensile yield strength, $\Delta\sigma$, caused by the precipitate particles is proportional to the shear stress, τ,

$$\Delta\sigma = \alpha GbV_f^{1/2}/r, \tag{11.7}$$

where α is a constant. This predicts that the strength increases with increasing volume fraction and decreasing particle size.

---

* Strictly d should be the diameter of the semicircle formed by the dislocation as it bows between the particles, so it should be $d - d_p$, where d_p is the particle diameter.

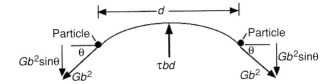

Figure 11.8. Force balance on a dislocation bowed between two particles.

EXAMPLE PROBLEM 11.2: A high-strength, low-alloy steel is strengthened by precipitates of VC. If the diameter of the particles is 50 nm, what volume percent particles would be required to increase the strength by 100 MPa? For iron, $G = 80$ GPa, and the atomic diameter is 0.248 nm.

Solution: Substituting into equation (11.10) and solving for V_f, $V_f = [r\Delta\sigma/(\alpha Gb)]^2 = [(25 \times 10^{-9}\,\text{m})(100 \times 10^6\,\text{Pa})/(4 \times 80 \times 10^9\,\text{Pa})(0.248 \times 10^{-9}\,\text{m})]^2 = 0.00099$, or 0.1%.

This picture is consistent with the dependence of strength on the number of precipitate particles. In the initial stages of precipitation hardening, strength increases as V_f increases. After virtually all solute has precipitated, the number of precipitate particles gradually decreases as the smaller ones dissolve and the larger ones grow (Ostwald ripening). This leads to a decreased strength. Throughout the precipitation, there is a loss of solid solution hardening so the net change of strength is a balance of these two effects. With overaging, the strength often drops below that of the unaged material.

The second possibility for dislocation motion in a particle-hardened matrix is that dislocations cut through the particles, even though they are harder than the matrix. Figure 11.10 is an illustration of the cutting Ni_3Al particles by dislocations. In this case, the overall strength is given by the rule of mixtures,

$$\tau = \tau_{\text{mat}}(1 - V_f) + \tau_{\text{part}} V_f, \tag{11.8}$$

where τ_{mat} and τ_{part} are the shear strengths of the matrix and particles, respectively. Because V_f is usually small,

$$\tau \approx \tau_{\text{mat}} + \tau_{\text{part}} V_f. \tag{11.9}$$

Equation (11.9) sets an upper limit on the value of τ in equation (12.7). The lower of the two values of τ is appropriate, as shown schematically in Figure 11.11.

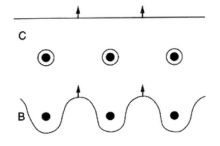

Figure 11.9. Recombination of dislocations after bowing between hard particles. As they pass on, loops are left around each particle.

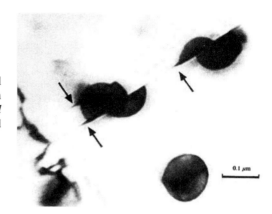

Figure 11.10. Photomicrograph showing Ni_3Al particles in a nickel-base alloy that have been sheared by dislocations. From Haasen, *Physical Metallurgy*, 2nd ed., Cambridge (1986). Attributed to H. Gleiter and E. Hornbogen.

During tempering of martensite, carbides precipitate. For most steels, the precipitation hardening effect is completely masked by the loss of solid solution strengthening as carbon comes out of solution to form the carbides. The net effect is a loss of hardness. However, in high-speed tool steels, which contain large amounts of strong carbide-forming elements (W, Mo, Cr, and V), the net effect can be an increase in hardness.

Yield Points and Strain Aging

The stress–strain curves of annealed low-carbon steels are characterized by an initial yield point. As soon as a very small amount of plastic deformation occurs, the stress necessary for continued deformation drops. Figure 11.12 illustrates this effect. The stress to initiate deformation is called the *upper yield strength*, whereas the stress to continue the deformation is called the *lower yield strength*. Measurements of the upper yield strength are very difficult to reproduce because upper yield strength is extremely sensitive to the alignment of the test specimen in a tension test. For steel, the lower yield strength is usually reported, instead of the upper yield strength, because it is more reproducible.

The yield point effect is explained by the affinity of interstitially dissolved atoms (C and N) for dislocations. The stress fields of edge dislocations attract the interstitials, which lower their energy and pin them. For slip to occur, the applied stress must be great enough to pull some dislocations away from the interstitial

Figure 11.11. Schematic illustration of how the yield strength of a dispersion should depend on volume fraction of dispersed phase. The yield strength should correspond to the lower of the two lines.

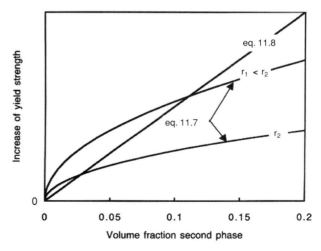

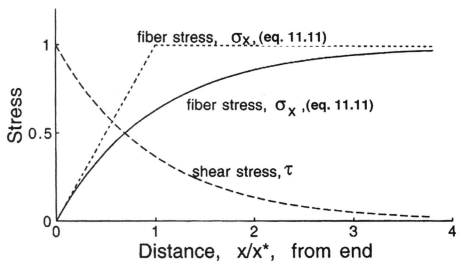

Figure 11.12. Stress–strain curve for a low-carbon steel, showing an upper and a lower yield point.

atoms that pin them. Once a dislocation breaks free from interstitials, it can move at a lower stress. Yielding starts at one location and spreads throughout the stressed regions. In a tension test, only a small region is deforming at any one instant. The deforming region is called a *Lüder's band*. It moves along the specimen causing the same strain wherever it passes. Strain hardening starts only after the Lüder's band has traversed the entire gauge length. Figure 11.13 shows Lüder's bands in a tensile specimen. During stamping of parts from sheet steels, this discontinuous yielding is undesirable. If the Lüder's band does not traverse the whole sheet, its boundaries will create an unsightly surface roughness, known as *stretcher strains* (Figure 11.14). To eliminate the yield point, sheet steel is usually deformed by bending or an extremely light rolling reduction to eliminate a yield point. It is not necessary to deform the entire cross-section of the sheet to remove the yield point effect. As long as the deformed regions are widely dispersed, they can initiate uniform deformation when the sheet is being formed (Figure 11.15).

If the steel is exposed to warm temperatures for a long enough period, the yield point will return (Figure 11.16). This return of the yield point after deformation is called *strain aging*. Carbon or nitrogen atoms diffuse to the dislocations that had been previously broken free from locking interstitials. Temperature has a large influence on how rapidly strain aging occurs and on the extent of strain aging (Figure 11.17). Strain aging increases the tensile strength and decreases the elongation, in addition to causing a new yield point. Most low-carbon steels are *aluminum killed*. Killing is deoxidation by the addition of a small amount of aluminum to molten steel just before ingot casting. In addition to reacting with the dissolved oxygen to form Al_2O_3, the aluminum also combines with nitrogen, making the steels more resistant to strain aging.

Combined Effects

The effects of the various hardening mechanisms are not additive in the mathematical sense that the overall increase in yield strength, $\Delta\sigma$, is the sum of the yield

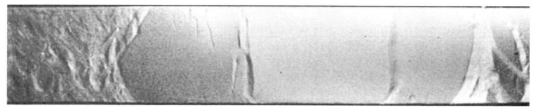

Figure 11.13. Lüder's bands on a low-carbon steel tensile specimen. From S. Kalpakjian, *Mechanical Processing of Materials*, Van Nostrand (1967). Courtesy of J. F. Butler, Jones and Laughlin Steel Corp.

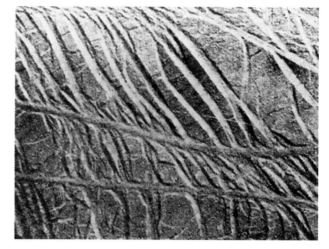

Figure 11.14. Stretcher strains on a low-carbon steel. From *Metals Handbook*, 8th ed., vol. 7 (1972), ASM.

Figure 11.15. Deformed regions created by a very light rolling reduction. From D. S. Blickwede, *Metals Progress* (July 1969).

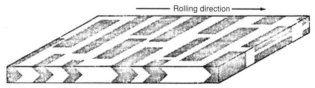

Figure 11.16. Strain aging of a low-carbon steel. From W. C. Leslie, *The Physical Metallurgy of Steels*, McGraw-Hill (1981).

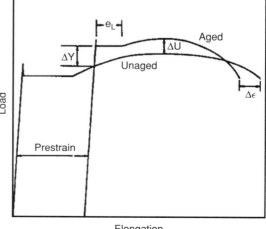

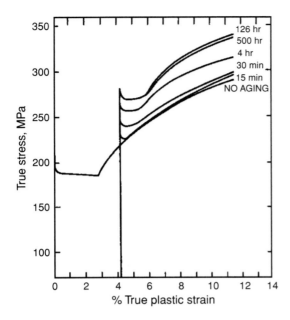

Figure 11.17. Strain aging of a 0.03%C steel at 60°C. From *Making, Shaping and Treating of Steels*, U.S. Steel (1971).

strength increased from the individual mechanisms, $\Delta\sigma_i$ (i.e., $\Delta\sigma \neq \Sigma\Delta\sigma_i$). For example, the effects of grain size and precipitation hardening are not additive. With a fine dispersion of hard particles as obstacles for dislocations, the additional effect of grain boundaries is negligible. The same is true of strain hardening. Strain hardening of a precipitation-hardened material is initially very high and rapidly decreases. This leads to a much lower strain hardening exponent.

Although not additive in a mathematical sense, strengthening mechanisms can be combined to achieve greater yield strengths. Combining strain hardening and age hardening is an example. The highest yield strengths in Be-Cu alloys are a result of combined strain and age hardening. The material is cold worked in the solution-treated condition and then aged. The aging temperatures are low enough so that annealing effects are negligible. Table 11.1 gives the properties of Cu-1.90% Be, 0.2% Ni after several treatments.

Table 11.1. *Strength of beryllium copper sheet*

Condition*	Tensile strength (MPa)
Annealed	415–540
1/4 H	515–605
1/2 H	585–690
H	690–825
HT	1,205
1/4 H & HT	1,275
1/2 H & HT	1,345
H & HT	1,380

* 1/4H, 1/2H, and H indicate increasing strengths by work hardening.
HT indicates age-hardening treatment. (Solution treatments were done prior to work hardening.)

EXAMPLE PROBLEM 11.3: Aluminum alloy 6061 is an age-hardenable alloy containing about 1% Mg and 0.6% Si. The precipitates that form on aging are Mg_2Si. The yield strength in the solution-treated and quenched condition is 150 MPa. After aging 1 hr at 230°C, the yield strength is about 250 MPa. In the completely overaged condition, the yield strength is about 75 MPa. In the aged condition, what is the strength increase, $\Delta\sigma$, attributable to particle hardening? What is the strength increase attributable to solid solution hardening by Mg and Si in the solution-treated condition?

Solution: If there is no solid solution hardening by Mg and Si in the aged condition, and there is no particle hardening or solid solution hardening by Mg and Si in the overaged condition, the strengthening due to the particles is $250 - 75 = 175$ MPa, and the solid solution strengthening in the solution-treated condition is $150 - 75 = 75$ MPa.

REFERENCES

G. I. Taylor, *J. Inst. Metals*, v. 62 (1938), and in *Timoshenko Anniv. Vol.*, Macmillan (1938).
J. F. W. Bishop and R. Hill, *Phil. Mag.*, ser. 7, v. 42 (1951).
F. McClintock and A. Argon, *Mechanical Behavior of Materials*, Addison-Wesley (1966).
W. C. Leslie, *The Physical Metallurgy of Steels*, McGraw-Hill (1981).
R. E. Reed-Hill and R. Abbaschian, *Physical Metallurgy Principles*, 3rd ed., PBS (1991).

Notes

Dr. Alfred Wilm, in the Netherlands, first recognized the phenomenon of age hardening in 1906. He made a series of experiments in an attempt to harden aluminum alloys by the same type of heat treatment used to harden steels by quenching them from a high temperature. His experiments on an aluminum alloy containing 4% copper, 0.5% magnesium were interrupted by a long weekend. When he returned, he found that the hardness readings were considerably higher than those taken before the weekend. Wilm did not publish this finding until 1911, and even then did not explain the reasons for the hardening.

Although this alloy had been used for aircraft parts for some time, apparently the first use of Wilm's findings to harden it was by the Germans during World War I. They used age-hardened aluminum on dirigible and airplane parts. This use led to the subsequent investigation by the U.S. Bureau of Standards. P. D. Merica, with the aid of R. G. Waltenberg and H. Scott, first explained the hardening by the precipitation of a compound. Their paper was published by the U.S. Bureau of Standards in 1919 and now is regarded as a classic.

In the manufacture of automobile bodies, a yield point in sheet steels is very undesirable because of surface markings that result when it is formed. After annealing, steel producers "temper roll" the steel, reducing its thickness a fraction of a percent. This removes the yield point effect. Plants that make automotive stampings want steel sheet that will not strain age before forming, so aluminum-killed steels are usually specified. However, steels that will strain age during paint baking cycles are desirable because this adds dent resistance to the already formed parts.

Many aluminum canoes are made from alloy 5052, which can be hardened only by deformation. However, one of the leading manufacturers of aluminum canoes uses aluminum alloy 6062, which is bought from the aluminum producer as solution-treated sheets. After the sheets are stamped to form canoe halves, they are precipitation hardened in a low-temperature oven.

Problems

1. For fcc crystals, the Schmid factors for tension parallel to [100] and [111] are reciprocals of the Taylor factors for these orientations. This is not true for any other orientation. Explain.

2. For a rod of an fcc metal with a fiber texture, what is the maximum amount of texture strengthening possible? (What is the largest possible ratio of yield strengths for textured material to that of randomly oriented material?) What is the greatest amount of texture softening possible?

3. The yield strength and tensile strength of copper increases with additions of zinc up to about 35% Zn. Further addition of zinc causes an abrupt drop in strength. Speculate as to the cause of this drop.

4. Using the data in Figure 11.7, determine the constants σ_o and K_Y in the Hall-Petch relation (equation 11.4).

5. Figure 11.18a shows how the yield strength of aluminum alloy 2014, which contains about 4.5% Cu, changes with aging after quenching from a solution treatment at 500°C. Figure 11.18b is the Al-rich end of the Al-Cu phase diagram.

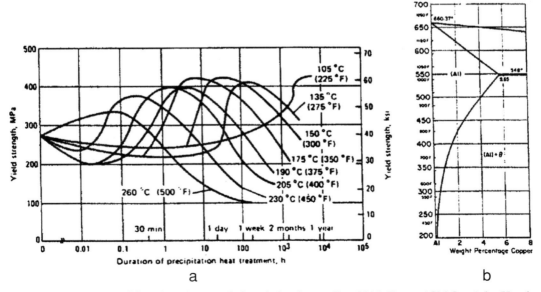

Figure 11.18. (a) Aging characteristics of aluminum alloy 2014. From *ASM Specialty Handbook: Aluminum and its Alloys*, ASM Inter. (1993). (b) The Al-Cu phase diagram. From *Metals Handbook*, vol. 8, 8th ed., ASM (1973).

A. Estimate the slope, $d\sigma_y/dc$, of a plot of yield strength, σ_y, versus atomic % copper, c, in solid solution. Assume that after aging for 1 week at 260°C, the copper has precipitated, and the dispersion strengthening is negligible. Also assume that the copper is in solid solution immediately after quenching.

B. Determine $\Delta\sigma_y$ attributable to precipitation, assuming complete precipitation for the curves reaching maximum yield strength (e.g., 1 day at 150°C). Note that $\Delta\sigma_y$ is the strengthening caused by the precipitates for the same matrix composition.

C. Estimate the distance between particles on the slip planes to achieve this strength increase, $\Delta\sigma_y$.

6. The maximum solubility of carbon in ferrite (bcc iron) is 0.022 wt % at 727°C.

A. Assuming that this carbon can be retained in solid solution during a quench to room temperature and then precipitated as Fe_3C, what volume fraction precipitate would be formed? Neglect any difference between the densities of Fe_3C and ferrite.

B. Estimate the increase of yield strength caused by the precipitation, assuming that the precipitate appears as circles of 50 nm measured under a microscope on a polished surface. Assume that $\Delta\sigma = 2.5\Delta\tau$.

C. Compare this with the loss of solid solution hardening that is indicated in Figure 11.2. For iron, $G = 208$ GPa, and the atomic diameter is 0.248 nm.

7. The yield strength of a low-carbon steel depends on its grain size. The following data have been reported:

Grain Size (mm)	Yield Strength (MPa)
0.025	205
0.15	125

Estimate the yield strength of the same steel treated to give a grain size of 0.0156 mm.

8. *Making, Shaping, and Treating of Steels* (U.S. Steel Co., Pittsburgh, 1971, p. 1127) provides a table of combinations of aging times and temperatures that give the same amount of strain aging in low-carbon steels.

Temperature (°C)	0	21	100	120	150
	1 yr	6 mo	4 hr	1 hr	10 min
	6 mo	3 mo	2 hr	30 min	5 min
	3 mo	6 wk	1 hr	15 min	2.5 min

A. Make a plot of time on a log scale (or $\ln t$) versus $1/T$ for equivalent aging.

B. Find the apparent activation energy for strain aging from the straight-line portion of the plot. (Note that the line does not go through the 0°C points.)

C. Can you offer any explanation for the deviation 0°C data from a straight line? (Think about how the data were probably determined.)

9. Consider a material strengthened by precipitation of fine particles. Suppose the shear yield strength of the particles is $G/10$ (theoretical), and the strength of matrix

is $G/500$. Also assume that the volume fraction of particles is 1 vol. % and that they have a diameter of 0.010 μm. Assume $b = 0.2$ nm.

    A. Calculate the applied shear stress, τ (relative to G), which is necessary to bow the dislocations between particles.

    B. Would dislocations bow between the particles or shear them?

10. The effect of grain size on flow stress decreases if the flow stress is taken at high strains, rather than low strains. The effect of grain size on precipitation-hardened metals is very low. Explain these two observations.

11. The following data have been obtained for the yield strength of nickel as a function of grain size:

Grain diameter (μm)	1.0	2.0	10.0	20.0	95.0	130.0
Yield strength (MPa)	250.0	185.0	86.0	95.0	33.0	25.0

A modified Hall-Petch expression of the form $YS = Kd^{1/a}$ has been suggested. Find the exponent, a, that best fits the data.

12 Discontinuous and Inhomogeneous Deformation

Stick-Slip Phenomena

Stick-slip phenomena involve intervals of motion separated by periods of rest. The model in Figure 12.1 illustrates the basic elements: a constantly moving driver (A), a spring (B), and a block (C) in frictional contact with a surface (D). If the static coefficient of friction between the block and the surface is greater than the coefficient of friction when the block is sliding, the motion of the block will be sporadic.

As the driver moves at a constant rate, the spring will elongate causing the force to increase. Once the force is great enough to overcome the static friction, the block will slide and will continue sliding until the spring shortens enough so that the force will drop below that needed to overcome sliding friction. At that point, the block will remain motionless until the force rises high again to overcome static friction. This is illustrated in Figure 12.2.

If plastic deformation behaves in this manner, the result is a serrated stress–strain curve. There are several possible causes of this phenomenon. One cause is dynamic strain aging, which results from the attraction of atoms in solid solution to dislocations. Because this attraction lowers the energy of the system, the solute atoms pin the dislocations. A larger force is required to break the dislocations free from the atoms pinning them than is required for them to continue moving. If the temperature is sufficiently high and the rate of deformation sufficiently slow, the solute atoms may diffuse to the new location of the dislocation and pin it again. This will result in a serrated stress–strain curve (Figure 12.3).

Another possible cause of serrated stress–strain curve is thermal instability. At very low temperatures, the heat capacity may be so low that the heat released by a small amount of plastic deformation will cause the flow stress to drop sufficiently below the applied stress. This will result in a burst of plastic deformation causing the stress to drop. The deforming region will then cool, and elastic reloading is necessary for the material to yield again.

Mechanical twinning is another cause of serrated stress–strain curves. The stress to initiate a twin is greater than that to make it grow because of the surface energy of the twin boundary. The ratio of surface area to twin volume decreases as the twin grows, so the stress drops. The growth of a twin is interrupted once it reaches a grain

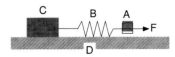

Figure 12.1. Model to explain a stick-slip phenomenon.

boundary or another twin. Then the specimen must elastically reload until the stress is sufficient to cause another twin to form.

Dynamic Strain Aging

Dynamic strain aging occurs as a metal is deformed. It is manifested by serrated stress–strain curves and stretcher strains. The stress–strain curves are characterized by sudden load drops that correspond to bursts of plastic deformation, followed by load drops and elastic reloading until another burst of deformation occurs.

Dynamic strain aging is the result of the attraction of solute atoms to dislocations. It is like static strain aging in that it takes a greater force to initiate plastic deformation by dislocations breaking free of solutes than to cause them to move once they are free. The difference is that dynamic strain aging occurs at a temperature and a strain rate that allows the solute atoms to pin the dislocations again while the material is deforming. The result is that the stress–strain curve consists of a series of load drops occurring during the deformation followed by elastic reloading until the stress is high enough to reinitiate deformation. Figure 12.4 shows the difference between static and dynamic strain aging in a low-carbon steel.

For low-carbon steels static strain from interstitial solutes (C and N), aging occurs in the temperature range of 100°C to 250°C, as shown in Figure 12.5. It may occur at lower temperatures if the amount of nitrogen in solution is high.

The temperature range for dynamics strain aging also depends on the strain rate as shown in Figure 12.6.

Substitutional solutes diffuse much less rapidly so they cause dynamic strain aging at a much higher temperature. Figure 12.7 shows the range of dynamic strain aging in iron-titanium alloys.

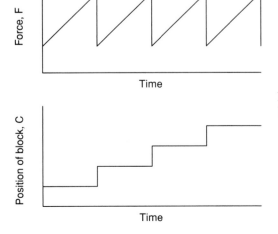

Figure 12.2. Changes of force (*top*) and position of block, C (*bottom*), with time.

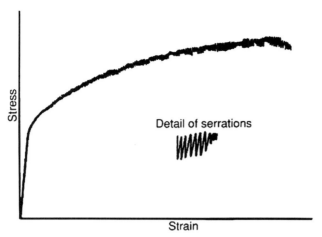

Figure 12.3. Stress–strain curve showing repeated load drops. From R. E. Reed-Hill, *Physical Metallurgy Principles*, Van Nostrand (1973).

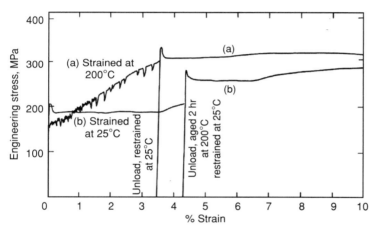

Figure 12.4. Dynamic and static stain aging of a 0.03% C rimmed steel. From A. S. Keh, *Materials Science Research*, v. 1 (1963).

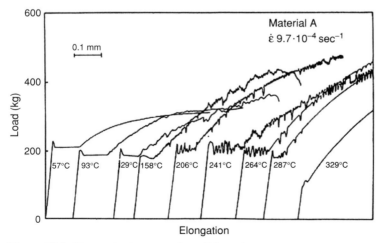

Figure 12.5. Stress–strain curves for mild steel over a temperature range. From Y. Bergstrom and W. Roberts, *Acta Met*, 19 (1971).

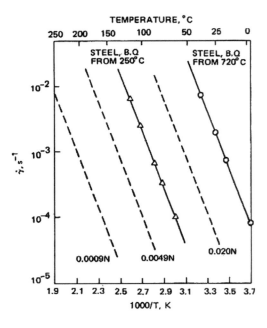

Figure 12.6. Temperature and strain rate dependence for the onset of dynamic strain aging in steels containing three different levels of nitrogen. From A. S. Keh, Y. Nakada, and W. C. Leslie, *Dislocation Dynamics*, McGraw-Hill Book Co. (1968).

Dynamic strain aging occurs in aluminum alloys containing magnesium in solid solution at room temperature. Figure 12.8 shows the stress–strain curve of aluminum alloy 5086.

The effects of temperature and strain rate are interrelated. Decreasing temperature has the same effect as increasing strain rate, as shown schematically in Figure 6.21. This effect occurs even in temperature regimes where the rate sensitivity is negative. In the temperature range where $d\sigma/dT$ is positive, σ is lower for a high strain than for a lower strain rate so the strain rate sensitivity is negative.

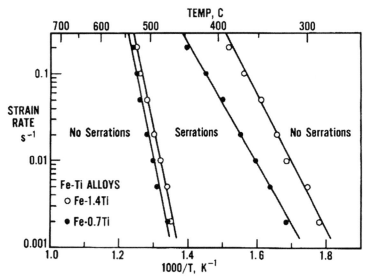

Figure 12.7. Temperature and strain rate dependence of dynamic strain aging in iron-titanium alloys. From W. C. Leslie, L. J. Cuddy, and R. J. Sober, *Proc. 3rd Intern. Conf. Strength of Metals and Alloys*, Institute of Metals, London (1973).

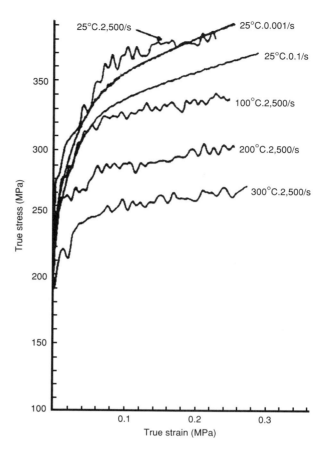

Figure 12.8. Stress–strain curves of aluminum alloy 5086 at different temperatures and strain rates. From M. Wagenhofer et al., *Scripta Mat.*, v. 41 (1999).

A positive $d\sigma/dT$ is associated with a negative strain rate sensitivity, $(d\sigma/d\dot{\varepsilon} < 0)$. In these materials, the stress necessary to deform a material at a fast strain rate is lower than the stress to deform it at a slow strain rate. Such materials undergo discontinuous yielding. If one tries to impose a slow strain rate on a specimen, the deformation will occur in such a way that one region deforms rapidly, whereas the rest of the material does not deform at all. Because the deforming region work hardens, the deforming region moves through the material as a Lüder's band. The stress–strain curves are serrated. Figure 12.9 shows how the yield strength of commercially pure titanium alloy varies with temperature.

The phenomenon is similar to the yield point effect in steels. The negative strain rate sensitivity can be ascribed to the pinning of dislocations by solute atoms. Once a region starts to deform and the dislocations are free of solute atoms, they can move at a lower stress than required to initiate the deformation. The deformation spreads to adjacent regions, forming a band, which traverses the entire specimen. While the band is propagating, the solutes can diffuse to the new dislocations, pinning them. An increased stress is necessary to reinitiate plastic deformation. This effect was first observed by Portevin and LeChatelier* in aluminum and is known as the *Portevin-LeChatier effect*. It is also called *dynamic strain aging*.

---

* A. Portevin and F. LeChatelier, *Acad. Sci. Compt. Rend*, v. 176 (1923).

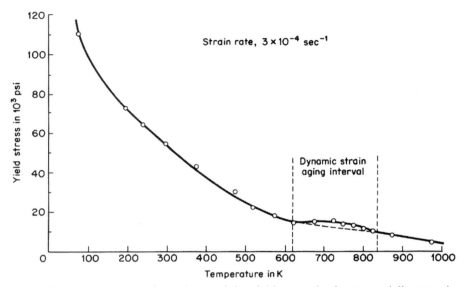

Figure 12.9. Temperature dependence of the yield strength of commercially pure titanium. Note that the range of dynamic strain aging corresponds to the range of a positive dependence of yield strength on temperature. From R. E. Reed-Hill, *Physical Metallurgy Principles*, 2nd ed., Van Nostrand (1983).

Figure 12.10 shows the surface of an aluminum alloy containing magnesium. Dynamic strain aging that causes this type of surface can be prevented with special tempers or by forming at temperatures above 150°C.

Dynamic strain aging has also been observed in copper, nickel zirconium, magnesium, and alloys. Table 12.1 shows the temperature range for dynamic strain aging in several metals.

The interrelation between strain rate and temperature dependence was treated in Chapter 6. For most metals dynamic strain aging and corresponding minimum strain rate, sensitivity in substitutional solutions occurs at about $0.3\ T_m$. In interstitial solutions, these occur at about $0.2\ T_m$.

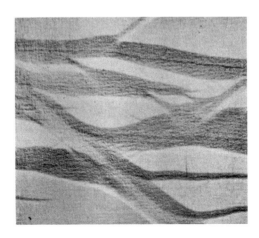

Figure 12.10. Stretcher strain on an aluminum alloy containing magnesium. From *Properties and Physical Metallurgy*, ASM (1984). Courtesy of ALCOA.

Table 12.1. *Temperature range for dynamic strain aging in several metals*

Metal		Temperature range for $d\sigma/dT > 0$
Ta	($\varepsilon = 0.01$, $\dot{\varepsilon} = 0.001/s$)	440°C–600°C
V	($\varepsilon = 0.01$, $\dot{\varepsilon} = 0.001/s$)	500°C–760°C
DH 36 steel	($\varepsilon = 0.01$, $\dot{\varepsilon} = 0.001/s$)	360°C–620°C
Ti	($\dot{\varepsilon} = 0.0003/s$)	620°C–720°C
Al 2024		−200°C–20°C

EXAMPLE PROBLEM 12.1: Derive an expression for how the temperature dependence of dynamic strain aging depends on strain rate, and apply that expression to the data in Figure 12.6.

Solution: In principle, one should be able to calculate how temperature range depends on the strain rate, from equation (6.17). Solving for Q,

$$Q/RT = -\ln(\dot{\varepsilon}) + \ln A \qquad (12.1)$$

Differentiating $d(1/T) - d\ln\dot{\varepsilon}/(Q/R)$, so

$$d(1/T)d\ln\dot{\varepsilon} = -1(Q/R) \qquad (12.2)$$

For diffusion of nitrogen in ferrite, $Q = 76.4$ kJ/mole so $d(1/T)/d\ln\dot{\varepsilon} = -1/[(8.314)(84,400)] = 1.6 \times 10^{-6}$ (1/K). Comparing this for the start of serrations with Figure 13.6, $d(1/T)/d\ln\dot{\varepsilon} = 0.7 \times 10^{-6}$ (1/K), and for the end of serrations, $d(1/T)/d\ln\dot{\varepsilon} = 1.1 \times 10^{-6}$ (1/K).

Other Causes of Serrated Stress–Strain Curves

Serrated stress–strain curves may also arise from adiabatic heating. At very low temperatures (e.g., 4.2 K), the specific heat of metals is very low so that the heat released by plastic deformation causes an appreciable temperature rise. This may be sufficient to cause a large enough decrease of flow stress to cause an avalanche of deformation. This phenomenon has caused serrated stress–strain curves of aluminum.[*]

Mechanical twinning is another cause of serrated stress–strain curves (see Chapter 10). The stress to initiate a twin is greater than that to make it grow because of the surface energy of the twin boundary. The ratio of surface area to twin volume decreases as the twin grows so the stress drops. The growth of a twin is interrupted once it reaches a grain boundary or another twin. Then the specimen must elastically reload until the stress is sufficient to cause another twin to form.

Strain Localization

Strain may localize into deformation bands or shear bands. When an aluminum single crystal is deformed in tension parallel to <110>, four slip systems are equally favored. Different regions of the crystal will deform by slip on different slip systems,

[*] Z. S. Basinski, *Proc. Roy. Soc. (London)*, A240 (1957).

Figure 12.11. Shear bands formed during cold rolling of magnesium sheet. Photomicrograph by S. L. Couling, *Metals Handbook*, v. 9, 9th ed. (1985).

with the result that the lattice in different regions will rotate in different directions, forming deformation bands. In grains of a polycrystal, different regions may deform on different slip systems to accommodate deformation in neighboring grains.

Shear bands often traverse many grains, particularly during plane–strain deformation. Figure 12.11 shows shear bands in cold-rolled magnesium.

REFERENCE

ASM, *Inhomogeneity of Plastic Deformation* (1972).

Notes

In 1909, Le Chatelier* reported serrated stress–strain curves in mild steel from 80°C to 250°C. Le Catelier was born in 1850 on the French-Italian border. His father was French and his mother Italian. He taught chemistry at the Collége de France. He is most known for his work on chemical equilibria. His principal can be summarized as "Any change in concentration, temperature or pressure on a system in equilibrium will shift the system in a way that minimizes that change."

In 1949, McReynolds[†] studied the stress–strain curves of 2S aluminum (modern aluminum alloy 2002) at room temperature and found that plastic deformation occurs as waves traveling along a specimen. He found that this phenomenon did not occur in pure aluminum.

Problems

1. Measurements of negative strain rate sensitivites underestimate how negative the rate sensitivity is. Explain why.

2. If strain aging occurs in aluminum alloys containing magnesium deformed at a strain rate of 100/s below 150°C, predict the temperature below which dynamic strain aging will occur during deformation at a strain rate of 1/s. The diffusivity of magnesium in aluminum is given by $1.2 \times 10^{-4}\exp[-131,000/(RT)]$ m^2/s.

* A. Le Chatelier, *Rev. de Met.*, v. 6 (1909).
[†] A. W. McReynolds, *Trans AIME*, v. 185 (1949).

3. For the stick-slip model in Figure 12.2, predict the frequency of load-drops if the weight of block A is 10 N; the sticking and sliding coefficients of friction are 0.20 and 0.10, respectively; the spring constant is 20 N/m; and the speed of travel of C is 10 cm/s.

4. From the slopes in Figure 12.7 of the temperature and strain rate dependence, find the activation energies for the onset of dynamic strain aging in steel containing 1.4% titanium.

13 Ductility and Fracture

Introduction

Throughout history, there has been a never-ending effort to develop materials with higher yield strengths. However, a higher yield strengths is generally accompanied by a lower ductility and a lower toughness. *Toughness* is the energy absorbed in fracturing. A high-strength material has low toughness because it can be subjected to higher stresses. The stress necessary to cause fracture may be reached before there has been much plastic deformation to absorb energy. Ductility and toughness are lowered by factors that inhibit plastic flow. As schematically indicated in Figure 13.1, these factors include decreased temperatures, increased strain rates, and the presence of notches. Developments that increase yield strength usually result in lower toughness.

In many ways, the fracture behavior of steel is like that of taffy candy. It is difficult to break a warm bar of taffy candy to share with a friend. Even children know that warm taffy tends to bend rather than break. However, there are three ways to promote its fracture. A knife may be used to notch the candy bar, producing a stress concentration. The candy may be refrigerated to raise its resistance to deformation. Finally, rapping it against a hard surface raises the loading rate, thus increasing the likelihood of fracture. Notches, low temperatures, and high rates of loading also embrittle steel.

There are two important reasons for engineers to be interested in ductility and fracture. The first is that a reasonable amount of ductility is required to form metals into useful parts by forging, rolling, extrusion, or other plastic working processes. The second is that a certain degree of toughness is required to prevent failure in service. Some plastic deformation is necessary to absorb energy.

This chapter treats the mechanisms and general observations of failure under a single application of load at low and moderate temperatures. Chapter 14 is a quantitative treatment of fracture mechanics. Chapter 16 covers failure under creep conditions at high temperatures, and fatigue failure under cyclic loading is covered in Chapter 17.

Fractures can be classified several ways. A fracture is described as *ductile* or *brittle*, depending on the amount of deformation that precedes it. Failures may also be described as *intergranular* or *transgranular*, depending on the fracture path. The

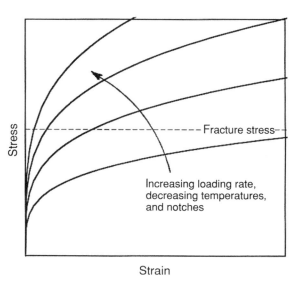

Figure 13.1. Lowered temperatures, increased loading rates, and the presence of notches all reduce ductility. These three factors raise the stress level required for plastic flow so the stress required for fracture is reached at lower strains.

terms *cleavage, shear,* and *void coalescence* are used to identify failure mechanisms. These descriptions are not mutually exclusive. A brittle fracture may be intergranular, or it may occur by cleavage.

The *ductility* of a material describes the amount of deformation that precedes fracture. Ductility may be expressed as percent elongation or as the percent reduction of area in a tension test. Failures in tension tests may be classified several ways (Figure 13.2). At one extreme, a material may fail by necking down to a vanishing cross-section. At the other extreme, fracture may occur on a surface that is more or less normal to the maximum tensile stress, with little or no deformation. Failures may also occur by shear.

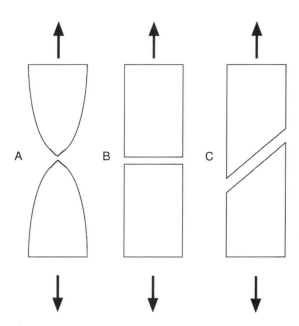

Figure 13.2. Several failure modes. (A) Rupture by necking down to a zero cross-section. (B) Fracture on a surface that is normal to the tensile axis. (C) Shear fracture.

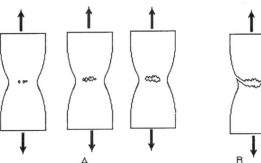

Figure 13.3. Development of a cup and cone fracture. (A) Internal porosity growing and linking up. (B) Formation of a shear lip.

Ductile Fracture

Tension test: Failure in a tensile test of a ductile material occurs well after the maximum load is reached and a neck has formed. In this case, fracture usually starts by nucleation of voids in the center of the neck where the hydrostatic tension is the greatest. As deformation continues, these internal voids grow and eventually link up by necking of the ligaments between them (Figures 13.3 to 13.5). Such a fracture starts in the center of the bar where the hydrostatic tension is greatest. With continued elongation, this internal fracture grows outward until the outer rim can no longer support the load and the edges fail by sudden shear. The final shear failure at the outside also occurs by void formation and growth (Figures 13.6 and 13.7). This overall failure is often called a *cup and cone* fracture. If the entire shear lip is on the same broken piece, it forms a cup. The other piece is the cone (Figure 13.8). More often, however, part of the shear lip is on one half of the specimen and part on the other half.

In ductile fractures, voids form at inclusions because either the inclusion–matrix interface or the inclusion itself is weak. Figure 13.9 shows the fracture surface formed by coalescence of voids. The inclusions can be seen in some of the voids. Ductility is strongly dependent on the inclusion content of the material. With increasing numbers of inclusions, the distance between the voids decreases so it is easier for them to link together and lower the ductility. Figure 13.10 shows the

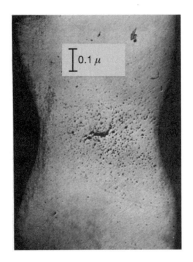

Figure 13.4. Section through a necked tensile specimen of copper, showing an internal crack formed by linking voids. From K. E. Puttick, *Phil Mag.*, v. 4 (1959).

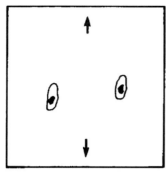 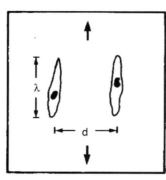 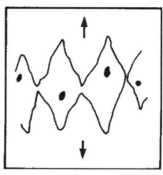

Figure 13.5. Schematic drawing showing the formation and growth of voids during tension, and their linking up by necking of the ligaments between them. From W. F. Hosford and R. M. Caddell, *Metal Forming: Mechanics and Metallurgy*, 3rd ed., Cambridge (2007).

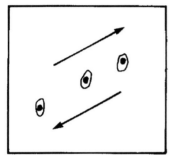 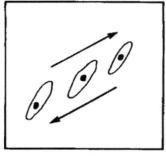 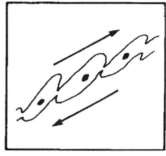

Figure 13.6. Schematic drawing illustrating the formation of and growth of voids during shear, and their linking up by necking of the ligaments between them. From W. F. Hosford and R. M. Caddell, *Ibid.*

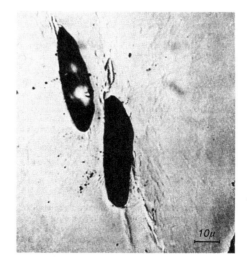

Figure 13.7. Large voids in a localized shear band in OFHC copper. From H. C. Rogers, *Trans. TMS-AIME*, v. 218 (1960).

Figure 13.8. A typical cup and cone fracture in a tension test of a ductile manganese bronze. From A. Guy, *Elements of Physical Metallurgy*, Addison-Wesley (1959).

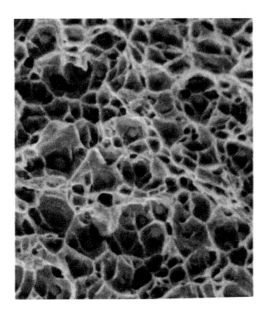

Figure 13.9. Dimpled ductile fracture surface in steel. Note the inclusions associated with about one-half the dimples. The rest of the inclusions are on the mating surface. Courtesy of J. W. Jones. From W. F. Hosford and R. M. Caddell, *Ibid*.

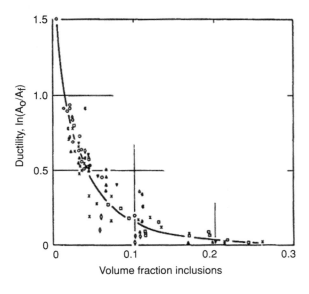

Figure 13.10. Effect of second-phase particles on the tensile ductility of copper. Data include alumina, silica, molybdenum, chromium, iron, and iron-molybdenum inclusions as well as holes. From B. I. Edelson and W. M. Baldwin, *Trans. Q. ASM.*, v. 55 (1962).

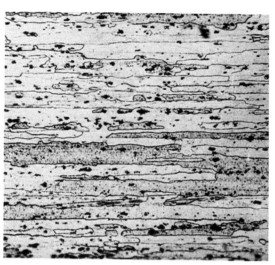

Figure 13.11. Directional features in the micro-structure of 2024-T6 aluminum sheet include grain boundaries, aligned inclusions, and elongated inclusions. From *Metals Handbook*, v. 7, 8th ed., ASM (1972).

decrease of ductility of copper with volume fraction inclusions. Ductile fracture by void coalescence can occur in shear as well as in tension testing.

Mechanical working tends to produce directional microstructures, as shown in Figures 13.11 to 13.13. Grain boundaries and weak interfaces are aligned by the working. Inclusions are elongated and sometimes broken up into strings of smaller inclusions. Often loading in service is parallel to the direction along which the interfaces and inclusions are aligned, so the alignment has little effect on the ductility. For example, wires and rods are normally stressed parallel to their axes, and the stresses in rolled plates are normally in the plane of the plate. In these cases, weak interfaces parallel to wire or rod axis and inclusions parallel to the rolling plane are not very important. For this reason, the anisotropy of fracture properties caused by mechanical fibering is often ignored. Sometimes, however, the largest stresses are normal to the aligned fibers, and in these cases, failures may occur by delamination parallel to the fiber axis. Welded T-joints of plates may fail this way by delamination. With severe bending, rods may splinter parallel to directions of prior working. Forged parts may fail along *flow lines* formed by the alignment of inclusions and

Figure 13.12. Microstructure of a steel plate consisting of bands of pearlite (dark regions) and ferrite (light regions). From *Metals Handbook*, v. 7, 8th ed., ASM (1972).

2% Nital   100×

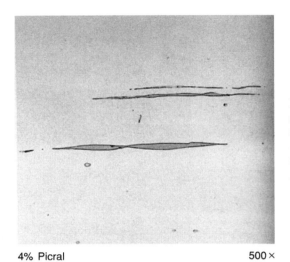

Figure 13.13. Higher magnification of the micro-structure of the steel plate in Figure 13.12 show-ing elongated sulfide inclusions. These inclusions are the major cause of directionality of fracture in steel. From *Metals Handbook*, v. 7, 8th ed., ASM (1972).

4% Picral   500×

weak interfaces. Annealing of worked parts does not remove such directionality. Even if recrystallization produces equiaxed grains, inclusion alignment is usually unaffected.

Figure 14.13 Microstructure of the steel plate in Figure 13.12 at a higher mag-nification showing elongated manganese sulfide inclusions. These inclusions are the major cause of the directionality of ductility in steel. From *Melals Handbook*, v. 7, 8th ed., ASM, 1972.

In steels, elongated manganese sulfide inclusions are a major cause of direction-ality. Reduced ductility perpendicular to the rolling direction can cause problems in sheet forming. This effect can be eliminated by reducing the sulfur content and adding small amounts of Ca, Ce, or Ti. These elements are stronger sulfide form-ers than Mn. At hot working temperatures, their sulfides are much harder than the steel and therefore do not elongate as the steel is rolled. Figure 13.14 shows the effect calcium on the through-thickness ductility. For applications requiring very

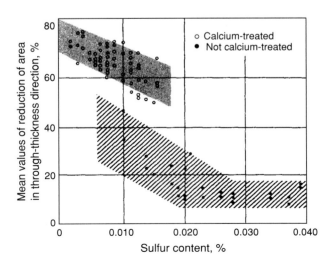

Figure 13.14. Effect of inclusion shape control on the through-thickness ductil-ity of an HSLA steel. From H. Pircher and W. Klapper, *Micro Alloying 75*, Union Carbide, New York (1977).

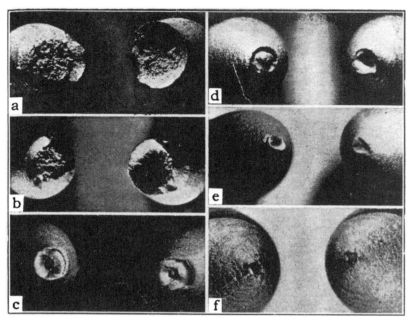

Figure 13.15. Effect of pressure on the area reduction in tension tests of steel. (a) atmospheric pressure, (b) 234 kPa, (c) 1 MPa, (d) 1.3 MPa, (e) 185 MPa, and (f) 267 MPa. From P. W. Bridgman in *Fracture of Metals,* ASTM (1947).

high ductility, a very low inclusion content can be achieved by vacuum or electroslag melting.

The level of hydrostatic stress plays a dominant role in determining the fracture strains. Hydrostatic tension promotes the formation and growth of voids, whereas hydrostatic compression tends to suppress void formation and growth. Photographs of specimens tested in tension under hydrostatic pressure (Figure 13.15), show that the reduction of area increases with pressure. Figure 13.16 shows how the level of hydrostatic stress affects ductility.

Figure 13.16. Effective fracture strain, $\bar{\varepsilon}_f$, increases as the ratio of the mean stress, σ_m, to the effective stress, $\bar{\sigma}$, becomes more negative (compressive) for two of the steels tested by Bridgman. Data from A. L. Hoffmanner, Interim Report, Air Force Contract 33615-67-C (1967).

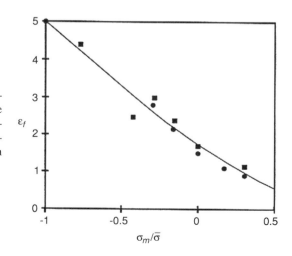

Table 13.1. *Cleavage planes of several crystal types*

Structure	Examples	Cleavage planes
bcc metals	Fe, W, Mo	$\{001\}$
hcp metals	Mg,	(0001), $\{10\bar{1}1\}$, $\{10\bar{1}0\}$
	Zn, Cd	(0001)
Rhombohedral	Bi	$\{111\}$, $\{110\}$
Rock salt	NaCl, MgO, . . .	$\{001\}$
Zinc blende	ZnS, CuCl, CuI	$\{110\}$
Cesium chloride	NH_4Cl, NH_4Br	$\{001\}$

Brittle Fracture

Cleavage: In some materials, fracture may occur by *cleavage*. Cleavage fractures occur on certain crystallographic planes (*cleavage planes*) that are characteristic of the crystal structure. Table 13.1 lists the cleavage planes for different crystal structures. In most cases, these are the most widely spaced planes. It is significant that fcc metals do not undergo cleavage. It is believed that cleavage occurs when the normal stress, σ_n, across the cleavage plane reaches a critical value, σ_c, as illustrated in Figure 13.17. The normal stress across a plane is $\sigma_n = \sigma_a / \cos^2 \phi$, where σ_a is the applied tensile stress, and ϕ is the angle between the tensile axis and the normal to the plane. Cleavage will occur when $\sigma_n = \sigma_c$, or

$$\sigma_a = \sigma_c / \cos^2 \phi. \tag{13.1}$$

In three dimensions, the cleavage planes in one grain of a polycrystal will not link up with cleavage planes in a neighboring grain, as indicated in Figure 13.18. Therefore, fracture cannot occur totally by cleavage. Some other mechanism must link up the cleavage fractures in different grains. Figure 13.19 shows a fracture surface in which there is cleavage of many grains.

Grain boundary fracture: Some polycrystals have brittle grain boundaries, which form easy fracture paths. Figure 13.20 shows such an *intergranular fracture* surface. The brittleness of grain boundaries may be inherent to the material or may be caused by segregation of impurities to the grain boundary or even by a film of a brittle second phase. Commercially pure tungsten and molybdenum fail by grain

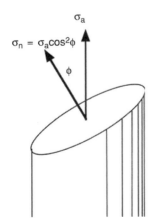

Figure 13.17. Cleavage plane and an applied stress. Cleavage occurs when the normal stress across the cleavage plane, $\sigma_n = \sigma_a \cos^2 \phi$, reaches a critical value, σ_c.

Figure 13.18. In polycrystalline material, cleavage planes in neighboring grains are tilted by different amounts relative to the plane of the paper. Therefore, they cannot be perfectly aligned with each other. Another mechanism is necessary to link up the cleavage fractures in neighboring grains.

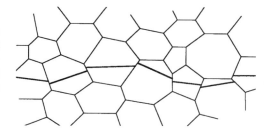

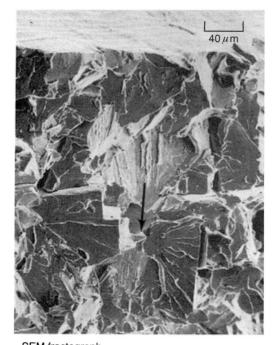

Figure 13.19. Cleavage fracture in a Fe-3.9 %Ni alloy. The arrow indicates the direction of crack propagation. From *Metals Handbook*, 8th ed., v. 9, ASM (1974).

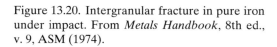

SEM fractograph

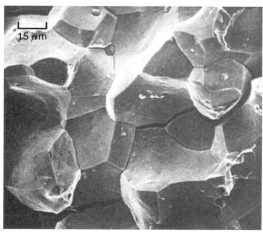

Figure 13.20. Intergranular fracture in pure iron under impact. From *Metals Handbook*, 8th ed., v. 9, ASM (1974).

boundary fracture. These metals are ductile only when the grain boundaries are aligned with the direction of elongation, as in tension testing of cold-drawn wire. Copper and copper alloys are severely embrittled by a very small amount of bismuth, which segregates to and wets the grain boundaries. Molten FeS in the grain boundaries of steels at hot working temperatures would cause failure along grain boundaries. Such loss of ductility at high temperatures is called *hot shortness*. Hot shortness is prevented in steels by adding Mn, which reacts with the sulfur to form MnS. Manganese sulfide is not molten at hot-working temperatures and does not wet the grain boundaries. Stress corrosion is responsible for some grain boundary fractures.

Role of grain size: With brittle fracture, toughness depends on grain size. Decreasing the grain size increases the toughness and ductility. Perhaps this is because cleavage fractures must reinitiate at each grain boundary, and with smaller grain sizes, there are more grain boundaries. Decreasing grain size, unlike most material changes, increases both yield strength and toughness.

Impact Energy

A material is regarded as being *tough* if it absorbs a large amount of energy in breaking. In a tension test, the energy per volume to cause failure is the area under the stress–strain curve and is the toughness in a tension test. However, the toughness under other forms of loading may be very different because toughness also depends on the degree to which deformation localizes. The total energy to cause failure depends on the deforming volume, as well as on energy per volume.

Charpy test: Impact tests are often used to assess the toughness of materials. The most common of these is the *Charpy test*. A notched bar is broken by a swinging pendulum. The energy absorbed in the fracture is measured by recording by how high the pendulum swings after the bar breaks. Figure 13.21 gives the details of the

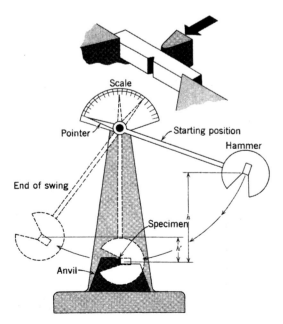

Figure 13.21. Charpy testing machine and test bar. A hammer on the pendulum breaks the bar. The height the pendulum swings after breaking the bar indicates the energy absorbed. From H. W. Hayden, W. G. Moffatt, and J. Wulff, *Structure and Properties of Materials, Vol. III, Mechanical Behavior*, Wiley (1965).

test geometry. The standard specimen has a cross-section of 10 mm by 10 mm. There is a 2-mm-deep V-notch with a radius of 0.25 mm. The pendulum's mass and height are standardized. Sometimes bars with U- or keyhole notches are employed instead. Occasionally, subsized bars are tested.

One of the principal advantages of the Charpy test is that the toughness can easily be measured over a range of temperatures. A specimen can be heated or cooled to the specified temperature and then transferred to the Charpy machine and broken quickly enough so that its temperature change is negligible. For many materials, there is a narrow temperature range, over which there is a large change of energy absorption and fracture appearance. It is common to define a *transition temperature* in this range. At temperatures below the transition temperature, the fracture is brittle and absorbs little energy in a Charpy test. Above the transition temperature, the fracture is ductile and absorbs a large amount of energy. Figure 13.22 shows typical results for steel.

The transition temperature *does not* indicate a structural change of the material. The ductile-brittle transition temperature depends greatly on the type of test being made. With less severe notches, as in keyhole or U-notched Charpy test bars, lower transition temperatures are measured (Figure 13.23). With decreased specimen width, there is less triaxiality, so the transition temperatures are lowered. With slow bend tests and unnotched tensile tests, even lower ductile-brittle transitions are observed. In discussing the ductile-brittle transition temperature of a material, one should specify not only the type of test, but also the criterion used. Specifications for ship steels area often require a certain *Charpy V-notch 15 ft-lb transition temperature* (the temperature at which the energy absorption in a V-notch Charpy test is 15 ft-lb). Some investigators define the transition temperature in a Charpy test as the temperature at which the energy is half of the plateau energy or in terms of fracture appearance. Occasionally, the amount of deformation beneath the notch is used as a criterion.

There are two principal uses for Charpy data. One is in the development of tougher materials. For example, it has been learned through Charpy testing, that the transition temperature of hot-rolled carbon steels can be lowered by decreasing the

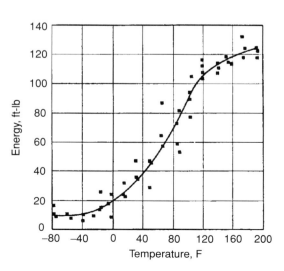

Figure 13.22. Ductile-brittle transition in a Charpy V-notch specimen of a low carbon, low alloy, hot-rolled steel. From R. W. Vanderbeck and M. Gensamer, *Welding J. Res. Suppl.* (Jan. 1950).

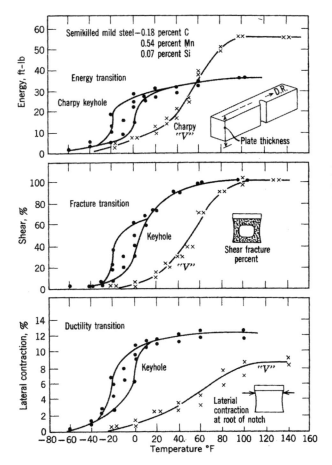

Figure 13.23. Charpy test results for one steel with two different notch geometries (standard V-notch and keyhole notch.) Three different criteria (energy absorption, fracture appearance, and deformation below the notch) are shown. Note that the "transition temperature" depends on the mode of testing and the criterion used. From Pellini, *Spec. Tech. Publ.*, 158 (1954).

carbon content (Figure 13.24). Increasing the manganese content and decreasing the ferrite grain size also lower transition temperatures. The beneficial and detrimental effects of other elements have also been documented. Alloying with nickel greatly lowers the transition temperature of steels, but nickel is too expensive for many

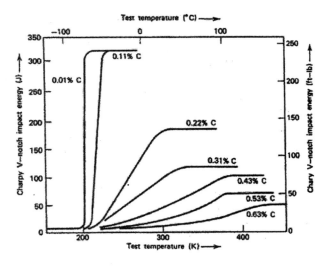

Figure 13.24. Effect of carbon content on Charpy V-notch impact energy. Decreasing carbon content lowers the ductile-brittle transition temperature and raises the shelf energy. From J. A. Rinebilt and W. J. Harris, *Trans. ASM*, v. 43 (1951).

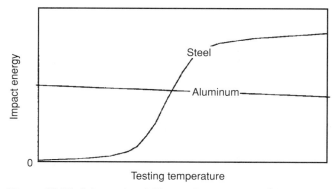

Figure 13.25. Schematic of Charpy impact energy for a steel alloy and an aluminum alloy. Note that like all fcc metals, the aluminum alloy has no ductile-brittle transition.

uses. For steels, decreasing the hot-rolling temperatures results in finer grain size and, therefore, a lower transition temperatures.

The other main use of Charpy data is for documenting and correlating service behavior. This information can then be used in the design and specifications for new structures. For example, a large number of ships failed by brittle fracture during World War II. Tests were conducted on steel from plates in which cracks initiated, plates through which cracks propagated, and a few plates in which cracks stopped. The results were used to establish specifications for the 15 ft-lb Charpy V-notch transition temperature for plates used in various parts of a ship.

In general, bcc metals and many hcp metals exhibit a ductile-brittle transition, but it is significant that fcc metals do not. For fcc metals, changes of impact energy with temperature are small, as shown for aluminum in Figure 13.25. Because of this, austenitic stainless steels or copper is frequently used in equipment for cryogenic applications.

There are ductile-brittle transitions in thermoplastics associated with the glass transition. See Chapter 20.

REFERENCES

ASM, *Ductility*, Metals Park (1967).
ASM, *Fracture of Engineering Materials*, Metals Park (1964).
E. Parker, *Brittle Fracture of Engineering Structures*, Wiley (1957).
F. McClintock and A. Argon, *Mechanical Behavior of Materials*, Addison-Wesley (1966).
W. Hosford and R. Caddell, *Metal Forming: Mechanics and Metallurgy*, 3rd ed. Cambridge (2007).

Notes

A catastrophic failure of a molasses tank in Boston on January 15, 1919 was an early brittle failure that received much attention. A full molasses tank on a hill in Boston suddenly burst. As a result of the flood of molasses, 12 people and several horses were drowned or otherwise killed. Part of the Boston elevated transit system was demolished, and houses were damaged. Engineers at the time did not understand how this failure could have occurred because the stresses, even after accounting for concentrations around rivet holes, were below the measured tensile strength of the steel.

From 1948 to 1951, there were many fractures of natural gas pipelines. Most occurred during testing and most started at welding defects but propagated through sound metal. One of the longest cracks was 3,200 feet long. Once started, cracks run at speeds greater than the velocity of sound in the pressurized gas. Therefore, there is no release of the gas pressure to reduce the stress at the tip of the crack. Figure 13.26 is a photograph of one of the cracked lines.

During World War II, there was a rapid increase of shipbuilding. Production of ships by welding of steel plates together (in contrast to the earlier procedure of joining them by riveting) became common. As a result, a large number of ships, particularly Liberty Ships and T-2 Tankers, failed at sea. More ships sunk as the result brittle fractures than by German U-boat activity. Recovery of some ships and half-ships allowed the cause of the failures to be investigated. There were three main factors: poor welds, ship design (cracks often started at sharp cornered hatchways that created stress concentrations), and high transition temperatures of the steels. Figure 13.27 is a photograph of a ship that failed in harbor.

Mica is one common material in which cleavage occurs readily. Pairs of covalently bonded planes of $(AlSi_3O_{10})^{-3}$ are ionically bonded by K^+, Al^{+3}, and OH^- ions to form electrically neutral plates. These are in turn bonded to each other by weak van der Waals bonds, so mica cleaves easily on these planes. Higher stresses are needed to cleave ionic bonds in alkali halide crystals. Cleavage in metals is still harder due to the nondirectional nature of the bonds.

Figure 13.26. Natural gas pipe line that failed during field testing. From E. Parker, *Brittle Fracture of Engineering Structures*, Wiley (1957).

Figure 13.27. Ship that fractured while in port. From C. F. Tipper, *The Brittle Fracture Story*, Cambridge University Press (1963).

Problems

1. Derive the relation between % El and % RA for a material that fractures before it necks. (Assume constant volume and uniform deformation.)

2. Consider a very ductile material that begins to neck in tension at a true strain of 0.20. Necking causes an additional elongation approximately equal to the bar diameter. Calculate the % elongation of this material if the ratio of the gauge length to bar diameter is 2, 4, 10, and 100. Plot % elongation versus L_o/D_o.

3. For a material with a tensile yield strength, Y, determine the ratio of the mean stress, $\sigma_m = $ to Y, at yielding in a

    A. Tension test,
    B. Torsion test,
    C. Compression test.

4. The cleavage planes in sodium chloride are the {100} planes. Assuming there is a critical normal stress, σ_c, for cleavage, what are the highest and lowest ratios of applied tensile stress, σ_a, to σ_c for cleavage? (In looking for the maximum, realize that if the angle between the tensile axis and the [100] direction gets too high, the angles between the tensile axis and the [010] or [001] directions will be smaller. You must decide at what of orientation the tensile axis, is the angle to the nearest <100> the greatest. It may help to refer to Figure AII.3.

5. Cleavage in bcc metals occurs more frequently as the temperature is lowered and as the strain rate is increased. Explain this observation.

6. It has been argued that the growth of internal voids in a material while it is being deformed is given by $dr = f(\sigma_H)d\bar\varepsilon$, where r is the radius of the void, and $f(\sigma_H)$ is a function of the hydrostatic stress. Explain, using this hypothesis, why ductile fracture occurs at higher effective strains in torsion than in tension.

7. Explain why voids often form at or near hard inclusions, both in tension and in compression.

8. Is it safe to say that brittle fracture can be avoided in steel structures if the steel is chosen so that its Charpy V-notch transition temperature is below the service temperature? If not, what is the value of specifying Charpy V-notch test data in engineering design?

9. Hold the upper left corner of a piece of newspaper with one hand and the upper right corner with the other, and tear it. Take another piece of newspaper, rotate it 90 degrees, and repeat. One of the tears will be much straighter than the other. Why?

14 Fracture Mechanics

Introduction

The treatment of fracture in Chapter 13 was descriptive and qualitative. In contrast, *fracture mechanics* provides a quantitative treatment of fracture. It allows measurements of the toughness of materials and provides a basis for predicting the loads that structures can withstand without failure. Fracture mechanics is useful in evaluating materials, in the design of structures, and in failure analysis.

Early calculations of strength for crystals predicted strengths far in excess of those measured experimentally. The development of modern fracture mechanics started when it was realized that strength calculations based on assuming perfect crystals were far too high because they ignored preexisting flaws. Griffith[*] reasoned that a preexisting crack could propagate under stress only if the release of elastic energy exceeded the work required to form the new fracture surfaces. However, his theory based on energy release predicted fracture strengths that were much lower than those measured experimentally. Orowan[**] realized that plastic work should be included in the term for the energy required to form a new fracture surface. With this correction, experiment and theory were finally brought into agreement. Irwin[†] offered a new and entirely equivalent approach by concentrating on the stress states around the tip of a crack.

Theoretical Fracture Strength

Early estimates of the theoretical fracture strength of a crystal were be made by considering the stress required to separate two planes of atoms. Figure 14.1 shows schematically how the stress might vary with separation. The attractive stress between two planes increases as they are separated, reaching a maximum that is the theoretical strength, σ_t, and then decaying to zero. The first part of the curve can be approximated by a sine wave of half wavelength, x^*,

$$\sigma = \sigma_t \sin(\pi x/x^*), \tag{14.1}$$

---

[*] A. A. Griffith, *Phil. Trans. Roy. Soc. London, Ser. A*, v. 221 (1920).
[**] E. Orowan, Fracture and Strength of Crystals, *Rep. Prog. Phys.*, v. 12 (1949).
[†] G. R. Irwin, *Fracturing of Metals*, ASM (1949).

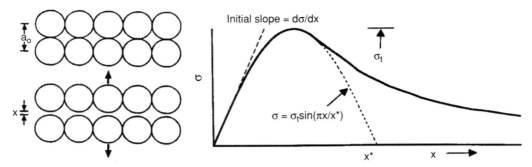

Figure 14.1. Schematic of the variation of normal stress with atomic separation and a sine wave approximation to the first part of the curve.

where x is the separation of the planes. Differentiating, $d\sigma/dx = \sigma_t(\pi/x^*)\cos(\pi x/x^*)$. At low values of x, $\cos(\pi x/x^*) \to 1$ so $d\sigma/dx \to \sigma_t(\pi/x^*)$. The engineering strain, $e = x/a_o$. Young's modulus, E, is the slope of the stress–strain curve, $d\sigma/d\varepsilon$, as $\varepsilon \to 0$

$$E = d\sigma/d\varepsilon = (a_o/x^*)\pi\sigma_t. \tag{14.2}$$

Solving for the theoretical strength,

$$\sigma_t = Ex^*/(\pi a_o). \tag{14.3}$$

If it is assumed that $x^* = a_o$,

$$\sigma_t \approx E/\pi. \tag{14.4}$$

Equation (14.4) predicts theoretical strengths that are much higher than those that are observed (65 GPa vs. about 3 GPa for steel and 20 GPa vs. about 0.7 GPa for aluminum).

The assumption that $x^* = a_o$ can be avoided by equating the work per area to create two fracture surfaces to twice the specific surface energy (surface tension), γ. The work is the area under the curve in Figure 14.1,

$$2\gamma = \int_0^{x^*} \sigma\,dx = \int_0^{x^*} \sigma_t \sin(\pi x/x^*)dx = -(x^*/\pi)\sigma_t[\cos\pi - \cos 0] = (2x^*/\pi)\sigma_t, \tag{14.5}$$

so

$$x^* = \pi\gamma/\sigma_t. \tag{14.6}$$

Substituting equation (14.6) into equation (14.3),

$$\sigma_t = [(\pi\gamma/\sigma_t)/a_o]E/\pi, \tag{14.7}$$

$$\sigma_t = \sqrt{(\gamma E/a_o)}. \tag{14.8}$$

The predictions of equations (14.8) and (14.4) are similar. Both are much too high.

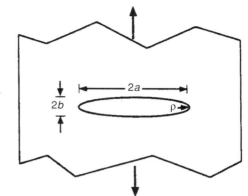

Figure 14.2. Internal crack in a plate approximated by an ellipse with major and minor radii of a and b.

Stress Concentration

The reason that the theoretical predictions are high is that they ignore flaws, and all materials contain flaws. In the presence of a flaw, an externally applied stress is not uniformly distributed within the material. Discontinuities such as internal cracks and notches are stress concentrators. For example, the stress at the tip of the crack, σ_{max}, in a plate containing an elliptical crack (Figure 14.2), is given by

$$\sigma_{max} = \sigma_a(1 + 2a/b), \qquad (14.9)$$

where σ_a is the externally applied stress. The term, $(1 + 2a/b)$, is called the *stress concentration* factor. The radius of curvature, ρ, at the end of an ellipse is given by

$$\rho = b^2/a, \qquad (14.10)$$

so $a/b = \sqrt{(a/\rho)}$. Substitution of equation (14.10) into equation (14.9) results in

$$\sigma_{max} = \sigma_a(1 + 2\sqrt{(a/\rho)}). \qquad (14.11)$$

Because a/r is usually very large ($a/\rho \gg 1$), equation (14.11) can be approximated by

$$\sigma_{max} = 2\sigma_a\sqrt{(a/\rho)}. \qquad (14.12)$$

It is possible to rationalize the difference between theoretical and measured fracture stresses in terms of equation (14.12). Fracture will occur when the level of σ_{max} reaches the theoretical fracture strength, so the average stress, σ_a, at fracture is much lower than the theoretical value.

EXAMPLE PROBLEM 14.1: Calculate the stress concentration at the tip of an elliptical crack having major and minor radii of $a = 100$ nm and $b = 1$ nm.

Solution: $\rho = b^2/a = 10^{-18}/10^{-7} = 10^{-11}$. $\sigma_{max}/\sigma_a = 1 + 2\sqrt{(a/\rho)} = 1 + 2\sqrt{(10^{-7}/10^{-11})} = 201$.

Griffith Theory

Griffith approached the subject of fracture by assuming that materials always have preexisting cracks. He considered a large plate with a central crack under a remote stress, σ, and calculated the change of energy, ΔU, with crack size (Figure 14.3). There are two terms. One is the surface energy associated with the crack,

$$\Delta U_{\text{surf}} = 4at\gamma, \tag{14.13}$$

where t is the plate thickness, and $2a$ is the length of an internal crack. The other term is the decrease of stored elastic energy due to the presence of the crack,[*]

$$\Delta U_{\text{elast}} = -\pi a^2 t \sigma^2 / E. \tag{14.14}$$

Combining equations (14.13) and (14.14),

$$\Delta U_{\text{total}} = 4at\gamma - \pi a^2 t(\sigma^2 / E). \tag{14.15}$$

This equation predicts that the energy of the system first increases with crack length and then decreases as shown in Figure 14.3. Under a fixed stress, there is a critical crack size above which crack growth lowers the energy.

This critical crack size can be found by differentiating equation 14.15 with respect to a and setting to zero. $d\Delta U_{\text{total}}/da = 4t\gamma - 2\pi at(\sigma^2 / E) = 0$,

$$\sigma = \sqrt{(2E\gamma / \pi a)}. \tag{14.16}$$

This is known as the Griffith criterion. A preexisting crack of length greater than $2a$ will grow spontaneously when equation (14.16) is satisfied. Griffith found reasonable agreement between this theory and experimental results on glass. However, the predicted stresses were much too low for metals.

In this development, the plate is assumed to be thin enough relative to the crack length that there is no stress relaxation in the thickness direction (e.g., plane–stress conditions prevail). If, on the other hand, the plate is thick enough that there is complete strain relaxation in the thickness direction (plane–strain conditions), equation (14.16) should be modified to

$$\sigma = \sqrt{(2E\gamma / [(1 - v^2)\pi a]}. \tag{14.17}$$

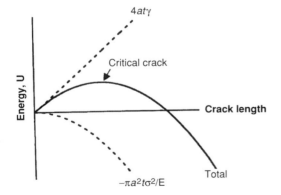

Figure 14.3. Effect of a crack's length on its energy. The elastic energy decreases, and the surface energy increases. There is a critical crack length at which growth of the crack lowers the total energy.

---

[*] This term will not be derived here. However, to check its magnitude, one can simplify the problem by assuming that all strain energy is lost inside a circle of diameter $2a$, and none outside. The energy/volume is $(1/2)\sigma\varepsilon = (1/2)\sigma^2 / E$ and the volume is $\pi a^2 t$, so the total elastic strain energy would be $\Delta U_{\text{elast}} = -(1/2)\pi a^2 t \sigma^2 / E$. The correct solution is just twice this. Note that the derivation of this assumes that plate width, w, is very large compared to a ($w \gg a$) and that the plate thickness, t, is small compared to a ($t \ll a$). For thick plates, ($t \gg a$), plane strain prevails, and equation (14.14) becomes $\Delta U_{\text{elast}} = -\pi a^2 t \sigma^2 (1 - v^2)/E$.

Orowan Theory

Orowan proposed that the reason that the predictions of equations (14.16) and (14.17) were too low for metals is because the energy expended in producing a new surface by fracture is not just the true surface energy. There is a thin layer of plastically deformed material at the fracture surface, and the energy to cause this plastic deformation is much greater than γ. To account for this, equation (14.16) is modified to

$$\sigma = \surd(EG_c/\pi a), \tag{14.18}$$

where G_c replaces 2γ and includes the plastic work in generating the fracture surface.

Fracture Modes

There are three different modes of fracture, each having a different value of G_c. These modes are designated I, II, and III, as illustrated in Figure 14.4. In mode I fracture, the fracture plane is perpendicular to the normal force. This is what is occurs in tension tests of brittle materials. Mode II fractures occur under the action of a shear stress, with the fracture propagating in the direction of shear. An example is the punching of a hole. Mode III fractures are also shear separations, but here the fracture propagates perpendicular to the direction of shear. An example is the cutting of paper with scissors.

Irwin's Fracture Analysis

Irwin noted that in a body under tension, the stress state around an infinitely sharp crack in a semiinfinite elastic solid is entirely described by*

$$\sigma_x = K_I/(2\pi r)^{1/2}\cos(\theta/2)[1 - \sin(\theta/2)\sin(3\theta/2)] \tag{14.19a}$$

$$\sigma_y = K_I/(2\pi r)^{1/2}\cos(\theta/2)[1 + \sin(\theta/2)\sin(3\theta/2)] \tag{14.19b}$$

$$\tau_{xy} = K_I/(2\pi r)^{1/2}\cos(\theta/2)\sin(\theta/2)\cos(3\theta/2) \tag{14.19c}$$

$$\sigma_z = \upsilon(\sigma_x + \sigma_y) \text{ for plane strain } (\varepsilon_z = 0) \text{ and } \sigma_z = 0 \text{ for plane stress} \tag{14.19d}$$

$$\tau_{yz} = \tau_{zx} = 0, \tag{14.19e}$$

Figure 14.4. Three modes of cracking.

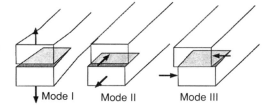

Mode I   Mode II   Mode III

* These equations are attributed to Westergaard, *Trans. ASME, J. Appl. Mech.,* v. 61 (1939).

where θ and r are the coordinates relative to the crack tip (Figure 14.5), and K_I is the stress intensity factor,* which in a semiinfinite body is given by

$$K_I = \sigma(\pi a)^{1/2}. \tag{14.20}$$

Here, σ is the applied stress. For finite specimens, $K_I = f\sigma(\pi a)^{1/2}$, where f depends on the specimen geometry and is usually a little greater than 1 for small cracks. Figure 14.6 shows how f varies with the ratio of crack length, a, to the specimen width, w.

The Westergaard equations predict that the local stresses, σ_x and σ_y, are infinite at the crack tip and decrease with distance from the crack, as shown in Figure 14.7. Of course, the infinite stress prediction is unrealistic. The material will yield wherever σ_y is predicted to be greater than the material's yield strength, so yielding limits the actual stress near the crack tip ($\sigma_y \leq Y$).

In Irwin's analysis, fracture occurs when K_I reaches a critical value, K_{Ic}, which is a material property. This predicts that the fracture stress, σ_f, is

$$\sigma_f = K_{Ic}/\left[f(\pi a)^{1/2}\right]. \tag{14.21}$$

Comparison with equation (14.18) shows that for $f = 1$,

$$K_{Ic} = \sqrt{(EG_c)}. \tag{14.22}$$

EXAMPLE PROBLEM 14.2: A 6-foot-long steel strut, 2.0 in. wide and 0.25 in. thick, is designed to carry an 80,000-lb tension load. Assume that the steel has a toughness of 40 ksi$\sqrt{}$in. What is the longest edge crack that will not cause failure?

Solution: According to equation (14.21), failure will occur if $a = (K_{Ic}/f\sigma)^2/\pi$. To solve this problem, we must first guess a value for f. As a first guess let $f = 1.2$. Substituting $f = 1.2$, $K_{Ic} = 40$ ksi$\sqrt{}$in. and $\sigma = 80,000$ lb/(2.0 × 0.25 in.^{2}) = 160 ksi, $a = 0.014$ in. Now checking to see whether the guess of $f = 1.2$ was reasonable, $a/w = 0.08/2 = 0.04$. From Figure 14.6a, $f = 1.12$. Repeating the calculation with $f = 1.12$, $a = 0.016$. The value $f = 1.12$ is reasonable for $a = 0.016$.

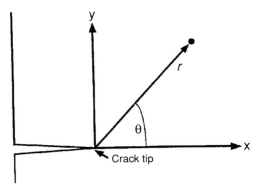

Figure 14.5. Coordinates for the stress state (equation 14.19) near a crack tip.

---

* The stress intensity factor should not be confused with the stress concentration factor

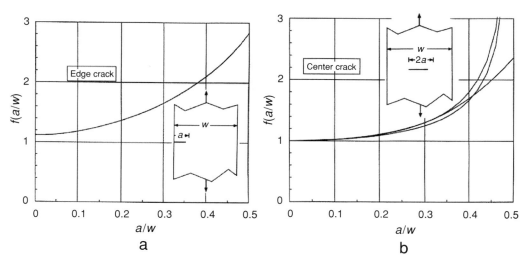

Figure 14.6. Variation of f with a/w for edge crack (a) and center crack (b). The three lines for the center crack are from three different mathematical approximations.

Plastic Zone Size

The plastic work involved in the yielding of material near the crack tip as the crack advances is responsible for the energy absorption. Figure 14.8 is a three-dimensional sketch of the plastic zone shape. Plane strain ($\varepsilon_z = 0$) is characteristic of the interior where adjacent material prevents lateral contraction. At the surface, where plane stress ($\sigma_z = 0$) prevails, the fracture corresponds to mode II rather than mode I. The energy absorption per area is greater here because of the larger volume of deforming material in the plane–stress region.

The sizes and shapes of the plastic zones calculated for plane–strain ($\varepsilon_z = 0$) and plane–stress ($\sigma_z = 0$) conditions are shown in Figure 14.9. The details of the calculation method are given in the Appendix of this chapter. It is customary to characterize the size of the plastic zone by a radius, r_p. For plane stress,

$$r_p = (K_{Ic}/Y)^2/(2\pi). \tag{14.23}$$

For plane strain,

$$r_p = (K_{Ic}/Y)^2/(6\pi). \tag{14.24}$$

Figure 14.7. Stress distribution ahead of a crack. The Westergaard equations predict an infinite stress at the crack tip, but the stress there can be no greater than the yield strength.

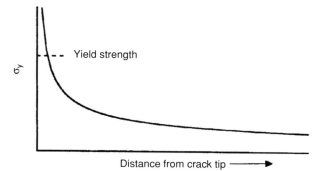

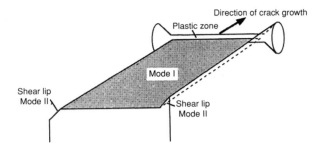

Figure 14.8. Three-dimensional sketch showing the shape of the plastic zone and the shear lip formed at the edge of the plate.

There is no stress, σ_z, normal to the surface of the specimen, so plane stress prevails at the surface with a characteristic toughness K_{IIc} that is higher than K_{Ic}. There is a transition from plane stress to plane strain with increasing distance from the surface. To obtain valid K_{Ic} data, the thickness of the specimen should be much greater than radius of the plastic zone for plane stress. Specifications require that

$$t \geq 2.5(K_{Ic}/Y)^2. \tag{14.25}$$

This assures that the fraction of the fracture surface failing under plane–stress conditions, $2r_p/t$, equals or is less than $[2(K_{Ic}/Y)^2/(2\pi)]/[2.5(K_{Ic}/Y)^2] = 12.7\%$.

EXAMPLE PROBLEM 14.3: Using stress field in equation (14.19) for plane stress ($\sigma_z = 0$), calculate the values of r at $\theta = 0°$ and $90°$ at which the von Mises yield criterion is satisfied. Express the values of r in terms of (K_{Ic}/Y).

Solution: According to equation (14.19), for $\theta = 0$, $\sigma_x = \sigma_y = K_{Ic}/\sqrt{(2\pi r)}$, $\sigma_z = 0$, $\tau_{yz} = \tau_{zx} = \tau_{xy} = 0$. Substituting into the von Mises criterion, $\sigma_x = \sigma_y = Y$ so $Y = K_{Ic}/\sqrt{(2\pi r)}$. Solving for r, $r = (K_{Ic}/Y)^2/(2\pi) = 0.159(K_{Ic}/Y)^2$.

For $\theta = 90°$, $\theta/2 = 45°$, $\cos(\theta/2) = \sin(\theta/2) = \sin(3\theta/2) = -\cos(3\theta/2) = 1/\sqrt{2}$. According to equation (14.19), $\sigma_x = K_{Ic}/\sqrt{(2\pi r)}[1/(2\sqrt{2})]$, $\sigma_y = K_{Ic}/\sqrt{(2\pi r)}[3/(2\sqrt{2})] = 3\sigma_x$, $\sigma_z = 0$, $\tau_{xy} = K_{Ic}/\sqrt{(2\pi r)}[-1/(2\sqrt{2})] = -\sigma_x$,

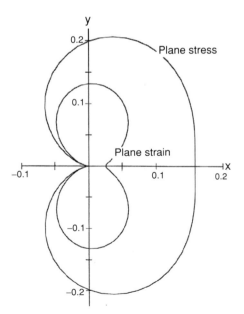

Figure 14.9. Plots of the plastic zones associated with plane stress and plane strain. Dimensions are in units of $(K_I/Y)^2$. The curve for plane strain was calculated for $\upsilon = 0.3$. Note that the "radii" of the plane–stress and plane–strain zones are conventionally taken as $(K_I/Y)^2/(2\pi) = 0.159$ and $(K_I/Y)^2/(6\pi) = 0.053$, respectively.

Figure 14.10. Dependence of K_c on thickness. For very thick plates, the effect of mode II in the surface region is negligible, so K_c approaches K_{Ic}. For thinner sheets, more of the fracture surface is characterized by plane–stress conditions so K_c approaches K_{IIc}. For even thinner sheets, K_c decreases with thickness because the plastic volume decreases.

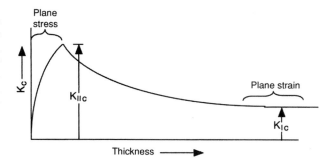

$\tau_{yz} = \tau_{zx} = 0$. Substituting into the von Mises criterion, $2Y^2 = (3\sigma_x - 0)^2 + (0 - \sigma_x)^2 + (\sigma_x - 3\sigma_x)^2 + 6(\sigma_x)^2 = 20\sigma_x^2$, so $\sigma_x^2 = 2Y^2/20 = Y^2/10$. Now substituting $\sigma_x^2 = K_{Ic}^2/(2\pi r)[1/(2\sqrt{2})]^2 = Y^2/10$, $r = [10/(16\pi)](K_{Ic}/Y)^2 = 0.199(K_{Ic}/Y)^2$. Compare these values with Figure 14.9 for $\theta = 0°$ and $\theta = 90°$ for plane–stress conditions.

The overall fracture toughness (critical stress intensity), K_c, depends on the relative sizes of the plane–stress and plane–strain zones, and therefore depends on the specimen thickness as sketched in Figure 14.10.

Thin Sheets

Figure 14.10 shows how toughness depends on the sheet or plate thickness. For thick sheets, K_c decreases with thickness because a lower fraction of the fracture surface is in mode II. However, if the sheet thickness is less than twice r_p for plane stress, the entire fracture surface fails in mode II. In this case, the failure is by through-thickness necking (Figure 14.11). The volume of the plastic region equals t^2L, where t is the specimen thickness and L is the length of the crack. Because the area of the fracture is tL, the plastic work per area is proportional to the thickness, t.

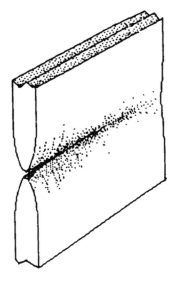

Figure 14.11. Sketch of a sheet failure by through-thickness necking.

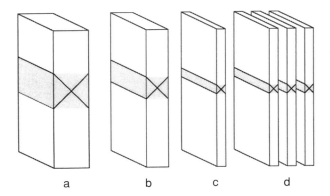

Figure 14.12. Volume of the plastic zone is proportional to t^2, so the energy absorbed per area of fracture surface decreases as t decreases. The toughness of a laminate (d) is less than that of a monolithic sheet of the same total thickness (a) because the plastic volume is smaller.

a   b   c   d

Therefore, K_c is proportional to $\sqrt{t}$. Very thin sheets tear at surprisingly low stresses. Plates made of laminating sheets have lower toughness than monolithic ones. See Figure 14.12.

EXAMPLE PROBLEM 14.4: Derive expressions that show how K_c and G_c depend on thickness for thin sheets that fail by plane–stress fracture.

Solution: The fracture energy per area, $G_c = U/A = U/(tL)$, where U is the plastic work and L is the length of the fracture. Assume that necks in different thickness specimens are geometrically similar. Then the necked volume $= Ct^2L$ so $U = C'Lt^2$, where C and C' are constants. Substituting, $G_c = U/(tL) = C'Lt^2/(tL) = C't$, so G_c is inversely proportional to t. $K_c = \sqrt{(G_cE)} = \sqrt{(C'tE)}$, so K_c is inversely proportional to the square root of thickness.

Temperature and Loading Rate

Because the size of the plastic zone depends on (K_{Ic}/Y), the factors that raise yield strengths, Y, decrease the size of the plastic zones and therefore lower fracture toughnesses. These include lowered temperatures and increased rates of loading. Figure 14.13 shows the dependence of temperature on G_c for steel forgings. An increased loading rate has an effect similar to decreased temperature.

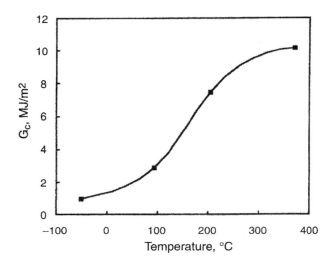

Figure 14.13. Dependence of G_c on temperature for steel forgings. Data from D. H. Winne and B. M. Wundt, *Trans. ASME*, v. 80 (1958).

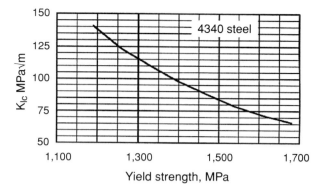

Figure 14.14. Inverse correlation of K_{Ic} with yield strength for 4340 steel.

Metallurgical Variables

Most metallurgical variations responsible for increasing strength, such as increased carbon content of steels, alloying elements in solid solution, cold working, and martensitic hardening, tend to decrease toughness. The sole exception seems to be decreased grain size, which increases both strength and toughness. Figure 14.14 shows the decrease of toughness with strength for 4340 steels that had been tempered to different yield strengths. There is a similar correlation between strength and toughness for aluminum alloys. Other factors such as the presence of inclusions also affect K_{Ic}.

Fracture Mechanics in Design

Engineering design depends on the size of the largest possible crack that might exist in a stressed part or component. Various techniques are used to inspect for preexisting cracks. These include ultrasonics, x-rays, magnetic inspection, and dye penetrants. For each technique, there is a limit to how small a crack can be reliably detected. Safe design is based on assuming the presence of cracks of the largest size that cannot be detected with 100% certainty. For example, if that inspection technique cannot assure detection of edge cracks smaller than 1 mm, the designer must assume $a = 1$ mm. Then the permissible stress is calculated using equation (14.21) and applying an appropriate safety factor. The design should also assure that the component does not yield plastically, so the permissible stress should not exceed the yield strength times the safety factor. If the stress calculated from equation (14.21) exceeds the yield strength, failure from accidental overload will occur by plastic yielding rather than fracture. Yielding, however, is usually regarded as less dangerous than fracture.

Even if there are no preexisting cracks larger than those assumed in the design, "safe failure" is still not assured because smaller cracks may grow by fatigue during service. In critical applications, such as cyclically loaded aircraft component, periodic inspection during the life of a part is required.

EXAMPLE PROBLEM 14.5: A support is to be made from 4340 steel. The steel may be tempered to different yield strengths. The correlation between strength and toughness is shown in Figure 14.14. Assume that $f = 1.1$ for the support

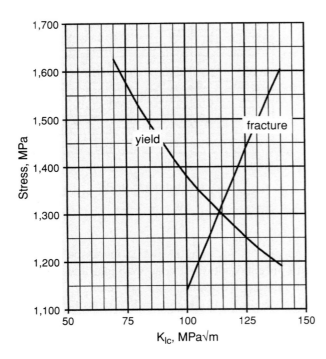

Figure 14.15. Highest possible load is reached when $\sigma_f = Y = 1,305$ MPa.

geometry and that nondestructive inspection can detect all edge cracks of size $a = 2$ mm or larger.

A. If a 4340 steel of $K_{Ic} = 120$ MPa$\sqrt{m}$ is used, what is the largest stress that will not cause either yielding or fracture? If the support is overloaded, will it fail by yielding or by fracture?

B. By selecting a 4340 steel of a different yield strength, find the yield strength that will allow the largest stress without either yielding or fracture.

Solution:

A. From Figure 14.14, for $K_{Ic} = 120$ MPa$\sqrt{m}$. $Y = 1275$ MPa. Using equation (14.21), $\sigma_f = K_{Ic}/[f\sqrt{(\pi a)}] = 120\,\text{MPa}/[1.1\sqrt{(0.002\,\text{m} \cdot \pi)} = 1,376\,\text{MPa}$. This is higher than the yield strength, so it will fail by yielding when $\sigma = 1,275$ MPa.

B. For the highest load-carrying capacity, yielding and fracture should occur at the same stress level. Figure 14.15 is a plot of calculated values of σ_f for several different levels of K_{Ic}. The yield strength versus K_{Ic} data from Figure 14.14 is replotted on the same axes. The intersection of two curves indicates that the optimum value of Y is about 1,305 MPa.

Compact Tensile Specimens

The *compact tensile specimen* (Figure 14.16) is often used for measuring K_{Ic}. The specimen is loaded by pins inserted in the holes. The dimensions are selected so that loading of the specimen will be entirely elastic except for the plastic zone at the crack tip. With the requirement that $a \geq 2.5(K_{Ic}/Y^2)$, the radius of the plane–stress

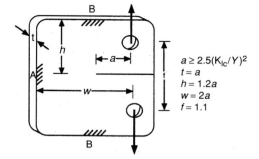

Figure 14.16. Compact tensile specimen used to measure K_{Ic}.

plastic zone at the surface, $r_p = (K_{Ic}/Y)^2/2\pi$ is $\leq a/5\pi$. Mode II extends inward from each surface a distance of r_p, so mode II will prevail over less than $a/(2.5\pi)$. If the thickness $= a$, this amounts to $1/(2.5\pi) = 12.7\%$ of the fracture surface. The requirement that $w/a = 2$ ensures that a plastic hinge will not develop at the back of the specimen (region A). Similarly, if h were less than $1.2a$, regions B would yield in bending.

The initial crack is made by machining. Then cyclic loading is applied to cause the crack to grow and sharpen by fatigue. Finally, the fracture toughness is measured by loading the specimen until the crack grows. The load at this point is noted, and K_{Ic} is calculated from equation (14.19), taking σ as the load divided by w and t.

EXAMPLE PROBLEM 14.6: A medium carbon steel has a yield strength of 250 MPa and a fracture toughness of 80 MPa$\sqrt{m}$. How thick would a compact tensile specimen have to be for valid plane–strain fracture testing?

Solution: The specimen thickness should be at least $2.5(K_{Ic}/Y)^2 = 2.5(80/250)^2 = 0.26$ m (or about 10 in.). (This would be impossible unless a very thick plate and a very large testing machine were available.)

Strain–Energy Release

Gurney and coworkers* suggested a still different approach to understanding brittle fracture. Every specimen and structure can be characterized by a compliance (inverse of stiffness), defined as $1/M = du/dP$, where u is displacement and P is the load. Figure 14.17 shows schematically the load displacement characteristics of a typical specimen. Initially, the loading is linear ($1/M$ is constant), but when the crack begins to grow, $1/M$ decreases. Under load, the energy stored elastically is

$$W_{elast} = (1/2)Pu. \tag{14.26}$$

Figure 14.18 illustrates how the energy changes as the crack grows from a length a_1 to a_2, increasing the crack area from $A_1 = a_1t$ to $A_2 = a_2t$. As the crack grows, the external work, ΔW, can be approximated by

$$\Delta W = \Delta u P_{ave} = (u_2 - u_1)(P_1 + P_2)/2. \tag{14.27}$$

* C. Gurney and J. Hunt, *Proc. Roy. Soc.*, v. 299A (1967), pp. 207–222 and C. Gurney and Y. W. Mai, *Engr. Fract. Mech.*, v. 4 (1972), pp. 853–863.

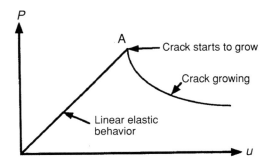

Figure 14.17. Load displacement curve for a cracked specimen. The initial loading is linear, but the compliance, $1/M = u/P$, increases when the crack begins to grow at A.

This external work must equal the change of energy of the system, which involves a change in elastic stored energy, and the work* in producing the increased crack area,

$$\Delta W = \Delta W_{elast} + G_c \Delta A. \tag{14.28}$$

Substituting $\Delta W_{elast} = (P_2 u_2 - P_1 u_1)/2$ and $\Delta A = t \Delta a$

$$G_c = [(1/2)(u_2 - u_1)(P_1 + P_2) - (1/2)(P_2 u_2 - P_1 u_1)]/(t \Delta a), \tag{14.29}$$

which simplifies to

$$G_c = (1/2)(P_1 u_2 - P_2 u_1)/(t \Delta a). \tag{14.30}$$

With a load displacement diagram like that in Figure 14.17, equation (14.30) can be used to determine G_c as long as the crack propagates in a stable manner so the crack length, a, can be measured during growth.

The J Integral

Linear elastic behavior was been assumed in the development of equation (14.19) that forms the basis for the treatment up to this point. For relatively tough materials, very large specimens are required to assure elastic behavior. Often materials are not available in the thicknesses needed. The J integral offers a method of evaluating the toughness of a material that undergoes a nonlinear behavior during loading. Strictly, it should be applied only for nonlinear elastic behavior, but it has been shown that the errors caused by a limited amount of plastic deformation are not large.

J is the work done on a material per area of fracture. It is the area between the loading path and the unloading path after the fracture area divided by the increase of fracture area, Δa. If there has been no plastic deformation, the unloading will return to the origin as shown in Figure 14.19a. This area can be measured by unloading or by testing a specimen with a longer initial crack. The value of J is equal to G_c.

$$J_{Ic} = G_{Ic} = K_{Ic}^2 / [E/(1 - u^2)]. \tag{14.31}$$

If, however, there has been plastic deformation accompanying the crack propagation, the unloading curve will not return to the origin, as shown in Figure 14.19b. In

---

* Gurney used the symbol, R, to designate the energy per area required to produce new crack surface. The symbol, G_c, is used here for consistency with the previous developments.

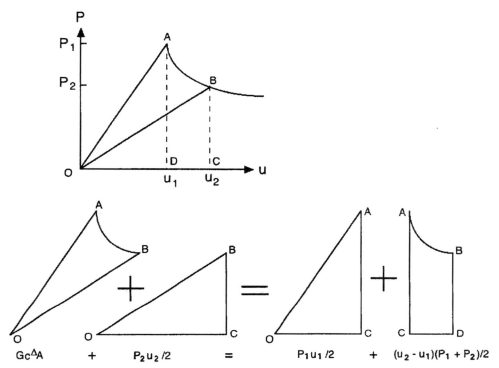

Figure 14.18. Load displacement diagram illustrating the several terms in the development of equations (14.26) through (14.30). The area of sector OAB equals $G_c\Delta A$.

this case, J is still taken as the area between the loading and unloading curves, and J should be somewhat larger than G_c. Figure 14.20 is a comparison of experimentally determined values of J_{Ic} and K_{Ic}.

There are other approaches to measuring the fracture toughness of tough materials, including measurements based on the crack opening displacement (COD). These methods will not be treated here.

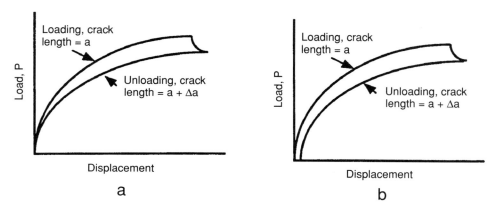

Figure 14.19. J is taken as the area between the loading and unloading curves. (a) For fully elastic behavior, the unloading curve returns to the origin. (b) If some plasticity accompanies the crack propagation, the unloading curve will not return to the origin.

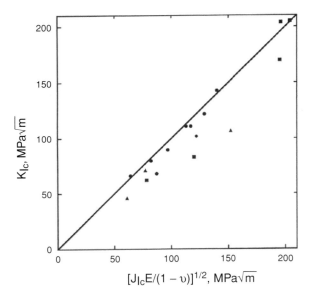

Figure 14.20. Comparison of experimentally determined values of J_{Ic} and K_{Ic} for several steels. Adapted from R. W. Hertzberg, *Deformation and Fracture and Fracture Mechanics of Engineering Materials*, 4th ed., John Wiley (1995).

REFERENCES

R. W. Hertzberg, *Deformation and Fracture Mechanics of Engineering Materials*, 4th ed., John Wiley (1996).

T. H. Courtney, *Mechanical Behavior of Materials*, McGraw-Hill (1990).

G. E. Dieter, *Mechanical Metallurgy*, 3rd ed., McGraw-Hill (1986).

N. E. Dowling, *Mechanical Behavior of Materials*, Prentice Hall (1993).

Notes

Alan A. Griffith is considered to be the father of fracture mechanics. He received his B.Eng., M.Eng., and D.Eng. degrees from the University of Liverpool. In 1915, he was employed by the Royal Aircraft Establishment (then the Royal Aircraft Factory), where he had a brilliant career. He did pioneering work on turbines, that led to the early development of the jet engine. In his work on aircraft structures, he became interested in the stress concentrations caused by notches and scratches. He realized that if the curvature at the end of a crack were of molecular dimensions, the stress concentration would reduce strengths far below those that were actually observed.

Griffith measured the fracture strength of glass fibers that naturally contained small cracks. He found reasonable agreement with his theory. Such agreement would not have been found, had he worked with metals or other materials in which there is energy absorbed by the plastic zone associated with crack propagation. His classic paper, in *Phil. Trans. Roy. Soc London*, ser. A, v. 221A (1923), has been reprinted with commentary in *Trans. ASM*, v. 61 (1968).

The basis of modern fracture mechanics was developed at the Naval Research Laboratories by George Irwin in 1947. He used the symbol G for plastic work per area to honor Griffith.

Problems

1. Using the two theoretical predictions of fracture strength, $\sigma_t = E/\pi$ (equation 14.4) and $\sigma_t = \sqrt{(\gamma E/a_o)}$ (equation 14.8), calculate the theoretical fracture strength of iron and MgO. For iron, take a_o as its atomic diameter (0.124 nm) and, for MgO, as the average of the ionic diameters of Mg^{+2} and O^{-2} (0.105 nm). The surface energies of iron and MgO are about 2.0 J/m^2 and 1.2 J/m^2, respectively. Young's moduli of iron and MgO are about 270 GPA and 300 GPa, respectively. How do these answers compare with each other? How do these answers compare with the actual fracture strengths of each?

2. Class 20 and class 60 gray cast irons have tensile strengths of about 20 ksi and 60 ksi, respectively. Assuming that the fractures start from graphite flakes and that the flakes act as preexisting cracks, use the concepts of the Griffith analysis to predict the ratio of the average graphite flake sizes of the two cast irons.

3. A wing panel of a supersonic aircraft is made from a titanium alloy that has a yield strength of 1,035 MPa and toughness of $K_{Ic} = 55$ MPa$\sqrt{m}$. It is 3.0 mm thick, 2.40 m long, and 2.40 m wide. In service, it is subjected to a cyclic stress of ±700 MPa, which is not enough to cause yielding but does cause gradual crack growth of a preexisting crack normal to the loading direction at the edge of the panel. Assume that the crack is initially 0.5 mm long and grows at a constant rate of $da/dN = 120$. Calculate the number of cycles to failure.

4. The support in Figure 14.21a is to be constructed from a 4340 steel plate tempered at 800°F. The yield strength of the steel is 228 ksi, and; its value of K_{Ic} is 51 ksi$\sqrt{(in.)}$. The width of the support, w, is 4 in.; the length, L, is 36 in.; and the thickness, t, is 0.25 in. Figure 14.21b gives $f(a/w)$ in the equation $\sigma = K_{Ic}/[f(a/w)\sqrt{(\pi a)}]$.

    A. If the crack length, a, is small enough, the support will yield before it fractures? What is the size of the largest crack for which this is true (i.e., what is the largest value of a for which general yielding will precede fracture)? Assume that any fracture would be in mode I (plane strain). Discuss critically the assumption of mode I.

    B. If it is guaranteed that there are no cracks longer than a length equal to 80% of that in Figure 14.21a, does this assure that failure will occur by yielding, rather than fracture? Explain briefly.

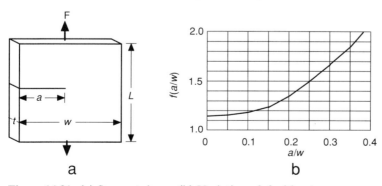

Figure 14.21. (a) Support shape. (b) Variation of f with a/w.

5. For 4340 steel, the fracture toughness and yield strength depend on the prior heat treatment, as shown in Figure 14.14. As yield strength is increased, the fracture toughness decreases. A pipeline is to be built of this steel, and to minimize the wall thickness of the pipe, the stress in the walls should be as high as possible without either fracture or yielding. Inspection techniques ensure that there are no cracks longer than 2 mm ($a = 1$ mm). What level of yield strength should be specified? Assume a geometric constant of $f = 1.15$.

6. A steel plate, 10 ft long, 0.25 in. thick, and 6 in. wide is loaded under a stress of 50 ksi. The steel has a yield strength of 95 ksi and a fracture toughness of 112 ksi$\sqrt{}$(in.). There is a central crack perpendicular to the 10-ft dimension.

    A. How long would the crack have to be for failure of the plate under the stress?

    B. If there is an accidental overload (i.e., the stress rises above the specified 50 ksi), the plate might fail by either yielding or by fracture, depending on the crack size. If the designer wants to be sure that an accidental overload would result in yielding rather than fracture, what limitations must be placed on the crack size?

7. A structural member is made from a steel that has a $K_{Ic} = 180$ MP$\sqrt{}$m and a yield strength of 1,050 MPa. In service, it should neither break nor deform plastically because either would be considered a failure. Assume $f = 1.0$. If there is a preexisting surface crack of $a = 4$ mm, at what stress will the structural member fail? Will they fail by yielding or fracture?

8. An estimate of the effective strain, ε, in a plane–strain fracture surface can be made in the following way. Assume that the material is not work-hardening and the effective strain is constant throughout the plastic zone so the plastic work per volume is $Y\varepsilon$. Assume the depth of the plastic zone is given by equation $r_p = (K_I/Y)^2/(6\pi)$ and the strained volume is $2r_pA$. Derive an expression for the plastic strain, ε, associated with running of a plane–strain fracture. (Realize that G_c is the plastic work per crack area and that K_{Ic} and G_c are related by equation (14.22).)

9. The data below were taken on a test material to determine its fracture toughness. The specimens were 25 mm thick.

    A. Make a plot of P versus u.

    B. Using this plot with Gurney's method, determine the value of G for several combinations of points.

Crack length, a (mm)	Displacement, u (mm)	Load, P (N)
51	0.17	2,370
76	0.31	1,495
100	0.54	1,170
125	0.81	980
155	1.15	825
180	1.52	750

10. Find the factor, f, for the compact tensile specimen in Figure 14.16. ASTM Standard E399-8 gives $f = (2 + a/b)[0.886 + 4.64(a/b) - 13.33(a/b)^2 + 14.72(a/b)^3 - 5.6(a/b)^4]$.

Appendix. Size and Shape of the Plastic Zone at the Crack Tip

The approximate size and shape of the plastic zone at the crack tip can be found by assuming that even where there is plastic deformation near the crack tip, equation (14.19) holds outside of the plastic zone. Then expressions for σ_x, σ_y, τ_{xy}, and σ_z can be substituted into a suitable yield criterion,

$$\sigma_x = A/(2\pi r)^{1/2} K_{Ic}/\sqrt{(2\pi r)}, \tag{14.32a}$$

$$\sigma_y = B/(2\pi r)^{1/2} K_{Ic}/\sqrt{(2\pi r)}, \tag{14.32b}$$

$$\sigma_z = C/(2\pi r)^{1/2} K_{Ic}/\sqrt{(2\pi r)}, \tag{14.32c}$$

$$\text{and} \quad \tau_{xy} = D/(2\pi r)^{1/2} K_{Ic}/\sqrt{(2\pi r)}, \tag{14.32d}$$

where $A = \cos(\theta/2)[1 - \sin(\theta/2)\sin(3\theta/2)]$,

$B = \cos(\theta/2)[1 + \sin(\theta/2)\sin(3\theta/2)]$,

$C = 2\upsilon\cos(\theta/2)$ for plane strain or $C = 0$ for plane stress,

$D = \cos(\theta/2)\sin(\theta/2)\cos(3\theta/2)$.

For example, substituting in the von Mises criterion,

$$\left[(\sigma_y - \sigma_z)^2 + (\sigma_z - \sigma_x)^2 + (\sigma_x - \sigma_y)^2 + 6\tau_{xy}^2\right] = 2Y^2 \tag{14.33}$$

$$[(B - C)^2 + (C - A)^2 + (A - B)^2 + 6(D)^2]K_{Ic}^2/(2\pi r) = 2Y^2.$$

Solving for r,

$$r/(K_{Ic}/Y)^2 = [(B - C)^2 + (C - A)^2 + (A - B^2) + 6(D)^2]/(4\pi). \tag{14.34}$$

Taking $(K_{Ic}/Y)^2 = 1$, the boundaries of the plastic zone then are found as $x = r\cos\theta$ and $y = r\sin\theta$, calculating for $0 \le \theta \le 360°$. Figure 14.9 is a plot of these loci. The calculations should be regarded as approximations because the stress state used in the calculations is based on an elastic solution, which would be altered by the plastic deformation in the zone.

15 Viscoelasticity

Introduction

In classic elasticity, there is no time delay between application of a force and the deformation that it causes. For many materials, however, there is additional time-dependent deformation that is recoverable. This is called *viscoelastic* or *anelastic* deformation. When a load is applied to a material, there is an instantaneous elastic response, but the deformation also increases with time. This viscoelasticity should not be confused with *creep* (Chapter 16), which is time-dependent plastic deformation. Anelastic strains in metals and ceramics are usually so small that they are ignored. In many polymers, however, viscoelastic strains can be very significant.

Anelasticity is responsible for damping of vibrations. A high damping capacity is desirable where vibrations might interfere with the precision of instruments or machinery and for controlling unwanted noise. A low damping capacity is desirable in materials used for frequency standards, in bells, and in many musical instruments. Viscoelastic strains are often undesirable. They cause the sagging of wooden beams, denting of vinyl flooring by heavy furniture, and loss of dimensional stability in gauging equipment. The energy associated with damping is released as heat, which often causes an unwanted temperature increase. Study of damping peaks and how they are affected by processing has been useful in identifying mechanisms. The mathematical descriptions of viscoelasticity and damping are developed in the first part of this chapter. Then several damping mechanisms are described.

Rheological Models

Anelastic behavior can be modeled mathematically with structures constructed from idealized elements representing elastic and viscous behavior, as shown in Figure 15.1.

A spring models a perfectly elastic solid. The behavior is described by

$$e_e = F_e/K_e, \qquad (15.1)$$

where e_e is the change of length of the spring, F_e is the force on the spring, and K_e is the spring constant.

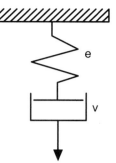

Figure 15.1. Spring and dashpot in series (Maxwell model)

A dashpot models a perfectly viscous material. Its behavior is described by

$$\dot{e}_v = de_v/dt = F_v/K_v, \tag{15.2}$$

where e_v is the change in length of the dashpot, F_v is the force on it, and K_v is the dashpot constant.

Series Combination of Spring and Dashpot

The *Maxwell model* consists of a spring and dashpot in series, as shown in Figure 15.1. Here and in the following, e and F, without subscripts, refer to the overall elongation and the external force. Consider how this model behaves in two simple experiments. First, let there be a sudden application of a force, F, at time, $t = 0$, with the force being maintained constant (Figure 15.2). The immediate response from the spring is $e_e = F/K_e$. This is followed by a time-dependent response from the dashpot, $e_v = Ft/K_v$. The overall response will be

$$E = e_e + e_v = F/K_e + Ft/K_v, \tag{15.3}$$

so the strain rate would be constant. The viscous strain would not be recovered on unloading.

Now consider a second experiment. Assume that the material is forced to undergo a sudden elongation, e, at time $t = 0$, and that this elongation is maintained for period of time as sketched in Figure 15.3. Initially, the elongation must be accommodated entirely by the spring ($e = e_e$), so that the force initially jumps to a level, $F_o = K_e e$. This force causes the dashpot to operate, gradually increasing

Figure 15.2. Strain relaxation predicted by the series model. The strain increases linearly with time.

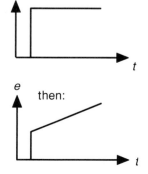

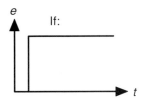

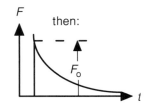

Figure 15.3. Stress relaxation predicted by the series model. The stress decays to zero.

the strain e_v. The force in the spring, $F = K_e e_e = (e - e_v)K_e$, equals the force in the dashpot, $F = K_v de_v/dt$, $(e - e_v)^{-1} de_v = -(K_e/K_v)dt$. Integrating, $\ln[(e - e_v)/e] = -(K_e/k_v)t$. Substituting $(e - e_v) = F/K_e$, and $K_e e = F_0$, $\ln(F/F_0) = t/(K_v/K_e)$. Now defining a relaxation time, $\tau = K_v/K_e$,

$$F = F_0 \exp(-t/\tau). \tag{15.4}$$

Parallel Combination of Spring and Dashpot

The *Voigt model* consists of a spring and dashpot in parallel, as sketched in Figure 15.4. For this model $F = F_e + F_v$ and $e = e_e = e_v$.

Now consider the behavior of the Voight model in the same two experiments. In the first, there is sudden application at time, $t = 0$, of a force, F, which is then maintained at that level (Figure 15.5). Initially, the dashpot must carry the entire force because the spring can carry a force only when it is extended. At an infinite time, the spring carries all the force, so $e_\infty = F/K_e$. Substituting $de_v = de$ and $F_v = F - F_e$ into $de_v/dt = F_v/K$, $de/dt = (F - F_e)/K_v$.

Now substituting $F_e = e_e K_e = e K_e$, $de/dt = (F - e K_e)/K_v = (K_e/K_v)(F/K_e - e)$. Denoting F/K_e by e_∞, defining the relaxation time as $\tau = K_v/K_e$ and rearranging, $\int (e_\infty - e)^{-1} de = \int dt/t$. Integrating $\ln[(e - e_\infty) - e_\infty] = t/\tau$. Rearranging, $(e_\infty - e)/e_\infty = \exp(-t/t)$ or

$$e = e_\infty[1 - \exp(-t/\tau)]. \tag{15.5}$$

Note that $e = 0$ at $t = 0$, and $e \to e_\infty$ as $t \to \infty$.

The experiment in which an extension is suddenly applied to the system (Figure 15.6) is impossible because the dashpot cannot undergo an sudden extension without an infinite force.

Combined Series Parallel Model

Neither the series nor the parallel model adequately describes both stress and strain relaxation. A combined series parallel, or *Voight-Maxwell model*, is much better. In

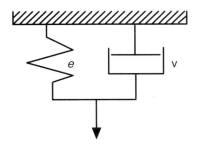

Figure 15.4. Spring and dashpot in parallel (Voigt model).

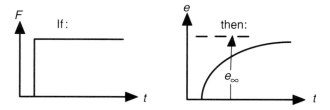

Figure 15.5. Strain relaxation predicted by the parallel model. The strain saturates at e_∞.

Figure 15.7, spring #2 is in parallel with a dashpot, and spring #1 is in series with the combination. The basic equations of this model are $F = F_1 = F_2 + F_v$, $e_v = e_2$ and $e = e_1 = e_2$.

Consider first the sudden application of a force, F, at time $t = 0$ (Figure 15.8). One can write $de_v/dt = F_v/K_v = (F - F_2)/K_v = (F - K_2 e_2)/K_v = (F/K_2 - e_v)(K_2/K_v)$. Rearranging, $\int (F/K_2 - e_v)^{-1} de_v = (1/\tau_e)\int dt$, where $\tau_e = K_v/K_2$ is the relaxation time for strain relaxation. Integration gives $\ln[(F/K_2 - e_v)/F/K_2)] = -t/\tau_e$ or $e_v = (1/K_2)F[1 - \exp(-t/\tau_e)]$. Substituting $F/K_2 = e_\infty - e_o$, where e_∞ and e_o are the relaxed and initial (unrelaxed) elongations, $e_v = (e_\infty - e_o)(1 - \exp(-t/\tau_e))$. The total strain, $e = e_v + e_1$, so

$$e = e_\infty - (e_\infty - e_o)\exp(-t/\tau_e). \tag{15.6}$$

Note that $e = e_o$ at $t = 0$ and $e \to e_\infty$ as $t \to \infty$.

Now consider the experiment in which an elongation, e, is suddenly imposed on the material. Immediately after stretching, the strain occurs in spring 1, so $e = F/K_1$, so the initial force, $F_o = eK_1$ (Figure 15.9). After an infinite time, the dashpot carries no load, so $e = (1/K_1 + 1/K_2)F$, or

$$F_\infty = [K_1 K_2/(K_1 + K_2)]e. \tag{15.7}$$

The force decaying from F_o to F_∞ is given by

$$F = F_o - (F_o - F_\infty)\exp(-t/\tau_\sigma). \tag{15.8}$$

where the relaxation time, τ_σ, is given by

$$\tau_\sigma = K_v/(K_1 + K_2). \tag{15.9}$$

Note that the relaxation time for stress relaxation is shorter than the relaxation time for strain relaxation,

$$\tau_e = K_v/K_2. \tag{15.10}$$

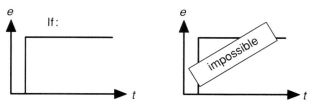

Figure 15.6. An instantaneous strain cannot be imposed on the parallel model.

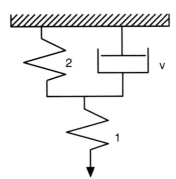

Figure 15.7. Combined series parallel model (Voight-Maxwell model).

EXAMPLE PROBLEM 15.1: For a given material, $F_o/F_\infty = 1.02$. Find the ratio of t_e/t_s for this material.

Solution: Combining equations (15.9) and (15.10), $\tau_e/\tau_\sigma = (K_1 + K_2)/K_2$, For strain relaxation, $e = F_o/K_1 = F_\infty(1/K_1 + 1/K_2)$, so $F_o/F_\infty = K_1(1/K_1 + 1/K_2) = (K_1 + K_2)/K_2$. Therefore, $\tau_e/\tau_\sigma = F_o/F_\infty = 1.02$.

More Complex Models

More complicated models may be constructed using more spring and dashpot elements or elements with nonlinear behavior. Nonlinear elasticity can be modeled by a nonlinear spring for which $F = K_e f(e)$. Non-Newtonian viscous behavior can be modeled by a nonlinear dashpot for which $F = K_v f(\dot{e})$.

Damping

Viscoelastic straining causes damping. Consider the cyclic loading of a viscoelastic material,

$$\sigma = \sigma_o \sin(wt). \tag{15.11}$$

The strain,

$$e = e_o \sin(wt - \delta), \tag{15.12}$$

lags the stress by δ, as shown in Figure 15.10. A plot of stress versus strain is an ellipse. The rate of energy loss per volume is given by $dU/dt = \sigma \, de/dt$, so the energy loss per cycle per volume, ΔU,

$$\Delta U = \oint \sigma \, d\varepsilon \tag{15.13}$$

or $\Delta U = \int \sigma[d\varepsilon/d(\omega t)]d(\omega t)$, where the integration limits are 0 and 2π. Substituting $\sigma = \sigma_o \sin(\omega t)$ and $d\varepsilon/d(\omega t) = \varepsilon_o \cos(\omega t - \delta)$, $\varepsilon = \varepsilon_o \sin(\omega t - \delta)$

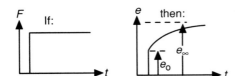

Figure 15.8. Strain relaxation predicted by the series parallel model. The strain saturates at e_∞.

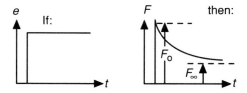

Figure 15.9. Stress relaxation with the series parallel model. The stress decays to F_∞.

$\Delta U = \sigma_o \varepsilon_o \int \sin(\omega t) \cos(\omega t - \delta) \, d(\omega t)$. But $\cos(\omega t - \delta) = \cos(\omega t) \cos \delta + \sin(\omega t) \sin \delta$, so $\Delta U/(\sigma_o \varepsilon_o) = \int \cos \delta \cos(\omega t) \sin(\omega t) \, d(\omega t) + \int \sin\delta \sin^2(\omega t) \, d(\omega t) = \pi \sin \delta$, or

$$\Delta U = \pi (\sigma_o \varepsilon_o) \sin \delta. \qquad (15.14)$$

Because the elastic energy per volume required to load the material to σ_0 and ε_0 is $u = (1/2)\sigma_o \varepsilon_o$, this can be expressed as

$$\Delta U / U = 2\pi \sin \delta. \qquad (15.15)$$

If δ is small,

$$\Delta U / U = 2\pi \delta. \qquad (15.16)$$

Natural Decay

During free oscillation, the amplitude will gradually decrease, as shown in Figure 15.11. It is usually assumed that the decrease between two successive cycles is proportional to the amplitude, e. A commonly used measure of damping is the logarithmic decrement, Λ, defined as

$$\Lambda = \ln(e_n / e_{n+1}). \qquad (15.17)$$

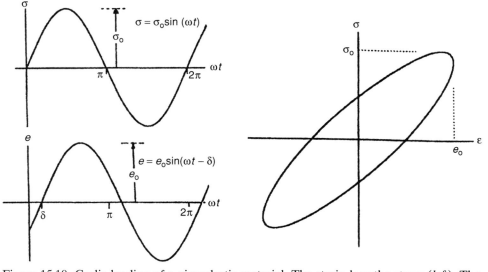

Figure 15.10. Cyclic loading of a viscoelastic material. The strain lags the stress (*left*). The plot of stress versus strain is an ellipse (*right*).

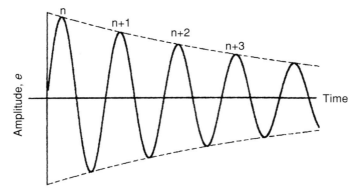

Figure 15.11. Decay of a natural vibration.

Λ can be related to δ by recalling that $U_n = \sigma_n e_n/2 = Ee_n^2/2$, so $e_n = [2u_n/E]^{1/2}$ and $e_{n+1} = [2U_{n+1}/E]^{1/2}$. Substituting, $\Lambda = \ln\{[2U_{n+1}/E]/[2U_n/E]\}^{1/2} = \ln(U_{n+1}/U_n)^{1/2} = \ln(1 + \Delta U/U)^{1/2}$, or

$$\Lambda = (1/2)\ln(1 + \Delta U/U). \tag{15.18}$$

Using the series expansion, $\Lambda = (1/2)[\Delta U/U - (\Delta U/U)^2/2 + (\Delta U/U)^3/3 - \ldots$ For small values of $\Delta U/U$, $\delta \approx (\Delta U/U)/2$, or

$$\Lambda \approx \pi\delta. \tag{15.19}$$

Sometimes the extent of damping is denoted by Q^{-1}, where $Q^{-1} = \tan\Lambda \approx \delta$.

> **EXAMPLE PROBLEM 15.2:** The amplitude of a vibrating member decreases so that the amplitude on the 100th cycle is 13% of the amplitude on the first cycle. Determine Λ.
>
> **Solution:** $\Lambda = \ln(e_n/e_{n+1})$. Because $(e_m/e_{m+1}) = (e_n/e_{n+1}) = (e_{n+1}/e_{n+2})$, etc., and $(e_n/e_{n+m}) = (e_n/e_{n+1})(e_{n+1}/e_{n+2})(e_{n+2}/e_{n+3})\ldots(e_{n+m-1}/e_{n+m}) = m(e_n/e_{n+m})$.
> Therefore, $(e_n/e_{n+1}) = (e_n/e_{n+m})^{1/m}$. $\Lambda = \ln(e_n/e_{n+1}) = \ln(e_n/e_{n+m})^{1/m} = (1/m)\ln(e_n/e_{n+m})$. Substituting m $= 100$ and $(e_n/e_{n+m}) = 1/0.13$, $\Lambda = 0.0204$.

Elastic Modulus – Relaxed versus Unrelaxed

If the frequency is high enough, there is no relaxation and therefore no damping, and the modulus, $E = \sigma/e$, is high. At low frequencies, there is complete relaxation with the result that there is no damping, but the modulus will be lower. At intermediate frequencies, there is partial relaxation, with high damping and a frequency-dependent modulus. The frequency dependence of the modulus is given by

$$E/E_r = \omega^2(\tau_e - \tau_\sigma)/(1 + \omega^2\tau_e^2). \tag{15.20}$$

The hysteresis curves for high, low, and intermediate frequencies are illustrated in Figure 15.12, and Figure 15.13 illustrates the corresponding interdependence of damping and modulus.

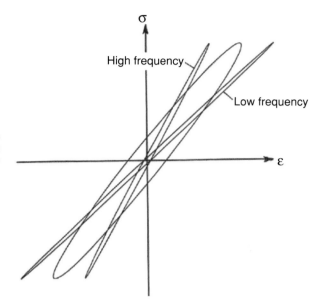

Figure 15.12. Hysteresis curves at a high, low, and intermediate frequency. The damping is low at both high and low frequencies.

Thermoelastic Effect

When a material is elastically deformed rapidly (adiabatically), it undergoes a temperature change,

$$\Delta T = -\sigma \alpha T / C_v, \tag{15.21}$$

where T is the absolute temperature, α is the linear coefficient of thermal expansion, and C_v is the volume heat capacity. For most materials, elastic stretching leads to a cooling because all terms are positive. Rubber is an exception because α is negative when it is under tension.

Consider a simple experiment illustrated by Figure 15.14. Let the material be elastically loaded in tension adiabatically. It undergoes an elastic strain, $e_a = \sigma/E_a$, where E_a is the Young's modulus under adiabatic conditions. At the same time, it will experience a cooling, $\Delta T = -\sigma \alpha T / C_v$. Now let it warm back up to ambient temperature while still under the stress, σ, changing its temperature by $\Delta T = +\sigma T \alpha / C_v$, so it will undergo a further thermal strain,

$$e_{\text{therm}} = \alpha \Delta T = \sigma T \alpha^2 / C_v. \tag{15.22}$$

Figure 15.13. Dependence of damping and elastic modulus on frequency. The peak damping occurs at the frequency that the modulus is rapidly changing. E/E_r was calculated from equation (15.20) with $\tau_e = 1.2\tau_\sigma$, and δ was calculated from $\tan \delta = \omega(\tau_e - \tau_\sigma)/(1 + \omega^2 \tau_e)$.

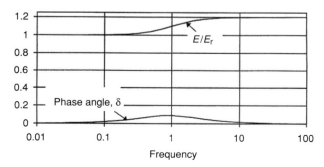

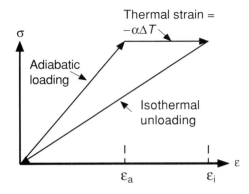

Figure 15.14. Schematic of loading adiabatically, holding until thermal equilibrium is reached and then unloading isothermally.

At room temperature the total strain will be

$$e_{tot} = \sigma(1/E_a + T\alpha^2/C_v). \tag{15.23}$$

This must be the same as the strain that would have resulted from stressing the material isothermally (so slowly that it remained at room temperature), so $e_{iso} = \sigma/E_i$, where E_i is the isothermal modulus. Equating $e_{iso} = e_{tot}$, $1/E_i = 1/E_a + T\alpha^2/C_p$, or

$$\Delta E/E = (E_a - E_i)/E_a = E_i T\alpha^2/C_p. \tag{15.24}$$

EXAMPLE PROBLEM 15.3: A piece of metal is subjected to a cyclic stress of 80 MPa. If the phase angle $\delta = 0.1°$ and no heat is transferred to the surroundings, what will be the temperature rise after 100,000 cycles? Data: $E = 80$ GPa, $C = 800$ J/kg·K, and $\rho = 4.2$ Mg/m³.

Solution: $U = (1/2)\sigma^2/E$. $\Delta U/U = 2\pi\sin\delta$. $\Delta U = 2\pi\sin\delta[(1/2)\sigma^2 E] = \sin(0.1°)$ $(80 \times 10^6 \text{ J/m}^3)^2/(80 \times 10^9 \text{ J/m}^3) = 219$ J/m³ per cycle. The heat released in 100,000 cycles would be 2.19×10^7 J/m³. This would cause a temperature increase of $Q/(\rho C) = 2.19 \times 10^7$ J/m³$/[(4,200 \text{ kg/m}^3)(800 \text{ J/kg·K}) = 6.5°$C.

The thermoelastic effect causes damping. The frequencies of the peaks depend on the time for thermal diffusion. The relaxation time, τ, is given by

$$\tau \approx x^2/D, \tag{15.25}$$

where x is the thermal diffusion distance and D is the thermal diffusivity. The relaxation time, τ, and therefore the frequencies of the damping peaks, depend strongly on the diffusion distance, x. There are thermal currents in a polycrystal between grains of different orientations because of their different elastic responses. In this case, the diffusion distance, x, is comparable to the grain size, d. For specimens in bending, there are also thermal currents from one side of the specimen to the other. These cause damping peaks with a diffusion distance, x, comparable to the specimen size.

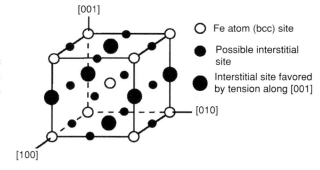

Figure 15.15. Interstitial sites in a bcc lattice. The sites indicated by the large dots are more favored when the cell is extended in the [001] direction.

Snoek Effect in bcc Metals

Interstitial atoms in bcc metals occupy positions that are at the centers of the edges and the centers of the face, as illustrated in Figure 15.15. When the lattice is under no external force, these sites are equivalent, each interstitial atom being midway between two iron atoms. However, with elastic extension along [001], one-third of the sites become more favorable. The Poisson contraction along [100] and [010] makes the other two-third of the sites less favorable. Given sufficient thermal energy and time, the interstitial atoms will jump to the favorable sites, causing a slight additional extension parallel to [001] and a slight elastic contraction perpendicular to it. The response is not immediate, so the strain lags the stress, causing a damping effect. The basic cause of this damping is the jumping interstitial atoms from one position to another, which is equivalent to diffusion. The damping is therefore frequency and temperature dependent, with the frequency, f^*, at maximum damping being given by an Arrhenius relation,

$$f^* = f_o \exp[(-Q/R)/T], \tag{15.26}$$

where the activation energy, Q, is the same as that for diffusion of the interstitial atom.

Figure 15.16 shows that as the frequency is increased, the damping peak occurs at higher temperatures.

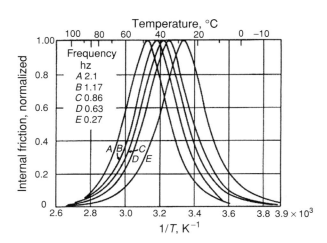

Figure 15.16. Damping of iron as a function of frequency and temperature. From C. Wert and Zener, *Phys. Rev.*, v. 76 (1949). Copyright 1949 by American Physical Society.

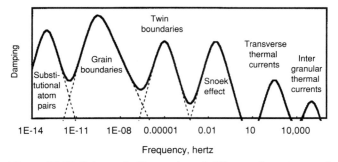

Figure 15.17. Schematic illustration of different damping mechanisms in metals as a function of frequency.

EXAMPLE PROBLEM 15.4: Assume that carbon interstitially dissolved in iron causes Snoek amping and that the frequency (cycles/s) at the peak amplitude is proportional to the diffusivity (m^2/s) of C in Fe. Use the data in Figure 15.16 to find the activation energy for diffusion of C in Fe.

Solution: Comparing the temperatures and frequencies of two damping peaks, $f_2/f_1 = D_2/D_1 = [D_o \exp(-Q/RT_2)]/[D_o \exp(-Q/RT_1)] = \exp[(+Q/R)(1/T_1 - 1/T_2)]$. $Q = R \ln(f_2/f_1)/(1/T_1 - 1/T_2)$. Now substituting data from extreme conditions, $f_2 = 2.1$ at $T_2 = 48°C = 321$ K and $f_1 = 0.27$ at $T_1 = 28°C = 301$ K, $Q = (8.314$ J/mole·K$)[\ln(2.1/0.27)]/(1/301 - 1/321) = 82.4$ kJ/mole.

Other Damping Mechanisms

Many other mechanisms contribute to damping in metals. These include bowing and unbowing of pinned dislocations, increasing and decreasing dislocation density in pile-ups, viscous grain-boundary sliding, opening and closing of microcracks, and possibly movement of twin boundaries and magnetic domain boundaries. Figure 15.17 shows the frequency range of different mechanisms in metals.

In thermoplastic polymers, the largest damping peak (α) is associated with the glass transition temperature, which is sensitive to the flexibility of the main backbone chain. Bulky side groups increase the stiffness of the backbone, which increases the glass transition temperature. With many polymers, there is more than one damping peak, as indicated in Figure 15.18.

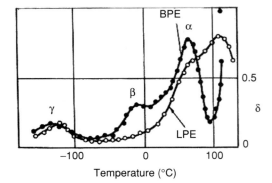

Figure 15.18. Damping peaks in linear polyethylene, LPE, and branched polyethylene, BPE. The α-peak is associated with motions of the backbone, and the β-peak with branches. Linear PE has no branches and hence no b-peak. From N. G. McCrum, C. P. Buckley, and C. B. Bucknell, *Principles of Polymer Engineering*, Oxford Science Pub. (1988).

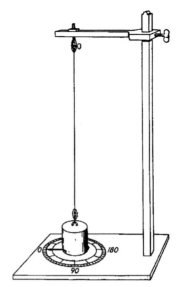

Figure 15.19. Coulomb's torsion pendulum for measuring damping. It is very similar in principle to modern torsion pendulums used for damping studies today. From S. Timoshenko, *History of the Strength of Materials*, McGraw-Hill (1953).

REFERENCES

C. M. Zener, *Elasticity and Anelasticity of Metals*, U. of Chicago Press, Chicago (1948).
N. G. McCrum, C. P. Buckley, and C. B. Bucknell, *Principles of Polymer Engineering*, Oxford Science Pub. (1988).

Notes

Rubber is unlike most other materials. On elastic stretching, it undergoes an adiabatic heating, $\Delta T = -\sigma \alpha T/C_v$ because under tension α is negative. If one stretches a rubber band and holds it long enough for its temperature to come to equilibrium with the surroundings and then releases it, the cooling on contraction can be sensed by holding it to the lips. The negative α of rubber under tension is the basis for an interesting demonstration. A wheel can be built with spokes of stretched rubber bands extending from hub to rim. When a heat lamp is placed so that it heats half the spokes, those spokes will shorten, unbalancing the wheel so that it rotates. The direction of rotation is just opposite of the direction that a wheel with metal spokes would rotate.

Bells should have a very low damping. Traditionally, they are made from bronze that has been heat treated to form a microstructure consisting of hard intermetallic compounds.

One of the useful characteristics of gray cast iron is the high damping capacity that results from the flake structure of graphite. Acoustic vibrations are absorbed by the graphite flakes. The widespread use of gray cast iron for bases of tool machines (e.g., beds of lathes, drill presses, milling machines) makes use of this damping capacity.

Coulomb (1736–1806) invented a machine for measuring damping with a torsion pendulum (Figure 15.19).

Problems

1. Consider a viscoelastic material whose behavior is adequately described by the combined series parallel model. Let it be subjected to the force versus time history shown in Figure 15.20. There is a period of tension followed by compression and tension again. After that, the stress is 0. Let the time interval, $\Delta t = K_v/K_{e2} = \tau_e$, and assume $K_{e1} = K_{e2}$. Sketch as carefully as possible the corresponding variation of strain with time.

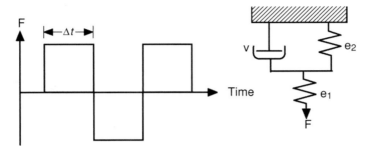

Figure 15.20. Loading of a viscoelastic material.

2. Consider a viscoelastic material whose behavior is adequately described by the combined series parallel model. Let it be subjected to the strain versus time history shown in Figure 15.21. There is a period of tension followed by compression and tension again. After that, the stress is returns to zero. Let the time interval, $\Delta t = K_v/K_{e2} = \tau_e$, and assume $K_{e1} = K_{e2}$. Sketch in as carefully as possible the corresponding variation of force with time.

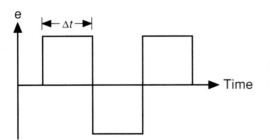

Figure 15.21. Strain cycles imposed on a viscoelastic material.

3. An elastomer was suddenly stretched in tension and the elongation was held constant. After 10 minutes, the tensile stress in the polymer dropped by 12%. After an extremely long time, the stress dropped to 48% of its original value.

    A. Find the relaxation time, τ, for stress relaxation.

    B. How long will it take the stress to drop to 75% of its initial value?

4. A certain bronze bell is tuned to middle C (256 Hz). It is noted that the intensity of the sound drops by one decibel (i.e., 20.56%) every 5 seconds. What is the phase angle, δ, in degrees?

5. A piece of aluminum is subjected to a cyclic stress of ±120 MPa. After 5,000 cycles, it is noted that the temperature of the aluminum has risen by 1.8°C. Calculate

Λ and the phase angle, δ, assuming that there has been no transfer of heat to the surroundings and the energy loss/cycle is converted to heat.

Misc. data for aluminum:

Crystal structure	fcc	Lattice parameter	0.4050 nm
Density	2.70 Mg/m^3	Young's modulus	62 GPa
Heat capacity	900 J/kg·K	Melting point	660°C

(*Hint:* ΔU can be found from the temperature rise, and U can be found from the applied stress and Young's modulus.)

6. Measurements of the amplitude of vibration of a freely vibrating beam are

Cycle	Amplitude
0	250
25	206
50	170
100	115
200	53

A. Calculate the log decrement, Λ.
B. Is Λ dependent on the amplitude for this material in the amplitude range studied? (Justify your answer.)
C. What is the phase angle δ?

(*Hint:* Assume Λ is constant so $e_n/e_{n+1} = e_{n+1}/e_{n+2} =$ etc. Then $e_n/e_{n+m} = [e_n/e_{n+1}]^m$).

7. A. A high-strength steel can be loaded up to 100,000 psi in tension before any plastic deformation occurs. What is the largest amount of thermoelastic cooling that can be observed in this steel at 20°C?
B. Find the ratio of adiabatic Young's modulus to isothermal Young's modulus for this steel at 20°C.
C. A piece of this steel is adiabatically strained elastically to 10^{-3} and then allowed to reach thermal equilibrium with the surroundings (20°C) at constant stress. It is then unloaded adiabatically and again allowed to reach thermal equilibrium with its surroundings. What fraction of the initial mechanical energy is lost in this cycle? (*Hint:* Sketch the $\sigma - \varepsilon$ path.) For iron, $\alpha = 11.76 \times 10^{-6}/°$C, $E = 29 \times 10^6$ psi, $C_v = 0.46$ J/g°C, and $\rho = 7.1$ g/cm^3.

8. For iron, the adiabatic Young's modulus is $(1 + 2.3 \times 10^{-3})$ times the isothermal modulus at room temperature. If the anelastic behavior of iron is modeled by a series parallel model, what is the ratio of K_1 to K_2?

9. Damping experiments on iron were made using a torsion pendulum with a natural frequency of 0.65 cycles/s. The experiments were run at various temperatures, and the maximum log decrement was found at 35°C. The activation energy for diffusion of carbon in α-iron is 78.5 kJ/mole. At what temperature would you expect the damping peak to occur if the pendulum were redesigned so that it had a natural frequency of 10 Hertz?

10. A polymer is subjected to a cyclic stress of 5 MPa at a frequency of 1 Hz for 1 minute. The phase angle is 0.05 degrees. Calculate the temperature rise, assuming no loss to the surroundings. $E = 2\,\text{GPa}$, $C = 1.0\,\text{J/kg} \cdot \text{K}$ and $\rho = 1.0\,\text{Mg/m}^3$.

16 Creep and Stress Rupture

Introduction

Creep is time-dependent plastic deformation that is usually significant only at high temperatures. Figure 16.1 illustrates typical creep behavior. As soon as the load is applied, there is an instantaneous elastic response, followed by period of transient creep (*stage I*). Initially, the rate is high, but it gradually decreases to a steady state (*stage II*). Finally, the strain rate may increase again (*stage III*), accelerating until failure occurs.

Creep rates increase with higher stresses and temperatures. With lower stresses and temperatures, creep rates decrease, but failure usually occurs at lower overall strains (Figure 16.2).

The acceleration of the creep rate in stage III occurs because the true stress increases during the test. Most creep tests are conducted under constant load (constant engineering stress). As creep proceeds, the cross-sectional area decreases so the true stress increases. Porosity develops in the later stages of creep, further decreasing the load-bearing cross-section.

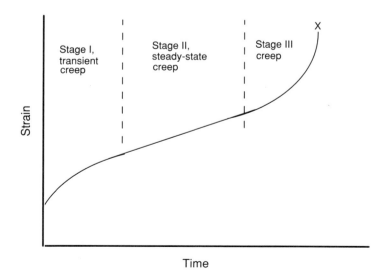

Figure 16.1. Typical creep curve showing three stages of creep.

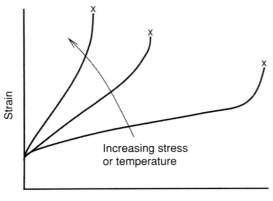

Figure 16.2. Decreasing temperature and stress lead to slower creep rates, but failure often occurs at a lower strains.

Creep Mechanisms

Viscous flow: Several mechanisms may contribute to creep. These include viscous flow, diffusional flow, and dislocation movement. Viscous flow is the dominant mechanism in amorphous materials. With Newtonian viscosity, the rate of strain, $\dot{\gamma}$, is proportional to the stress, τ,

$$\dot{\gamma} = \tau/\eta, \tag{16.1}$$

where η is the viscosity. For tensile deformation, this may be expressed as

$$\dot{\varepsilon} = \sigma/\eta', \tag{16.2.}$$

where $\eta' = 3\eta$.

In polycrystalline materials, grain boundary sliding is viscous in nature. The sliding velocity on the boundary is proportional to the stress and inversely proportional to the viscosity, η. The rate of extension depends on the amount of grain boundary area per volume and is therefore inversely proportional to the grain size, d, $\dot{\varepsilon} = C(\sigma/\eta)/d$. Viscous flow is thermally activated, so $\eta = \eta_o \exp(Q_v/RT]$. The strain rate attributable to grain boundary sliding can be written as

$$\dot{\varepsilon}_V = A_V(\sigma/d) \exp(-Q_V/RT). \tag{16.3}$$

If grain boundary sliding were the only active mechanism, there would be an accumulation of material at one end of each boundary on which sliding occurs and a deficit at the other end, as sketched in Figure 16.3. This incompatibility must be relieved by another deformation mechanism, one involving dislocation motion, diffusion, or grain boundary migration.

Diffusion-controlled creep: A tensile stress increases the separation of atoms on grain boundaries that are normal to the stress axis, and the Poisson contraction decreases the separation of atoms on grain boundaries that are parallel to the stress axis. The result is a driving force for diffusional transport of atoms from grain boundaries parallel to the tensile stress to boundaries normal to the tensile stress. Such diffusion produces a plastic elongation, as shown in Figure 16.4. The specimen elongates as atoms are added to grain boundaries perpendicular to the stress.

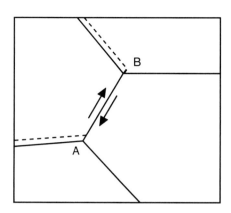

Figure 16.3. Grain boundary sliding causes incompatibilities at both ends of the planes, A and B, on which sliding occurs. This must be relieved by another mechanism for sliding to continue.

If the creep occurs by diffusion through the lattice, it is called *Nabarro-Herring creep*. The diffusional flux, J, between the boundaries parallel and perpendicular to the stress axis is proportional to the stress, σ, and the lattice diffusivity, D_L, and it is inversely proportional to the diffusion distance, $d/2$, between the diffusion source and sink. Therefore, $J = CD_L\sigma/(d/2)$, where C is a constant. The velocity, v, at which the diffusion source and sink move apart is proportional to the diffusional flux, so $v = CD_L\sigma/(d/2)$. Because the strain rate equals $v/(d/2)$,

$$\dot{\varepsilon}_{\text{N-H}} = A_L(\sigma/d^2)D_L \tag{16.4}$$

where A_L is a constant.

On the other hand, if creep occurs by diffusion along the grain boundaries, it is called *Coble creep*. The driving force for Coble creep is the same as for Nabarro-Herring creep. The total number of grain boundary diffusion paths is inversely proportional to the grain size, so now J is proportional to $d^{-1/3}$, and the creep rate is given by

$$\dot{\varepsilon}_{\text{C}} = A_G(\sigma/d^3)D_{\text{gb}}, \tag{16.5}$$

where D_{gb} is the diffusivity along grain boundaries and A_G is a constant.

The ratio of lattice diffusion to grain boundary diffusion increases with temperature because the activation energy for grain boundary diffusion is always lower

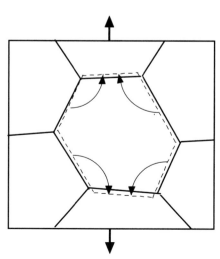

Figure 16.4. Creep by diffusion between grain boundaries. As atoms diffuse from lateral boundaries to boundaries normal to the tensile stress, the grain elongates and contracts laterally.

than that for lattice diffusion. Therefore, Coble creep is more important at low temperatures and Nabarro-Herring at high temperatures.

Dislocation motion: Slip is another mechanism of creep. The creep rate, in this case, is controlled by how rapidly the dislocations can overcome obstacles that obstruct their motion. At high temperatures, the predominant mechanism for overcoming obstacles is dislocation climb (Figure 16.5). With climb, the creep rate is not dependent on grain size, but the rate of climb does depend very strongly on the stress,

$$\dot{\varepsilon} = A_S \sigma^m. \tag{16.6}$$

The value of m is approximately 5 for climb-controlled creep.* Because climb depends on diffusion, the constant A_S has the same temperature dependence as lattice diffusion. At lower temperatures, creep is not entirely climb controlled, and higher exponents are observed. Equations (16.3) through (16.6) predict creep rates that depend only on stress and temperature and not on strain. Thus, they apply only to stage II or steady-state creep.

Multiple mechanisms: More than one creep mechanism may be operating. There are two possibilities: either the mechanisms operate independently, or they act cooperatively. If they operate independently, the overall creep rate, $\dot{\varepsilon}$, is the sum of the rates due to each mechanism,

$$\dot{\varepsilon} = \dot{\varepsilon}_A + \dot{\varepsilon}_B + \cdots \tag{16.7}$$

The result is like two mechanisms acting in series. The overall strain rate depends primarily on the most rapid mechanism.

On the other hand, two mechanisms may be required to operate simultaneously, as in the case of grain boundary sliding requiring another mechanism. Where two parallel mechanisms are required, both must operate at the same rate

$$\dot{\varepsilon} = \dot{\varepsilon}_A = \dot{\varepsilon}_B. \tag{16.8}$$

The overall rate is determined by the potentially slower mechanism. These two possibilities (equations (16.7) and (16.8)) are illustrated in Figure 16.6.

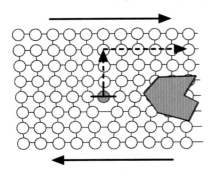

Figure 16.5. Climb-controlled creep. An edge dislocation can climb by diffusion of atoms away from the dislocation (vacancies to the dislocation), thereby avoiding an obstacle.

---

* This notation widely used in the creep literature is the inverse of that used in Chapter 6, where we wrote $\sigma \propto \dot{\varepsilon}^m$, or equivalently, $\dot{\varepsilon} \propto s^{1/m}$. The m used in the creep literature is the reciprocal of the m used previously. The value of m = 5 here corresponds to a value of m = 0.2 in the notation of Chapter 6.

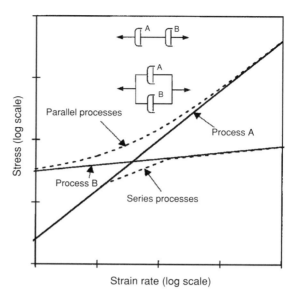

Figure 16.6. Creep by two mechanisms, A and B. If the mechanisms operate independently (series), the overall creep rate is largely determined by the faster mechanism. If creep depends on the operation of both mechanisms (parallel), the potentially slower mechanism will control the overall creep rate.

Temperature Dependence of Creep

Creep can be regarded as a rate process that depends on thermal activation. The simple approach taken by Sherby and Dorn was to assume that for any mechanism

$$\dot{\varepsilon} = f(\sigma)\exp(-Q/RT), \tag{16.9}$$

so that the stress dependence is incorporated into the preexponential term. This is equivalent to the Zener-Hollomon approach (equation (6.20)). The value of Q depends on whether creep is controlled by lattice diffusion or by grain boundary diffusion.

Sometimes the temperature dependence of creep rates are expressed in terms of preexponential term that is inversely proportional to the temperature, T, for example,

$$\dot{\varepsilon} = (A/T)\exp(-Q/RT)[f(\sigma, d)]. \tag{16.10}$$

However, the T in the preexponential term does not have a great affect on the temperature dependence. This is illustrated in example problem 16.1.

EXAMPLE PROBLEM 16.1: During creep under constant stress, the strain rate increased by a factor of 1.43 when the temperature was increased for 800 K to 805 K. Calculate the activation energy assuming (1) $\dot{\varepsilon} = (A/T)\exp(-Q/RT)$. (2) Repeat assuming $\dot{\varepsilon} = B\exp(-Q/RT)$.

Solution:

(1) $Q = R \ln[(T_2\dot{\varepsilon}_2)/(T_1\dot{\varepsilon}_1)]/(1/T_1 - 1/T_2) = 8.314 \ln[805 \times 1.43/800]/(1/800 - 1/805) = 390\,\text{kJ/mole}.$

(2) $Q = R \ln(\dot{\varepsilon}_2/\dot{\varepsilon}_1)/(1/T_1 - 1/T_2) = 8.314 \ln(1.43)/(1/800 - 1/805) = 383\,\text{kJ/mole}.$

Note that the values of Q differ by less than 2%.

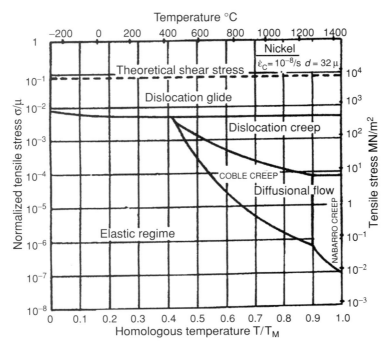

Figure 16.7. Deformation mechanism map for pure nickel with a grain size of $d = 32$ μm. The strain rate is a function of stress and temperature. Different mechanisms are dominant in d regimes. Coble creep is controlled by grain boundary diffusion and Nabarro creep by lattice diffusion. From M. F. Ashby, *Acta Met.*, v. 20 (1972).

Deformation Mechanism Maps

The controlling mechanisms of deformation change with temperature and stress. Figure 16.7 is a typical deformation mechanism map. At high stresses, slip by dislocation motion predominates. At lower stresses, different diffusion-controlled creep mechanisms are important. The nature of the dominant mechanism shifts with grain size. With decreased grain size, grain boundary diffusion (Coble creep) becomes important.

> **EXAMPLE PROBLEM 16.2:** For creep in Ni of grain size, $d = 200$ μm, determine the boundary between Coble and Nabarro-Herring creep on a deformation mechanism map using the data in Figure 16.7. The activation energy for lattice diffusion and grain boundary diffusion in Ni is $Q_L = 286$ kJ/mole and $Q_{gb} = 115$ kJ/mole.
>
> **Solution:** The boundary is where the two mechanisms contribute equally to the strain rate. Writing equations (16.4) and (17.5) as $\dot{\varepsilon}_{N\text{-}H} = A_L(\sigma/d^2)\exp(-Q_L/RT)$ and $\dot{\varepsilon}_C = A_{gb}(\sigma/d^3)\exp(-Q_{gb}/RT)$, and equating
>
> $$A_L(\sigma/d^2)\exp(-Q_L/RT) = A_{gb}(\sigma/d^3)\exp(-Q_{gb}/RT).$$
>
> So $A_L/A_{gb} = d\exp[(-Q_L + Q_{gb})/(RT)] = d\exp(-171{,}000/RT)$
> Substituting from Figure 16.7, $d = 32$ μ, and $T = 1{,}280 + 273 = 1{,}553$ K
>
> $$A_L/A_{gb} = (32\,\mu\text{m})\exp[-171{,}000/(8.314 \times 1553) = 5.667 \times 10^{-5}\,\mu\text{m}.$$

Now solving for T, $T = (-171,000)/\{R\ln[(5.667 \times 10^{-5}/200]\} = 1,364\,\text{K} = 1,090°\text{C}$.

The boundary for 200 μm, is at $T = 1,364\,\text{K} = 1,090°\text{C}$. Note that the boundary is vertical because both the Nabarro-Herring and Coble creep rates are proportional to stress.

Cavitation

Cavitation can lead to fracture during creep. Cavitation occurs by nucleation and growth of voids, particularly at grain boundaries and second-phase particles. A void will grow if its growth lowers the energy of the system. Consider the growth of a spherical void in a cubic element with dimensions x as illustrated in Figure 16.8. Let the radius be r and the stress, σ. The surface energy of the void is $E_S = 4\pi\gamma r^2$. If the radius increases by dr, the increase of the surface energy is

$$dU_S = 8\pi\gamma r\, dr. \tag{16.11}$$

The volume of the void is $(4/3)\pi r^3$. Growth by dr will change the volume of the sphere by $4\pi r^2\, dr$. If the atoms diffuse to positions that cause lengthening, the element will lengthen by dx so that its volume increases, $x^2\, dx = 4\pi r^2\, dr$. The strain, $d\varepsilon$, associated with the growth will be $d\varepsilon = dx/x = 4\pi r^2\, dr/x^3$. The energy per volume expended by the applied stress, σ, is $dU_S = \sigma d\varepsilon = 4\sigma\pi r^2\, dr/x^3$. The total energy associated with the element is

$$dU_\sigma = x^3 4\sigma\pi r^2 dr/x^3 = 4\sigma\pi r^2\, dr. \tag{16.12}$$

Equating (16.11) and (16.12), the critical condition for void growth is

$$r^* = 2\gamma/\sigma. \tag{16.13}$$

Voids with radii less than r^* should shrink. Ones larger than r^* should grow.

Failure may also occur by grain boundary fracture. Unless grain boundary sliding is accompanied by another mechanism, grain boundary cracks must form as illustrated in Figure 16.9. Figure 16.10 is a photomicrograph of a grain boundary fracture.

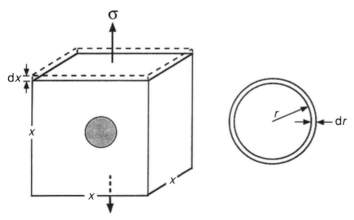

Figure 16.8. Spherical void in a cubic cell. Growth of the void by dr causes an elongation of the cell by dx.

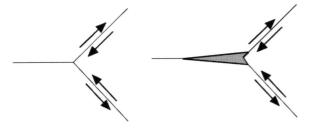

Figure 16.9. Schematic views of grain boundary cracks caused by grain boundary sliding.

Rupture versus Creep

The term *rupture* is used in the creep literature to mean fracture. Up to this point, the discussion has centered on creep deformation. Creep may cause component parts to fail in service by excessive deformation, so they no longer can function satisfactorily. However, after long times under load at high temperature, it is more common that parts fail by rupture than by excessive deformation. As the service temperatures and stress level are lowered to achieve lower rates of creep, rupture usually occurs at lower strains. Figure 16.11 shows this for a Ni-Cr-Co-Fe alloy. Successful design for high temperature service must ensure against both creep excessive deformation and against fracture (creep rupture).

Extrapolation Schemes

For many applications, such as power-generating turbines, boilers, engines for commercial jets, and furnace elements, components must be designed for long service at high temperatures. Some parts are designed to last 20 years (175,000 hr) or more. In such cases, it is not feasible to test alloys under creep conditions for times as long as the design life. Often tests are limited to 1,000 hr (42 days). To ensure that neither rupture nor excessive creep strain occurs, results from shorter time tests at higher

Figure 16.10. Grain boundary cracking in UPH Ni-16Cr-9Fe after 35% elongation at 350°C. From R. W. Herzberg, *Deformation and Fracture Mechanics of Engineering Materials*, 4th ed., Wiley (1996).

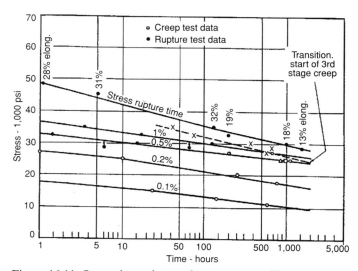

Figure 16.11. Stress dependence of stress rupture life and time to reach several creep strains for a Ni-Cr-Co-Fe alloy tested at 650°C. Note that as the stress level is reduced to increase the time to a given strain, rupture occurs at lower strains. From N. J. Grant in *High Temperature Properties of Metals*, ASM (1950).

temperatures and/or higher stresses must be extrapolated to the service conditions. Several schemes have been proposed for such extrapolation.

Sherby-Dorn: Sherby and Dorn proposed plotting creep strain as a function of a *temperature-compensated time, θ*,

$$\theta = t \, \exp(-Q/RT). \tag{16.14}$$

In creep studies, $\log_{10}\theta$ is often referred to as the Sherby-Dorn parameter, P_{SD}. Note that θ is the same as the Zener-Hollomon parameter discussed in Chapter 6. The basic idea is that the creep strain is plotted against the value of θ for a given stress, as shown in Figure 16.12. For a given material and stress, the creep strain depends only on θ. Alternatively, the stress to rupture or to achieve a given amount of creep strain is plotted against the value of P_{SD}, as shown in Figure 16.13. From such plots, one can predict long-time behavior.

> **EXAMPLE 16.3:** The chief engineer needs to know the permissible stress on the alloy in Figure 16.13 to achieve a rupture life of 20 years at 500°C.
>
> **Solution:** For these conditions $\theta - (20 \times 365 \times 24 \, \text{hr}) \exp -\{85,000/[1.987 (500 + 273)]\} = 1.62 \times 10^{-19}$, so $\log_{10}(1.62 \times 10^{-19}) = -18.8$. Figure 16.13 indicates that the permissible stress would be about 400 MPa.

Larson-Miller parameter: A different extrapolation parameter was proposed by Larson and Miller,[*]

$$P_{LM} = T(\log_{10} t_r + C). \tag{16.15}$$

---

[*] F. R. Larson and J. Miller, *Trans ASME*, v. 74 (1952).

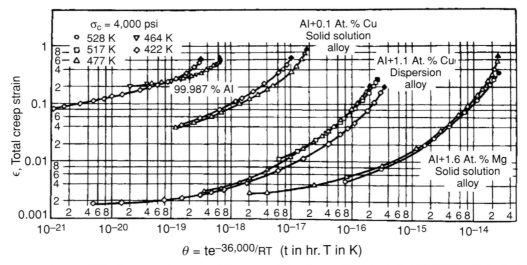

$$\theta = te^{-36,000/RT} \quad \text{(t in hr. T in K)}$$

Figure 16.12. Creep strain versus θ for several aluminum alloys tested at 27.6 MPa. Note that the test data for several temperatures fall nearly on the same line. From R. L. Orr, O. D. Sherby, and J. E. Dorn, *Trans ASM*, v. 96 (1954).

As they originally formulated this parameter, the temperature, T, was expressed as $T + 460$, where T is in degrees Fahrenheit and the rupture time, t_r, in hours. They found agreement for most high-temperature alloys with values of C of about 20. However the parameter can be expressed with temperature in Kelvin, keeping t_r in hours. Figure 16.14 is a Larson–Miller plot of stress for rupture versus P_{LM}. Again, one can use such plots for prediction of long-term behavior from shorter time tests at higher temperatures.

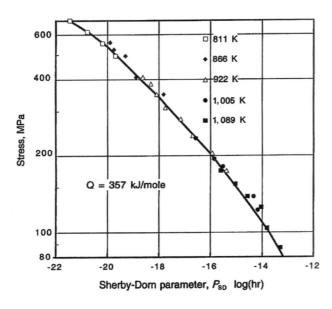

Figure 16.13. Stress to cause creep rupture of S-590 alloy as a function of $P_{SD} = \log_{10}\theta$. Q was taken as 357 J/mole. Note that the test data for several temperatures fall on the same line. Data taken from R. M. Goldoff, *Materials in Engineering Design*, v. 49 (1961).

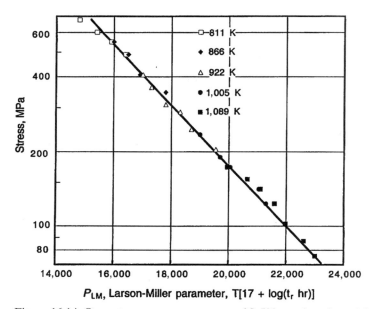

Figure 16.14. Stress to cause creep rupture of S-590 as a function of the Larson-Miller parameter P. This figure was constructed with the same data as used in Figure 16.13, with $C = 17$ and temperature in Kelvin. Note that the test data for several temperatures fall on the same line.

EXAMPLE 16.4: The chief engineer needs to know the permissible stress that can be applied to the S-590 alloy for a life of 20 years at 500°C.

Solution: The Larson-Miller parameter with $C = 17$ is $P_{LM} = 773[17 + \log_{10}(20 \times 365 \times 24\,\text{hr})] = 17{,}190$. From Figure 16.14, $P_{LM} = 17{,}190$ corresponds to $\sigma \approx 400$ MPa.

Alloys for High-Temperature Use

Alloys used for high-temperature applications fall into several classes. Because of their extremely high melting points, tungsten and molybdenum have a very high resistance to creep. Unfortunately, they oxidize rapidly at elevated temperatures. Their oxides are volatile at high temperatures and thus offer no protection. Therefore, tungsten and molybdenum can be used only where they are protected from exposure to air such as in incandescent lamps and vacuum systems. *Superalloys* are Ni- or Co-base alloys that have both good oxidation and good creep resistance at elevated temperatures. They contain substantial amounts of Cr for oxidation resistance. The carbon contents of all are low. The nickel-base superalloys are fcc solid solutions (γ) with enough Ti and Al to form a fcc γ' precipitate, Ni3(Ti,Al). It is the precipitation of γ' that is responsible is for the good high temperature strength. Coherency strains develop between γ' and the γ matrix which is a Ni-rich fcc solid solution. A typical microstructure of a nickel-base superalloy is shown in Figure 16.15.

Figure 16.15. Microstructure of a Ni-base $\gamma-\gamma'$ alloy. The precipitates are $Ni_3(Al,Ti)$ γ'. From *ASM Metals Handbook*, v. 7, 8th ed. (1972).

Marble's reagent   5,000×

In addition to composition, microstructure is importance in determining high temperature properties. Large grain size material is superior to fine-grain material. Turbine blades made by directional solidification are superior to conventionally cast blades because the grain boundaries are aligned parallel to the major stress axis so grain boundary sliding cannot occur. Still better are single-crystal blades that have no grain boundaries at all. Figure 16.16 shows these.

For less severe creep conditions, stainless steels and titanium-base alloys may be used. Aluminum or magnesium alloys may find service at still lower temperatures. Figure 16.17 gives a rough indication of the useful temperature range for various alloys.

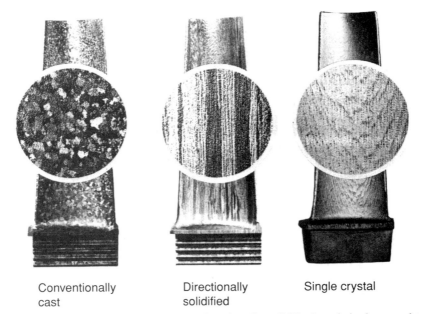

Conventionally   Directionally   Single crystal
cast   solidified

Figure 16.16. Conventionally cast, directionally solidified, and single-crystal turbine blades. From *Monocrystalloys A New Concept in Gas Turbine Materials*, PWA 1409, Pratt and Whitney.

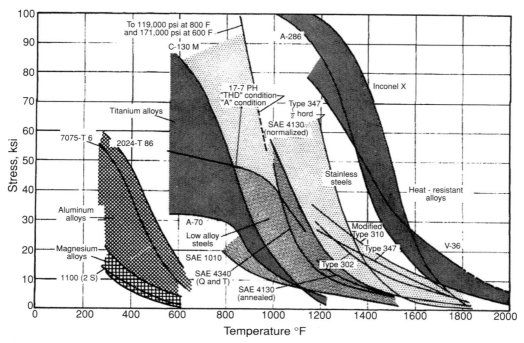

Figure 16.17. Strength of various alloys at high temperatures. From J. A. van Echo, *Short-Time High Temperature Testing*, ASM International (1958).

REFERENCES

N. E. Dowling, *Mechanical Behavior of Materials*, 2nd ed., Prentice-Hall (1999).
M. A. Meyers and K. K. Chawla, *Mechanical Behavior of Materials*, Prentice-Hall (1999).
T. H. Courtney, *Mechanical Behavior of Materials*, 2nd ed., McGraw-Hill (2000).
R. W. Hertzberg, *Deformation and Fracture Mechanics of Engineering Materials*, 4th ed., John Wiley (1995).

Notes

Larson and Miller,[*] in the paper that proposed their extrapolation scheme, rationalized the value of about 20 for the constant, C, in their parameter. They noted that if T were infinite, P_{LM} would be infinite unless $\log_{10} t_r + C = 0$, so $C = -\log_{10} t_r$. Furthermore, they assumed that at an infinite temperature, the pieces of a breaking specimen should fly apart with the speed of light, c. Taking the fracture to occur when the pieces are separated by an atomic diameter, d, the time, t, for fracture is d/c. Assuming $d = 0.25$ nm as a typical value, $t = (0.25 \times 10^{-9}\,\text{m}/3 \times 10^8\,\text{m/s})(3,600\,\text{s/hr}) = 2.3 \times 10^{-22}$ hr. $C = -\log_{10} t_r = 21.6$ or approximately 20.

In 1910, E. M. da C. Andrade[**] proposed a scheme for maintaining constant true stress during a creep experiment by using dead weight loading with a shaped weight partially suspended in water (Figure 16.18). As the specimen elongates, more of the weight is immersed, decreasing the force on the specimen.

[*] F. A. Larson and J. Miller, *Trans AIME*, v. 74 (1952), p. 765.
[**] E. M. daC. Andrade, *Proc. Roy. Soc. A London*, v. 84 (1910), pp. 1–12.

Figure 16.18. Andrade's constant true stress creep apparatus. As the specimen elongates, the buoyancy of the water lowers the force on the specimen.

Problems

1. Stress versus rupture life data for a super alloy are listed here. The stresses are given in MPa and rupture life is given in hours.

    A. Make a plot of stress (log scale) versus P_{LM}, where $P_{LM} = (T)(C + \log_{10} t)$. T is the temperature in Kelvin, t is in hours, and $C = 20$.

    B. Predict from the plot what stress would cause rupture in 100,000 hr at 450°C.

Stress	Rupture time (hr)			
MPa	500°C	600°C	700°C	800°C
600	2.8	0.018	0.0005	–
500	250.0	0.720	0.0040	–
400	–	12.100	0.0820	0.00205
300	–	180.000	0.8700	0.01100
200	–	2,412.000	11.0000	0.19800
100	–	–98.000	1.1000	

2. A. Using the data in problem 1, plot the Sherby-Dorn parameter, $P_{SD} = \log \theta$, where $\theta = t \exp(-Q/RT)$ and $Q = 340 \, \text{kJ/mole}$.

    B. Using this plot, predict from the plot what stress would cause rupture in 100,000 hr at 450°C.

3. For many materials, the constant C in the Larson-Miller parameter, $P_{LM} = (T + 460)(C + \log_{10} t)$ (where T is in Fahrenheit and t in hours) is equal to 20. However, the Larson-Miller parameter can also be expressed as $P' = T(C' + \ln t)$ with t in seconds and T in Kelvin, using the natural logarithm of time. In these cases, what is the value of C'?

4. Stress rupture data are sometimes correlated with the Dorn parameter, $\theta = i \exp[-Q/(RT)]$, where t is the rupture time, T is absolute temperature, and θ is assumed to depend only on stress. If this parameter correctly describes a set of data, then a plot of $\log(t)$ versus $1/T$ for data at a single level of stress would be a straight line. If the Larson-Miller parameter correctly correlates data, a plot of data at constant stress (therefore, constant P_{LM}) of $\log(t)$ versus $1/T$ would also be a straight line.

  A. If both parameters predict straight lines on $\log(t)$ versus $1/T$ plots, are they really the same thing?

  B. If not, how do they differ? How could you tell from a plot of $\log(t)$ versus $(1/T)$ which parameter better correlates a set of stress rupture data?

5. Sketch how the boundaries in Figure 16.7 for the creep mechanisms in nickel would change if the grain size were 1 mm. instead of 32 μm.

6. Data for the steady-state creep of a carbon steel are plotted in Figure 16.19.

  A. Using the linear portions of the plot, determine the exponent m in $\dot{\varepsilon}_{SC} = B\sigma^m$ at 538°C and 649°C.

  B. Determine the activation energy, Q, in the equation $\dot{\varepsilon} = f(\sigma) \exp[-Q/(RT)]$.

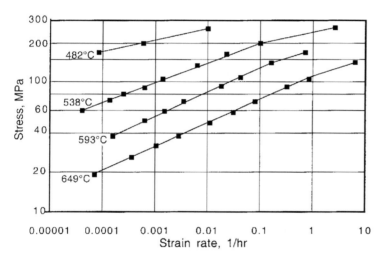

Figure 16.19. Creep data for a carbon steel. Data from P. N. Randall, *Proc. ASTM*, v. 57 (1957).

7. The following data were obtained in a series of stress rupture tests on a material being considered for high temperature service.

At 650°C	Stress (ksi)	80	65	60	40
	Rupture life (hr)	0.08	8.5	28	483
At 730°C	Stress (ksi)	60	50	30	25
	Rupture life (hr)	0.20	1.8	127	1,023
At 815°C	Stress (ksi)	50	30	20	
	Rupture life (hr)	0.30	3.1	332	
At 925°C	Stress (ksi)	30	20	15	10
	Rupture life (hr)	0.08	1.3	71	123
At 1,040°C	Stress (ksi)	20	10	5	
	Rupture life (hr)	0.03	1.0	28	211

  A. Make a Larson-Miller plot of the data.

  B. Predict the life for an applied stress of 30 ksi at 600°C.

8. Figure 16.17 shows how service temperature affects the usable stress levels for various metals.

9.   A.   Tungsten has a melting point of 3,400°C. Why is it not considered for use in jet engines?

  B.   What advantages do aluminum alloys have over more refractory materials at operating temperatures of 400°F?

10. Consider the creep rate versus stress curves for an aluminum alloy plotted in Figure 16.20. Calculate the stress exponent, m, at 755 K.

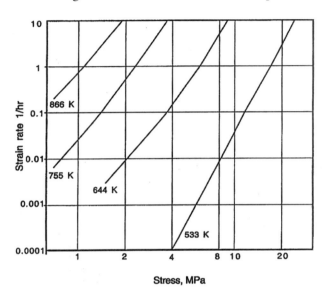

Figure 16.20. Steady-state creep rate of an aluminum alloy at several temperatures. Data from O. D. Sherby and P. M. Burke, in *Prog. Mater. Sci.*, v. 13 (1968).

17 Fatigue

Introduction

It has been estimated that 90% of all service failures of metal parts are caused by *fatigue*. A fatigue failure is one that occurs under a cyclic or alternating stress of an amplitude that would not cause failure if applied only once. Aircraft are particularly sensitive to fatigue. Automobile parts such as axles, transmission parts, and suspension systems may fail by fatigue. Turbine blades, bridges, and ships are other examples. Fatigue requires cyclic loading, tensile stresses, and plastic strain on each cycle. If any of these are missing, there will be no failure. The fact that a material fails after a number of cycles indicates that some permanent change must occur on every cycle. Each cycle must produce some plastic deformation, even though it may be very small. Metals and polymers fail by fatigue. Fatigue failures of ceramics are rare because there seldom is plastic deformation.

There are three stages of fatigue. The first is nucleation of a crack by small amounts of inhomogeneous plastic deformation at a microscopic level. The second is the slow growth of these cracks by cyclic stressing. Finally, sudden fracture occurs when the cracks reach a critical size.

Surface Observations

Often visual examination of a fatigue fracture surface will reveal *clamshell* or *beach markings*, as shown in Figure 17.1. These marks indicate the position of the crack front at some stage during the fatigue life. The initiation site of the crack can easily be located by examining these marks. The distance between these markings *does not* represent the distance that the crack propagated in one cycle. Rather, each mark corresponds to some change during the cyclic loading history, perhaps a period of time that allowed corrosion or perhaps a change in the stress amplitude. When the crack has progressed far enough, the remaining portion of the cross-section may fail in one last cycle by either brittle or ductile fracture. The final fracture surface may represent the major part or only a very small portion of the total fracture surface, depending on the material, its toughness, and the loading conditions.

Microscopic examination of a fracture surface often reveals markings on a much finer scale. These are called *striations*, and they do represent the position of the crack

Figure 17.1. Typical clamshell markings on a fatigue fracture surface of a shaft. The fracture started at the left side of the bar and progressed to the right, where final failure occurred in a single cycle. Courtesy of W. H. Durrant.

front at each cycle (Figure 17.2). The distance between striations is the distance advanced by the crack during one cycle. Sometimes striations cannot be observed because they are damaged when the crack closes.

A careful microscopic examination of the exterior surface of the specimen after cyclic stressing will usually reveal a roughening even before any cracks have formed. Under high magnification, intrusions and extrusions are often apparent, as shown in Figure 17.3. These intrusions and extrusions are the result of slip on set of planes during the compression half-cycle and slip on a different set of planes during the tensile half-cycle (Figure 17.4). *Persistent slip bands* beneath the surface are associated with these intrusions and extrusions. Fatigue cracks initiate at the intrusions and grow inward along the persistent slip bands.

Nomenclature

Most fatigue experiments involve alternate tensile and compressive stresses, often applied by cyclic bending. In this case, the mean stress is zero. In service, however,

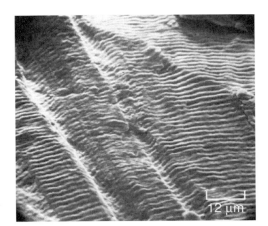

Figure 17.2. SEM picture of fatigue striations on a fracture surface of type 304 stainless steel. From *Metals Handbook*, v. 9, 8th ed., ASM (1974).

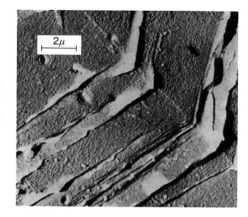

Figure 17.3. Intrusions and extrusions at the surface formed by cyclic deformation. These correspond to persistent slip bands beneath the surface. From A. Cottrell and D. Hull, *Proc. Roy. Soc. (London)*, v. A242 (1957).

materials may be subjected to cyclic stresses that are superimposed on a steady-state stress. Figure 17.5 shows this together with various terms used to define the stresses. The mean stress, σ_m, is defined as

$$\sigma_m = (\sigma_{max} + \sigma_{min})/2. \tag{17.1}$$

The amplitude, σ_a, is

$$\sigma_a = (\sigma_{max} - \sigma_{min})/2, \tag{17.2}$$

and the range, $\Delta\sigma$, is

$$\Delta\sigma = (\sigma_{max} - \sigma_{min}) = 2\sigma_a. \tag{17.3}$$

Also, the ratio of maximum and minimum stresses is

$$R = \sigma_{min}/\sigma_{max}, \tag{17.4}$$

Figure 17.4. Sketch showing how intrusions and extrusions can develop if slip occurs on different planes during the tension and compression portions of the loading.

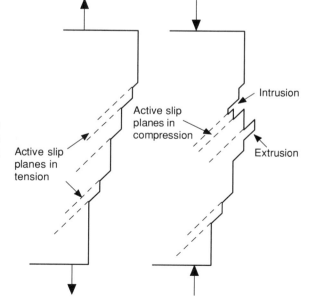

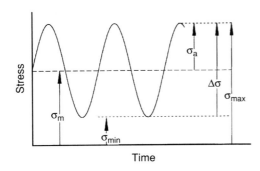

Figure 17.5. Schematic of cyclic stresses illustrating several terms.

which can be expressed as

$$R = (\sigma_m - \Delta\sigma/2)/(\sigma_m + \Delta\sigma/2). \qquad (17.5)$$

For completely reversed loading, $R = -1$, and for tension release, $R = 0$. Static tensile loading corresponds to $R = 1$.

Although Figure 17.5 is drawn with sinusoidal stress waves, the actual wave shape is of little or no importance. Frequency is also unimportant unless it is so high that heat cannot be dissipated and the specimen heats up or so low that creep occurs during each cycle.

S-N curves

Most fatigue data are presented in the form as *S-N* curves, which are plots of the cyclic stress amplitude ($S = \sigma_a$) versus the number of cycles to failure (N), with N conventionally plotted on a logarithmic scale. Usually *S-N* curves are for tests in which the mean stress (σ_m) is zero. Figures 17.6 and 17.7 show the *S-N* curves for a 4340 steel and a 7075 aluminum alloy. The number of cycles to cause failure decreases as the stress amplitude increases.

For low-carbon steels and other materials that strain age, there is a stress amplitude *(endurance limit* or *fatigue limit)* below which failure will never occur. The

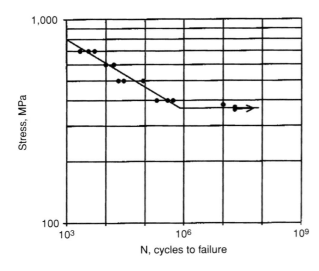

Figure 17.6. The *S-N* curve for annealed 4340 steel. Typically, the break in the curve for a material with a fatigue limit occurs at about 10^6 cycles. The points with arrows are for tests stopped before failure.

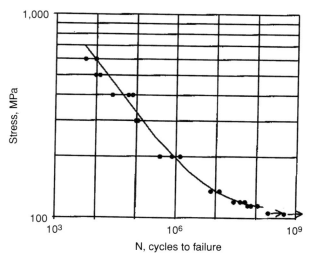

Figure 17.7. The *S-N* curve for an aluminum alloy 7075 T-6. Note that there is no true fatigue limit.

break in the *S-N* curve occurs at about 10^6 cycles, as shown in Figure 17.6. However, many materials, such as aluminum alloys (Figure 17.7), have no true fatigue limit. The stress amplitude for failure continues to decrease, even at a very large number of cycles. In this case, the *fatigue strength* is often defined as the stress amplitude at which failure will occur in 10^7 cycles.

If the *S-N* curve is plotted as $\log(S)$ against $\log(N)$, as in Figures 17.6 and 17.7, a straight line often results for $N < 10^6$. In this case, the relation may be expressed as

$$S = AN^{-b}, \tag{17.6}$$

where N is the number of cycles to failure. The constant, A, is approximately equal to the tensile strength.

EXAMPLE PROBLEM 17.1: Use the initial linear portion of the *S-N* curve for aluminum alloy 7075 T-6 in Figure 17.7 to find the values of A and b in equation (17.6).

Solution: For two points on the linear section, $S_1/S_2 = (N_1/N_2)^{-b}$, so $-b = \ln(S_1/S_2)\ln(N_1/N_2)$. Substituting $S_1 = 600\,\text{MPa}$ at $N_1 = 10^4$ and $S_2 = 200\,\text{MPa}$ at $N_2 = 10^6$, $-b = \ln(3)/\ln(10^{-2}) = 0.24$
$A = (S)/(N)^{-b}$. Substituting $S_2 = 200\,\text{MPa}$ at $N_2 = 10^6$, and $b = 0.24$, $A = 5,400$ MPa.

Effect of Mean Stress

Most *S-N* curves are from experiments in which the mean stress was zero. Under service conditions, however, the mean stress is usually not zero. Several simple engineering approaches to predicting fatigue behavior when the stress cycles about a mean stress have been proposed. Goodman[*] suggested that

$$\sigma_a = \sigma_e[1 - \sigma_m/UTS], \tag{17.7}$$

[*] A. Goodman, *Mechanics Applied to Engineering*, Longmans, London (1899).

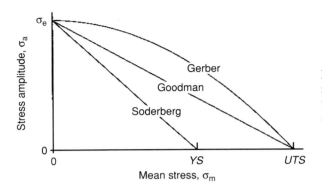

Figure 17.8. A plot representing the Goodman, Soderberg, and Gerber relations for the effect of mean stress on the stress amplitude for fatigue failure.

where σ_a is stress amplitude corresponding to a certain life, σ_m is the mean stress, and σ_e is the stress amplitude that would give the same life if σ_m were zero. UTS is the ultimate tensile strength.

Soderberg proposed a more conservative relation,

$$\sigma_a = \sigma_e[1 - \sigma_m/YS]. \qquad (17.8)$$

where YS is the yield strength.

Gerber proposed a less conservative relation,

$$\sigma_a = \sigma_e[1 - (\sigma_m/UTS)^2]. \qquad (17.9)$$

These are plotted in Figure 17.8. Any combination of σ_m and σ_a outside this region will result in fatigue failure.

These relations may also be represented as plots of σ_{min} and σ_{max} versus σ_m. Figure 17.9 is such a representation for the Goodman relation. In this plot, cycling between the σ_{min} and σ_{max} will not result in fatigue. Note that the permissible cyclic stress amplitude, σ_a, decreases as the mean stress, σ_m, increases reaching zero at the tensile strength.

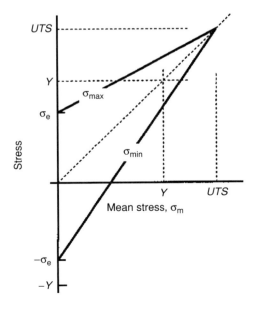

Figure 17.9. An alternative representation of the Goodman relation. Cycling outside of the lines σ_{min} and σ_{max} will result in failure.

Figure 17.10. Modified Goodman diagram showing the effect of mean stress on failure by fatigue
and yielding. Combinations of σ_a and σ_m above
the lines $-YS$ to YS and YS to YS will result
in yielding, whereas combinations of σ_a and σ_m
above the line σ_e to UTS will result in eventual
fatigue failure.

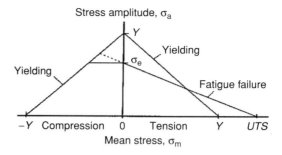

The conditions leading to yielding may be added to the Goodman diagram, as
shown in Figure 17.10.

EXAMPLE PROBLEM 17.2: A bar of steel having a yield strength of 40 ksi, a tensile
strength of 65 ksi and an endurance limit of 30 ksi is subjected to a cyclic loading.
Using a modified Goodman diagram (Figure 17.10), predict whether the material has an infinite fatigue life, or whether it will fail by yielding or fatigue for
the following cases. The cyclic stress is (A) between 0 and 36 ksi, (B) between
−27 ksi and +37, and (C) between 14 ksi ±32 ksi.

Solution: Draw a Goodman diagram and plot each case on the diagram. For A,
$\sigma_m = 18$, $\sigma_a = 18$, predict an infinite life; for B, $\sigma_m = 5$, $\sigma_a = 32$, predict fatigue
failure without yielding; for C, $\sigma_m = 14$, $\sigma_a = 32$, predict yielding and fatigue
failure.

The Palmgren-Miner Rule

The *S-N* curve describes fatigue behavior at constant stress amplitude, but often in
service the cyclic amplitude varies during the life of a part. There may be periods
of high-stress amplitude followed by periods of low amplitude or vice versa. This
is certainly true of the springs of an automobile that sometimes drives on smooth
roads and sometimes over potholes. A. Palmgren[*] and M. A. Miner[**] suggested a
simple approximate rule for analyzing fatigue life under these conditions. The rule
is that fatigue failure will occur when $\sum (n_i/N_i) = 1$, or

$$n_1/N_1 + n_2/N_2 + n_3/N_3 + \cdots = 1, \qquad (17.10)$$

where n_i is the number of cycles applied at an amplitude, σ_{ai}, and N_i is the number
of cycles that would cause failure at that amplitude. The term n_i/N_i represents the
fraction of the life consumed by n_i cycles at σ_{ai}. When $\sum (n_i/N_i) = 1$, the entire
life is consumed. This rule predicts that the fraction of the fatigue life consumed
by n_i cycles at a given stress amplitude depends on the total life, N_i, at that stress
amplitude.

According to this approximate rule, the order of cycling is of no importance.
Yet, experiments have shown that the life is shorter than predicted by equation
(17.10) if the amplitudes of the initial cycles are larger than the later ones. Likewise,
if the initial cycles are of lower amplitude than the later ones, the life will exceed

* A. Palmgren, *Z. Verien Deutscher Ingenieur*, v. 68 (1924).
** M. A. Miner, *Trans. ASME., J. Appl. Mech.*, v. 67 (1945).

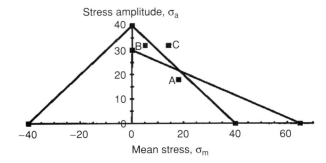

Figure 17.11. Modified Goodman diagram for example problem 17.2.

the predictions of the Palmgren-Miner rule. For steels, cycling at stresses below the endurance limit will promote longer lives. This practice is called *coaxing*.

EXAMPLE PROBLEM 17.3: A part made from the 7075-T4 aluminum alloy in Figure 17.7 has been subjected to 200,000 cycles of 250 MPa and 40,000 cycles of 300 MPa stress amplitude. According to Miner's rule, how many additional cycles of at 200 MPa can it withstand before failing?

Solution: Using the results of example problem 17.1, $N = (S/5,400)^{-1/0.24}$. $N_{250} = 3.63 \times 10^5$, $N_{300} = 1.70 \times 10^5$, $N_{200} = 10^6$. The remaining life at 200 MPa is $n = N_{200}(1 - N_{250}/N_{250} - n_{300}/N_{300}) = 10^6[1 - (2 \times 10^5/3.63 \times 10^5) - (4 \times 10^4/1.70 \times 10^5)] = 0.21 \times 10^6$ cycles.

Stress Concentration

At an abrupt change in cross-section, the local stress can be much higher than the nominal stress. The *theoretical stress concentration factor,* K_t, is the ratio of the maximum local stress to the nominal stress, calculated by assuming elastic behavior. In Chapter 14, the stress concentration factors for elliptical holes were given by equations (14.11) and (14.12) in semiinfinite plates. Figure 17.12 shows calculated values of K_t for circular holes and round notches in finite plates. Stress concentrators reduce fatigue strengths. Therefore, avoidance of such stress risers greatly lowers the likelihood of fatigue failure. However, the effect of notches on fatigue strength is not as great as would be expected by assuming that the actual stress was K_t times the nominal stress. Plastic deformation at the base of a notch reduces the actual stress there. How much the stress is reduced varies from material to material. The role of the material can be accounted for by a *notch sensitivity factor,* q, defined as

$$q = (K_f - 1)/(K_t - 1), \qquad (17.11)$$

where K_f is the *notch fatigue factor* defined as K_f = unnotched fatigue strength/notched fatigue strength. If a notch causes no reduction in fatigue strength, $K_f = 1$, $q = 0$. The value of q increases with strength level and notch radius, ρ. Several empirical equations for calculating proposed. H. Neuber[*] suggested for steels that

$$q = 1/[1 + \sqrt{(\beta/\rho)}] \qquad (17.12)$$

where β (in mm) is given by

$$\log \beta = -(\sigma_u - 134\,\text{MPa})/586 \qquad (17.13)$$

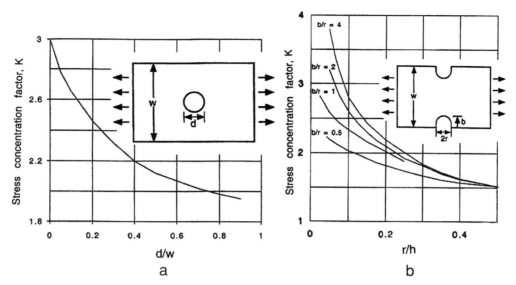

Figure 17.12. Theoretical stress concentration factors. Adapted from G. Neugebauer, *Production Engineering*, v. 14 (1943).

Here, σ_u is the tensile strength. Figure 17.13 values of the notch sensitivity factor, q, calculated from equations (17.12) and (17.13). The notch sensitivity increases with strength level and decreases with increasing notch sharpness.

EXAMPLE PROBLEM 17.4: Calculate the stress concentration factor for fatigue, K_f, for a plate of steel 2 in. wide and 0.25 in. thick with a hole 0.5 in. in diameter in the center. The steel has a tensile strength of 600 MPa.

Solution: $d/w = 0.125$. From Figure 17.12a, $K_t = 2.6$. The notch radius is 0.125 in. = 3.18 mm. From Figure 17.13, for 600 MPa, $q = 0.96$, $K_f = 0.96 \times 2.6 = 2.5$.

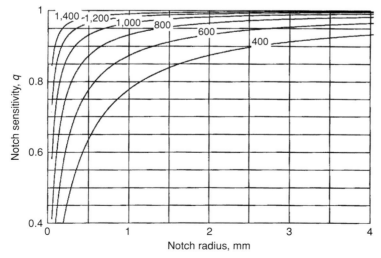

Figure 17.13. Values of notch sensitivity, q, for steels calculated from equations (17.12) and (17.13). The numbers are the tensile strengths in MPA. Note that q increases with tensile strength and larger notch radii.

Surfaces

Fatigue cracks usually start on the surface.* This is because most forms of loading involve some bending or torsion, so the stresses are highest at the surface. Surface defects also play a role. Therefore, the nature of the surface strongly affects fatigue behavior. There are three important aspects of the surface: hardness, roughness, and residual stresses.

In general, increased surface hardness increases fatigue limits. Carburizing, nitriding, flame, and induction hardening are used to harden surfaces and increase the fatigue strengths. Different finishing operations influence surface topography. Valleys of rough surfaces act as stress concentrators so fatigue strength decreases with surface roughness. Surfaces produced by machining are generally smoother than cast or forged surfaces. Grinding and polishing further increase smoothness. The use of polished surfaces to improve fatigue behavior is not warranted where exposure to dirt and corrosion during service may deteriorate the polished surface. Figure 17.14 shows the effects of surface condition. The effects of corrosive environment are also clear. Indeed, laboratory studies have shown considerable improvement in fatigue behavior when the cyclic stressing was done under vacuum instead of dry air.

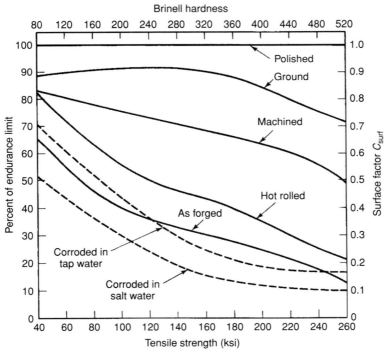

Figure 17.14. Effect of surface finish on fatigue. From C. Lipson and R. C. Juvinall, *Application of Stress, Analysis to Design and Metallurgy*, The University of Michigan Summer Conference, Ann Arbor, 1961.

* The most common exception is where the cyclic loading is from contact with a ball or a cylinder. In this case, the highest stress occurs some distance below the surface. This sort of loading occurs in ball and roller bearings.

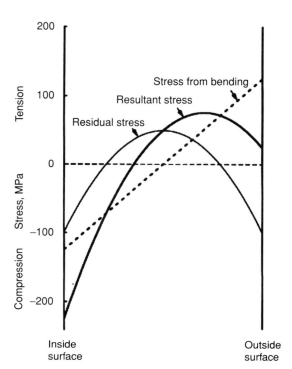

Figure 17.15. Schematic drawing showing the effect of residual stresses. With residual compression in the surface, a larger tensile stress can be applied by bending before there is tension in the surface.

Residual stresses play an important role in fatigue. When a part is subjected to a load, as in fatigue, the stress at any location is the sum of the residual stress at that point and the stress resulting from the external load (Figure 17.15). Because failures are tensile in nature and start at the surface, residual tension at the surface lowers the resistance to fatigue, while residual compression raises the fatigue strength. Note that this effect is in accord with the prediction of the Goodman diagram. Sometimes critical parts are shot peened to produce residual compression in the surface. In this process, the surface is indented with balls that produce local plastic deformation that does not penetrate into the interior of the part. The indentation would expand the surface laterally, but the undeformed interior prevents this leaving the surface under lateral compression.

Some investigators have found that the fatigue strength of a material decreases as the specimen size increases. However, the effect is not large (10% decrease for a diameter increase from 0.1 in. to 2 in.). The size effect is probably related to the increased amount of surface area where fatigue cracks initiate.

Design Estimates

Shigley* suggested that the endurance limit can be estimated, taking into account the various factors by

$$\sigma_e = \sigma_{eb} C_s C_d (1 - \sigma_m/\text{UTS})/K_f, \qquad (17.14)$$

* J. E. Shigley, *Mechanical Engineering Design*, 3rd ed., McGraw-Hill (1977).

where σ_{eb} is the base endurance limit (polished unnotched specimen of small diameter cycled about a mean stress of zero), C_s is the correction factor for the surface condition (Figure 17.14), and C_d is the correction factor for specimen size ($C_d = 1$ for $d < 7.6$ mm and 0.85 for $d > 7.6$ mm). The term $(1 - \sigma_m/UTS)$ accounts for the effect of mean stress, σ_m, and $K_f = 1 + q(K_t - 1)$. Equation (17.12) can be used to obtain a first estimate of the endurance limit, but it is always better to use data for the real conditions than to apply corrections to data for another condition.

> **EXAMPLE PROBLEM 17.5:** A round bar of a steel has a yield strength of 40 ksi, a tensile strength of 60 ksi, and an endurance limit of 30 ksi. An elastic analysis indicates that $K_t = 2$. It is estimated that $q = 0.75$. The bar is to be loaded under bending such that a cyclic bending moment of 1,500 in.-lb is superimposed on a steady bending moment of 1,000 in.-lb. The surface has a ground finish. What is the minimum diameter bar that would give an infinite life?
>
> **Solution:** For a round bar under elastic loading, the stress at the surface is given by $\sigma = Mc/I$, where $c = d/2$ and $I = \pi d^4/64$, so $\sigma = 32M/(\pi d^3)$. $\sigma_m = 10{,}186/d^3$ ksi and $\sigma_a = 15{,}279/d^3$ ksi, where M is the bending moment and d is the bar diameter.
> $K_f = 1 + q(K_t - 1) = 1.75$. Assuming that $d < 7.6$ mm, $C_d = 1$. From Figure 17.14, $C_s = 0.89$, so the endurance limit is estimated as $\sigma_e = C_s C_d \sigma_{eb}$ $(1 - \sigma_m/UTS)/K_f = (0.89)(1)(30)[1 - 10.86/60d^3]/1.75 = 15.25(1 - 0.181/d^3)$. Now equating this to the stress amplitude, $15.279/d^3 = 15.25(1 - 0.181/d^3)$. $d^3 = 1.182$, $d = 1.058$ in. Because this is larger than 7.6 mm, we should use $C_d = 0.85$. Recalculation with $C_d = 0.85$ instead of 1, results in $d = 1.11$ in.

Metallurgical Variables

Because fatigue damage occurs by plastic deformation, increasing the yield strength and hardness generally raises the endurance limit. For steels and titanium alloys, there is a rough rule of thumb that the fatigue limit is about half of the ultimate tensile strength. Figure 17.16 shows the correlation of fatigue limits with hardness in steels. For aluminum alloys, the ratio of the fatigue limit at 10^7 cycles to the tensile strength is between 0.25 and 0.35.

Nonmetallic inclusions lower fatigue behavior by acting as internal notches. Alignment of inclusions during mechanical working causes an anisotropy of fatigue properties. Fatigue strength for loading in the transverse direction (normal to the rolling or extrusion direction) is usually much poorer than for loading in the rolling direction. The number of inclusions can be greatly reduced by vacuum melting or electroslag melting. This decreases the directional effect. The addition of Ce or rare earth elements to steels produces sulfides that are much harder than MnS and are not elongated during rolling. Therefore, Ce additions reduce the directionality of fatigue behavior.

Strains to Failure

Cyclic loading in service sometimes subjects materials to imposed forces or stresses. Just as often, however, materials are subjected to imposed deflections or strains.

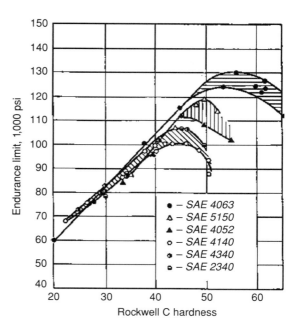

Figure 17.16. Correlation of the endurance limit with hardness for quenched and tempered steels. The endurance generally rises with hardness. From Garwood, Zurburg, and Erickson, *Interpretation of Tests and Correlation with Service*, ASM (1951).

These are not equivalent if the material strain hardens or strain softens during cycling.

Fatigue could not occur if the deformation during cyclic loading were entirely elastic. Some plastic deformation, albeit very little, must occur during each cycle. This probably accounts for the difference between steel and aluminum alloys. For steels, there seems to be a stress, below which no plastic deformation occurs. If fatigue data are analyzed by plotting the plastic strain amplitude, $\Delta\varepsilon_p/2$, versus the number of reversals to failure, $2N_f$, a straight line results. This was first noted by L. F. Coffin[*] and indicates a relationship of the form:

$$\Delta\varepsilon_p/2 = \varepsilon_f'(2N_f)^{-c}, \tag{17.15}$$

where ε_f' is the true strain at fracture in a tension test ($N = 1$) and $-c$ is the slope. Figure 17.17 is such a plot.

The total strain, including the elastic portion, is $\Delta\varepsilon/2 = \Delta\varepsilon_p/2 + \Delta\varepsilon_e/2$. The elastic term can be expressed as

$$\Delta\varepsilon_e/2 = \sigma_p/E = (B/E)(2N_f)^{-b}, \tag{17.16}$$

where B is a constant that increases with the tensile strength and the exponent, b, is related to the stain hardening exponent. Combining equations (17.15) and (17.16),

$$\Delta\varepsilon/2 = \varepsilon_f'(2N_f)^{-c} + (B/E)(2N_f)^{-b}. \tag{17.17}$$

Figure 17.18 is a plot of the total strain, $\Delta\varepsilon/2$, as well as the elastic and plastic strains versus $2N_f$. Although $\ln(\Delta\varepsilon_e/2)$ and $\ln(\Delta\varepsilon_p/2)$ vary linearly with $\ln(2N_f)$, $\ln(\Delta\varepsilon/2) = \ln(\Delta\varepsilon_e/2 + \Delta\varepsilon_p/2)$ does not.

The term, *low cycle fatigue*, is commonly applied to conditions in which the life, N_f, is less than 10^3 cycles. The term, *high cycle fatigue*, is applied to conditions in

[*] L. F. Coffin, *Trans. ASME*, v. 76 (1954).

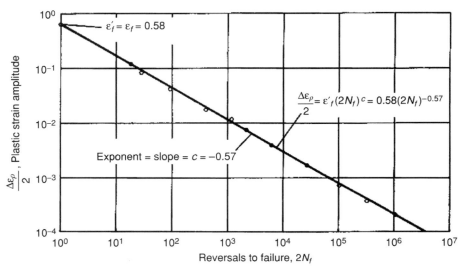

Figure 17.17. A plot of fatigue life, $2N_f$, of an annealed 4340 steel as a function of plastic strain per cycle. The slope is -c and the intercept at one cycle is ε'_f. From *Metals Handbook*, v. I, 9th ed., ASM (1978).

which the life, N_f, is greater than 10^4 cycles. Note that the range 10^3 to 10^4 cycles corresponds to the range, where $\Delta\varepsilon_p \approx \Delta\varepsilon_e$. For low cycle fatigue, $\Delta\varepsilon_p > \Delta\varepsilon_e$, whereas for high cycle fatigue, $\Delta\varepsilon_e > \Delta\varepsilon_p$. The total life, N_f, can be divided into the period necessary for crack initiation, N_i, and that necessary for the crack to propagate to failure, N_p.

$$N_f = N_i + N_p. \tag{17.18}$$

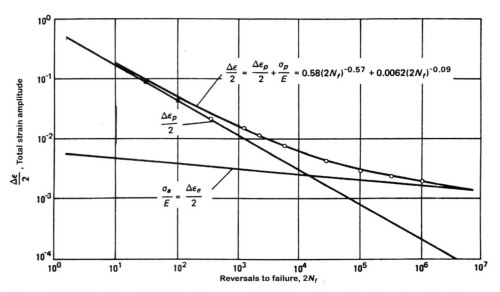

Figure 17.18. The fatigue life, $2N_f$, of an annealed 4340 steel as a function of total strain per cycle. The relation is no longer linear. From *Metals Handbook*, v. I, 9th ed., ASM (1978).

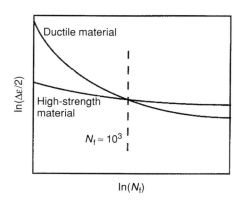

Figure 17.19. Fatigue life as a function of strain amplitude. At high strain amplitudes (low cycle fatigue), the softer, more ductile material will have the longer life, whereas at low strain amplitudes (high cycle fatigue), the stronger, less ductile material will have the longer life.

For low cycle fatigue, crack initiation is rapid and crack propagation accounts for most of the life ($N_p > N_i$), whereas for high cycle fatigue most of the life is spent in crack initiation ($N_i > N_p$).

This leads to the conclusion that under constant strain-amplitude cycling, a ductile material is desirable for low cycle fatigue and a high-strength material is desirable for high cycle fatigue. High-strength materials generally have low ductility (low value of ε_f' in equation (17.17)), and ductile materials generally have low strength (low B in equation (17.17)). This is shown schematically in Figure 17.19.

Crack Propagation

In the laboratory, crack growth rates may be determined during testing either optically with a microscope or by measuring the electrical resistance. It has been found that the crack growth rate depends on the range of stress intensity factor, ΔK, $\Delta K = K_{max} - K_{min} = \sigma_{max} f \sqrt{(\pi a)} - \sigma_{min} f \sqrt{(\pi a)} = (\sigma_{max} - \sigma_{min}) f \sqrt{(\pi a)}$ and the f in equation 17.19, or more simply,

$$\Delta K = f \Delta\sigma \sqrt{\pi a}, \tag{17.19}$$

where $\Delta\sigma = (\sigma_{max} - \sigma_{min})$.

Figure 17.20 shows schematically the variation of da/dN with ΔK. Below a threshold, ΔK_{th}, cracks do not grow. There are three regimes of crack growth. One is a region where da/dN increases rapidly with ΔK. Then there is large region where $\log(da/dN)$ is proportional to $\log(\Delta K)$, and finally at large values of ΔK_I, the crack growth rate accelerates more rapidly. Failure occurs when $K_I = K_{Ic}$. In viewing this figure, one should remember ΔK_I increases with growth of the crack, so that under constant load, a crack will progress up the curve to the right.

In stage II,

$$da/dN = C(\Delta K_I)^m, \tag{17.20}$$

where C is a material constant that increases with the stress ratio, $R = \sigma_{min}/\sigma_{max}$ (i.e., with the mean stress). The exponent, m, is also material dependent and is usually in the range of 2 to 7. Equation (17.20) is often called the *Paris* law.

* P. Paris and F. Erdogan, *Trans. ASME, J. Basic Engr.*, v. D 85 (1963).

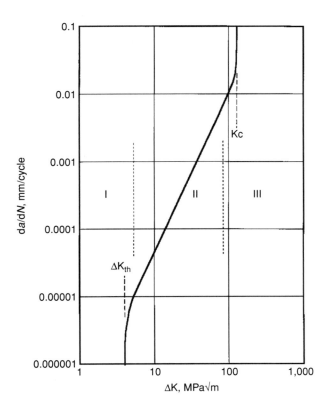

Figure 17.20. A schematic plot of the dependence of crack growth rate, da/dN on the stress intensity factor, ΔK_I. Because ΔK_I increases with crack length, a, crack growth will progress up the curve. Below the threshold stress, cracks do not grow. In general, there are three regions: an initial period (region I), a linear region (II) in which the rate of crack growth is given by equation (17.19), and a final region (III) of acceleration to failure. The slope of the linear region on the log-log plot equals m.

Note that the crack length at any stage can be found by substituting $(\Delta K_I)^m = [f\Delta\sigma\sqrt{(\pi a)}]^m$ into equation (17.20) and rearranging,

$$da/a^{m/2} = C(f\Delta\sigma\sqrt{\pi})^m dN. \tag{17.21}$$

Integration, neglecting the dependence of f on a, gives

$$a^{(1-m/2)} - a_o^{(1-m/2)} = (1 - m/2)C(f\Delta\sigma\sqrt{\pi})^m N. \tag{17.22}$$

The number of cycles to reach a crack size a is then

$$N = [a^{(1-m/2)} - a_o^{(1-m/2)}]/[(1 - m/2)C(f\Delta\sigma\sqrt{\pi})^m]. \tag{17.23}$$

The constant, C, in equations (17.20) through (17.23), depends on the stress ratio, $R = (\sigma_m - \Delta\sigma/2)/(\sigma_m + \Delta\sigma/2)$, as shown in Figure 17.21. This dependence is in accord with the effect of σ_m, predicted by the Goodman diagram.

Figure 17.22 shows how a crack length, a, depends on N for two different values of the initial crack length, a_o, and two different values of $\Delta\sigma$.

EXAMPLE PROBLEM 17.6: Find the exponent, m, and the constant C in equation (17.20) for aluminum alloy 7076-T6 using the data for $R = 0$ in Figure 17.21.

Solution: $m = \ln[(da/dN)_2/[(da/dN)_1]/[\ln(\Delta K_2/(\Delta K_1)]$. Substituting $(da/dN)_1 = 7 \times 10^{-7}$ m at $\Delta K_1 = 10$ MPa$\sqrt{m}$ and $(da/dN_2) = 6 \times 10^{-5}$ m at $\Delta K_1 = 30$ MPa$\sqrt{m}$, $M = \ln 85.7 \ln 3 = 4.05$. $C = (da/dN)_1/(\Delta K_1)m = 7 \times 10^{-7}/10^{4.05} = 6.2 \times 10^{-11}$.

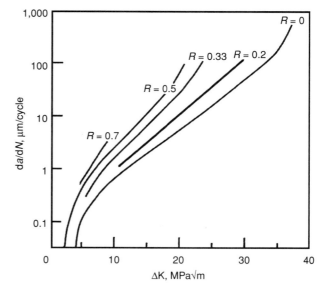

Figure 17.21. Crack growth rate in aluminum alloy 7076-T6 as a function of ΔK for several levels of R. Note that da/dN increases with R. Data rom C. A. Martin, NASA TN D5390 (1969).

For low cycle fatigue, the life can be found by assuming the initial crack size and knowing $\Delta\sigma$, which gives an initial value of ΔK. Integration under the da/dN versus ΔK curve allows life prediction.

EXAMPLE PROBLEM 17.7: Consider crack growth in 7076-T6 aluminum.

A. Determine the crack growth rate, da/dN, if $\Delta\sigma = 200$ MPa and the crack is initially 1 mm in length. Assume $R = 0$ and $f = 1$.

B. Calculate the number of cycles needed for the crack to grow from 1 mm to 10 mm if $\Delta\sigma = 200$ MPa.

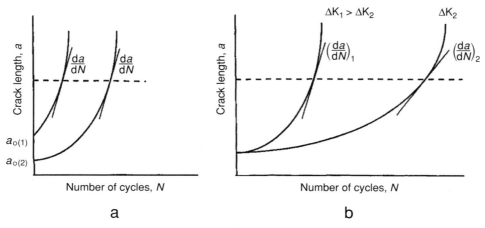

Figure 17.22. Crack growth during cycling. (a) The effect of the initial crack length with constant ΔK. Note that da/dN is the same at the same current crack length, a. (b) The effect of ΔK on crack growth. Note that da/dN increases with ΔK.

Solution:

A. $\Delta K = f\Delta\sigma\sqrt{(\pi a)} = 2\,\text{MPa}\sqrt{(0.001\pi)} = 11.2\,\text{MPa}\sqrt{m}$. From Figure 17.21, $da/dN \approx 10^{-3}$ mm.

B. Using equation (17.22) with m = 4.05 and C = 6.2×10^{-11} from example problem 17.6, N = $[(10^{-5})^{-1.025} - (10^{-2})^{-1.025}]/[(-1.0250)(6.2 \times 10^{-8})(200\sqrt{\pi})^{4.05}] = 99$ cycles.

Cyclic Stress–Strain Behavior

Materials subjected to cyclic loading in the plastic range may undergo strain hardening or strain softening in the case of heavily cold-worked material. If the cycling is done at constant strain amplitude, strain hardening causes an increase of stress amplitude during testing. On the other hand, for a material that strain softens, $\Delta\sigma$ decreases during cycling. Figure 17.23 illustrates both possibilities for constant plastic-strain amplitude cycling. Cyclic strain hardening can be measured by cycling at one strain amplitude until stress level saturates and then repeating this at increasing strain amplitudes. In Figure 17.24, the cyclic stress–strain hardening is compared to strain hardening under monotonic loading.

For a material that work hardens, the strain amplitude will decrease during constant stress amplitude cycling. On the other hand, for material that work softens, the strain amplitude will decrease during constant stress amplitude cycling. In contrast, if a material is cycled under constant strain amplitude, the stresses will increase in a material that work hardens and decrease in a material that work softens. These behaviors are illustrated in Figures 17.25a and b.

Temperature and Cycling Rate Effects

Increased temperatures lower flow stresses, and therefore, there is more plastic flow per cycle in constant stress amplitude cycling. Near room temperature, this effect

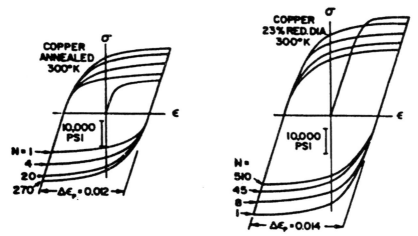

Figure 17.23. Cyclic hardening of annealed copper (a) and cyclic softening of a cold-worked copper (b) tested under constant plastic strain cycling. From C. E. Feltner and C. Laird, *Acta Met.*, v. 15 (1967).

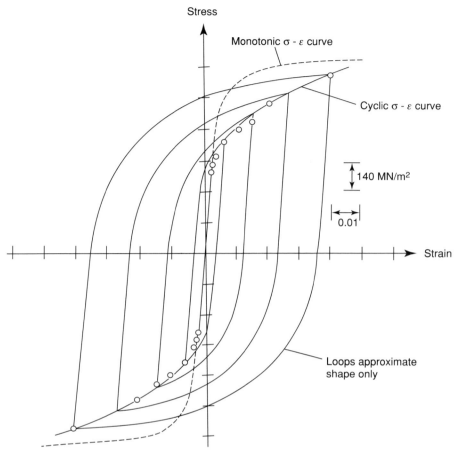

Figure 17.24. Comparison of strain hardening under cyclic and monotonic loading. The cyclic stress–strain curve is obtained by connecting the saturation stresses corresponding to various strain amplitudes. From R. W. Landgraf, *ASTM STP 467*, Philadelphia (1970).

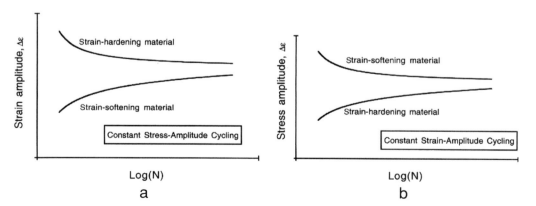

Figure 17.25. (a) Change of strain amplitude during constant stress amplitude cycling. The strain amplitude should decrease for a strain-hardening material and increase for a strain-softening material. (b) Change of stress range during constant strain amplitude cycling. The stress amplitude should increase for a strain-hardening material and decrease for a strain-softening material.

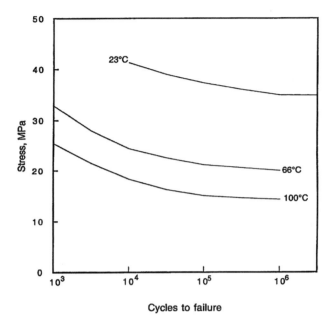

Figure 17.26. Effect of temperature on the fatigue behavior of an acetal polymer. Loading is complete stress reversals. Increased temperatures lower the *S-N* curves. Cycle frequency was 30 cycles/s. Data from *Design Handbook for du Pont Engineering Plastics.*

is more pronounced in polymers than in metals. Figure 17.26 shows the decreased fatigue strength of an acetal polymer with increasing temperature.

With higher frequencies, there is less plastic deformation per cycle, so the crack growth rate is lower. Figure 17.27 shows this effect for PVC.

On the other hand, with high cycling rates, the material's temperature may rise unless the heat generated by the mechanical hysteresis is dissipated. This is much

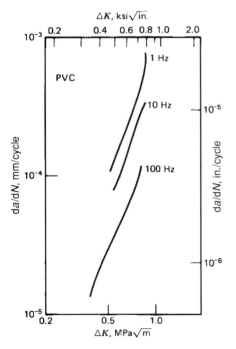

Figure 17.27. Effect of test frequency on the rate of crack propagation, da/dN, for PVC. From M. D. Skibo, PhD dissertation, Lehigh University (1977).

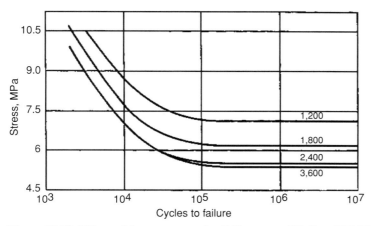

Figure 17.28. Effect of frequency on the *S-N* curves of Teflon (PTFE). Increased frequency lowers the *S-N* curve because of more heating of the specimen. From *Engineered Materials Handbook, Engineering Plastics*, vol. II, ASM International (1998).

more important with polymers than with metals because they tend to have larger hysteresis losses and lower thermal conductivities. With higher temperatures, there is more plastic strain per cycle under constant stress cycling so cracks grow more rapidly and the life will decrease (Figure 17.28).

Thus, increased frequency may either shorten or lengthen fatigue life. The net effect of frequency, for constant stress amplitude if any, will depend on which effect predominates.

The effects of frequency at constant strain amplitude cycling are different. Heating of the material and lower frequencies both cause lower stress amplitudes. However, the cumulative strain to failure should remain unaffected.

Unless the mean stress is zero ($R = 0$), creep during cyclic loading will cause the material to permanently elongate because the stress on the tensile half-cycle will be larger than on the compressive half-cycle.

Fatigue of Polymers

Figure 17.29 shows the *S-N* curves for several polymers. Many of the factors affecting fatigue behavior are similar to those affecting metals. Notches and surface roughness lower the levels of the *S-N* curves. Higher yield strengths promote higher *S-N* curves.

Cracks propagate slower in material of higher molecular weight than low molecular weight (Figure 17.30). This effect is probably related to ease of chain disentanglement. The fracture surfaces in fatigue often show striations similar to those found in metals that fail by fatigue (Figure 17.31).

However, there are differences. Crack initiation is often associated with crazes. Slip is not important in polymers, so there are no persistent slip bands. Yet, localized deformation plays a similar role. The spacing of fatigue striations is sometimes much greater than da/dN, indicating that cracks may progress by jumps. Perhaps damage ahead of the crack tip accumulates until the cracks suddenly advance.

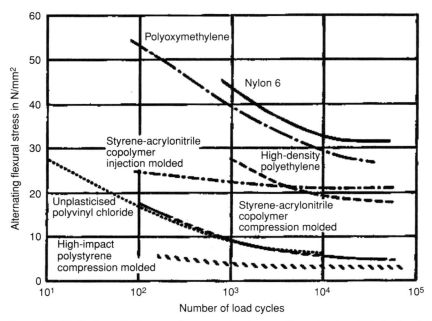

Figure 17.29. Fatigue *S-N* curves for several polymers From L. Engel et al., *An Atlas of Polymer Damage*, Prentice Hall (1978). After W. Läis, *Einfurhrung in die Werkstoffkunde der Kunstoffe* (1974).

Thermal effects are much more important in polymers than in metals. For most polymers, room temperature is about half of the absolute melting point, T_M, and thermal softening becomes important at $(0.4$ to $0.5)T_M$. Likewise, the effect of cyclic frequency is more important.

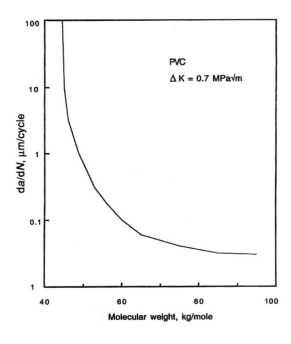

Figure 17.30. Effect of molecular weight on fatigue crack growth rate in PVC. From C. M. Rimac, J. A. Manson, R. W. Hertzberg, S. M. Webler, and M. D. Skibo, *J. Macromol. Sci. Phys.*, v. B19 (1981).

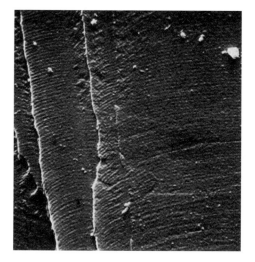

Figure 17.31. Striations on a fatigue failure surface of styrene acrylonnitrile (SAN). From Engel et al., *Atlas of Polymer Damage*, Prentice Hall (1978).

Fatigue Testing

Fatigue tests are often made using round bars rotated under a bending load (Figure 17.32) so that the entire surface is subjected to alternating tension and compression. Flat specimens from sheet or plate may be tested in fatigue by bending. In either case, the testing machine may be made to impose a constant deflection or a constant bending force. Fatigue testing with axial loading is normally much slower and requires great care. In this case, tension release is often used instead of tension compression. In special cases, other forms of loading may be employed. These include torsion and combined stresses as may be achieved by internal pressurization and release.

Design Considerations

There are two methods of designing to prevent fatigue failure: designing for an infinite life without inspection, or designing for periodic inspection. For many parts,

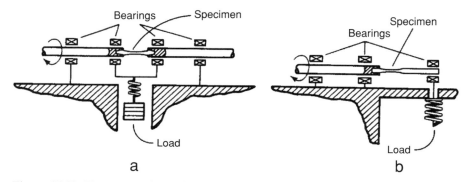

Figure 17.32. Two types of rotating beam fatigue testing machines. (a) Four-point bending imposed by an R. R. Moore machine. (b) Cantilever loading. In both cases, the specimens are machined so as to have a region of constant surface stress. From *Metals Handbook*, v. 8, 9th ed., ASM (1985).

inspection is not a possibility. An automobile axle is an example. It must be designed with enough of a safety factor so that it will not fail within the life of the car. In these cases, it is usual to assume initiation-controlled fatigue life, where endurance limit and life are improved by increased hardness.

The other possibility is to design for periodic inspection. This is done in the aircraft industry where safety factors must be lower to minimize weight. The interval between inspections must be short enough in terms of service so that a crack of the largest undetectable length cannot grow to failure before the next inspection. In this case, the design and the inspection schedule are based on crack growth rate, da/dN.

Summary

Cyclic application of loads can lead to failure at stresses lower than those for yielding. Surface roughness, corrosive environments, increased mean stress, and section size decrease the cyclic stress that a material can withstand without fatiguing. Higher hardness levels increase it. Residual compressive stresses in the surface layers are beneficial. Growth of cracks by fatigue eventually lead to a crack size that is unstable and sudden failure occurs.

REFERENCES

S. Suresh, *Fatigue of Materials*, Cambridge University Press (1991).
L. F. Coffin, Jr., *Trans ASME*, v. 76 (1954).
P. C. Paris, *Fatigue–An Interdisciplinary Approach*, Proc. 10th Sagamore Conf., Syracuse University Press (1964).
R. W. Hertzberg, *Deformation and Fracture of Engineering Materials*, 4th ed., Wiley (1995).
T. H. Courtney, *Mechanical Behavior of Materials*, 2nd ed., McGraw-Hill (2000).
J. E. Shigley, *Mechanical Engineering Design*, 3rd ed., McGraw-Hill (1977).
R. O. Ritchie, *Mater. Sci & Engrg.*, v. 103 (1988).
R. W. Hertzberg & J. A. Manson, *Fatigue of Engineering Plastics*, Academic Press (1980).

Notes

Railroad accidents in the middle of the nineteenth century stimulated the first research on the nature of fatigue. Poncelet, in France, first used the term "fatigue" in lectures as early as 1839. Inspection of railroad cars for fatigue cracks was recommended in France as early as 1853. In about 1850, a number of papers on fatigue were written in England. A. Wohler (1819–1914), a German railroad engineer, did the first systematic research on fatigue. He recognized the effects of sharp corners. He also noted that the maximum stress allowable depended on the stress range, that the maximum allowable stress increased as the minimum stress increased.

The Comet, built by de Haviland, was the first commercial passenger jet airliner. It was put into service in May 1952, after years of design, production, and testing. Almost 2 years later, in January and May 1954, two Comets crashed into the Mediterranean near Naples after taking off from Rome. The cause of the failures was discovered only after parts of one plane that crashed into the Mediterranean were retrieved by the British Navy from a depth of more than 300 ft. It was found that fatigue was responsible. Cracks originated from rivet holes near the windows. This conclusion was confirmed by tests on fuselage that was repeatedly pressurized and depressurized. The sharp radii of curvature at the corners acted as stress concentrators. Apparently, the engineers had overlooked the stresses to the fuselage caused by cycles of pressurization and depressurization during take-off and landing. Up to this time, the main fatigue concern of aircraft designers was with engine parts that experience a large number of stress cycles, and low cycle fatigue of the fuselage had not been seriously considered. As a result of the investigations, the Comets were modified and returned to service a year later.

In principle, fatigue cracks can be arrested by drilling holes at their tips. This requires the crack to reinitiate.

Resistance to fatigue can be improved by autofretage. The idea is that if a part is overstressed once in tension so that a small amount of plastic deformation occurs, it will be left in residual compression. In the past, cast bronze cannons were overpressurized so that the internal surface of the bore would be left in compression. It has been suggested (with tongue in cheek) that the underside of aircraft wings could be overstressed by having the plane sharply pull out of a steep dive, causing the wing tips to bend upward.

Problems

1. For steels, the endurance limit is approximately half of the tensile strength, and the fatigue strength at 10^3 cycles is approximately 90% of the tensile strength. The S-N curves can be approximated by a straight lines between 10^3 and 10^6 cycles when plotted as $\log(S)$ versus $\log(N)$. Beyond 10^6 cycles, the curves are horizontal.

   A. Write a mathematical expression for S as a function of N for the sloping part of the S-N curve, evaluating the constants in terms of the previous approximations.

   B. A steel part fails in 12,000 cycles. Use the previous expression to find what percent decrease of applied (cyclic) stress would be necessary to increase in the life of the part by a factor of 2.5 (to 30,000 cycles).

   C. Alternatively, what percent increase in tensile strength would achieve the same increase in life without decreasing the stress.

2. A. Derive an expression relating the stress ratio, R, to the ratio of cyclic stress amplitude, σ_a, to the mean stress, σ_m.

   B. For $\sigma_a = 100$ MPa, plot R as a function of σ_m over the range $0 \leq \sigma_m \leq 100$ MPa.

3. A steel has the following properties:

Tensile strength	460 MPa
Yield strength	300 MPa
Endurance limit	230 MPa.

   A. Plot a modified Goodman diagram for this steel showing the lines for yielding as well as fatigue failure.

B. For each of the cyclic loading given here, determine whether yielding, infinite fatigue life, or finite fatigue life is expected.

    i. $\sigma_{max} = 250\,$MPa, $\sigma_{min} = 0$.

    ii. $\sigma_{max} = 280\,$MPa, $\sigma_{min} = -200\,$MPa.

    iii. $\sigma_{mean} = 280\,$MPa, $\sigma_a = 70\,$MPa.

    iv. $\sigma_{mean} = -70\,$MPa, $\sigma_a = 140\,$MPa.

4. The notch sensitivity factor, q, gray cast iron is very low. Offer an explanation in terms of the microstructure.

5. A 1040 steel has been heat treated to a yield stress of 900 MPa and a tensile yield strength of 1,330 MPa. The endurance limit (at 10^6 cycles for cyclical loading about a zero mean stress) is quoted as 620 MPa. Your company is considering using this steel, with the same heat treatment for an application in which fatigue may occur during cyclic loading with $R = 0$. Your boss is considering shot peening the steel to induce residual compressive stresses in the surface. Can the endurance limit be raised this way? If so, by how much? Discuss this problem with reference to the Goodman diagram using any relevant calculation(s).

6. Low cycle fatigue was the cause of the Comet failures. Estimate how many pressurization–depressurization cycles the planes may have experienced in the 2 years of operations. An exact answer is not possible, but by making a reasonable guess of the number of landings per day and the number of days of service, a rough estimate is possible.

7. Frequently the *S-N* curves for steel can be approximated by a straight line between $N = 10^2$ and $N = 10^6$ cycles when the data are plotted on a log-log scale, as shown in the figure for SAE 4140 steel. This implies $S = AN^b$, where a and b are constants.

A. Find b for the 4140 steel for a certain part made from 4140 steel (Figure 17.33).

B. Fatigue failures occur after 5 years. By what factor would the cyclic stress amplitude have to be reduced to increase the life to 10 years? Assume the number of cycles of importance is proportional to time of service.

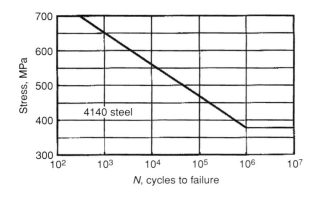

Figure 17.33. *S-N* curve for a SAE 4140 steel.

8. Figure 17.34 shows the crack growth rate in aluminum alloy 7075-T6 as a function of ΔK for $R = 0$. Find the values of the constants C and m in equation (17.19) that describe the straight-line portion of the data. Give units.

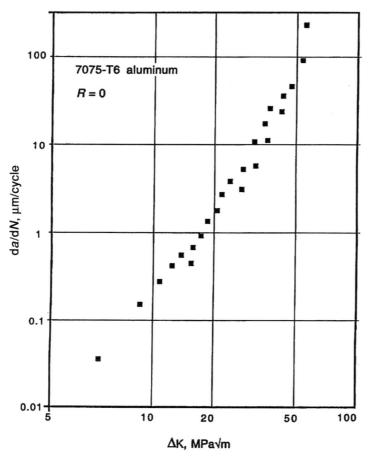

Figure 17.34. Crack growth rate of 075-T6 aluminum for $\Delta K = 0$. Data from C. M. Hudson, *NASA TN D-5300* (1969).

9. Find the number of cycles required for a crack to grow from 1 mm to 1 cm in 7075-T6 (problem 8) if $f = 1$ and $\Delta\sigma = 10$ MPa. Remember that $\Delta K = f \Delta\sigma \sqrt{(\pi a)}$.

10. The fatigue lives of a certain steel was found to be 10^4 cycles at ± 70 MPa and 10^5 cycles at ± 50 MPa. A part made from this steel was given 10^4 cycles at ± 61 MPa. If the part were then cycled at ± 54 MPa, what would be its expected life?

11. For another steel, the fatigue limits are 10,000 cycles at 100 ksi, 50,000 cycles at 75 ksi, and 200,000 cycles at 62 ksi. If a component of this steel had been subjected to 5,000 cycles at 100 ksi and 10,000 cycles at 75 ksi, how many additional cycles at 62 ksi would cause failure?

12. In fatigue tests on a certain steel, the endurance limit was found to be 1,000 MPa for $R = 0$ (tensile release) ($\sigma_a = \sigma_m = 500$ MPa) and 1,000 MPa for $R = -1$ (fully reversed cycling). Calculate whether a bar of this steel would fail by fatigue if it were subjected to a steady stress of 600 MPa and a cyclic stress of 500 MPa.

18 Residual Stresses

Introduction

There are often *residual stresses* in materials even when there are no external stresses. Residual stresses may arise because one region of a body has been heated or cooled differently than another region. Variations of orientation or composition cause different responses to temperature changes. Residual stresses may also arise from inhomogeneous mechanical working. They are responsible for warping of parts during machining, for spontaneous splitting of formed parts, for increased suscepti-bility to stress corrosion cracking, and for lowering (or raising) endurance limits in fatigue. Residual stresses have the same effect on materials and their performance as externally applied stresses.

The level of residual stress in a material varies with location. Tensile stresses in one region must be balanced by compressive stresses in another. Internal stresses must be balanced in the sense that the sum of forces acting across any imaginary cut through the body must be zero (Figure 18.1). Across a cut, $\overline{AB}$, perpendicular to x, this can be expressed mathematically as

$$\int \sigma_x w(z) \mathrm{d}z = 0, \tag{18.1}$$

where the width, $w(z)$, in the y-direction is a function of z.

There must also be moment balances within the material. About any axis in the y-direction, this may be expressed by

$$\int \sigma_x w(z) z \mathrm{d}z = 0. \tag{18.2}$$

Small-Scale Stresses

Residual stresses may occur on microscopic or macroscopic scales. Stresses can vary within a grain because of dislocation pile-ups near precipitates or other obstacles. On the other hand, there may be variation from one grain to another, or from the center of a part to its surface. In polycrystals, orientation differences cause grain-to-grain variations. Under tension, grains favorably oriented for slip will deform at

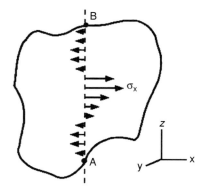

Figure 18.1. There must be a force balance and a moment balance on any cut through a body.

lower stresses than other grains that are less favorably oriented. During unloading, the elastic contractions must be the same in all grains. Those grains that deformed under the lower tensile stress will be left under residual compression, whereas grains less favorably oriented for slip will be left under residual tension. See Figure 18.2.

The orientation dependence of elastic moduli is another source of grain-to-grain residual stress patterns. During unloading, grains with higher elastic moduli undergo greater stress changes than grains of low moduli.

EXAMPLE PROBLEM 18.1: A bar of copper with randomly oriented grains was deformed in tension to a stress, σ_{av}, which caused plastic deformation and then the load was released. Determine whether a grain oriented with [100] parallel to the tensile axis would be left in a state of residual tension or residual compression. For [100], the Taylor factor, $M = 2.449$ and $E = 66$ GPa. For randomly oriented copper, $M = 3.067$ and $E = 128$ GPa.

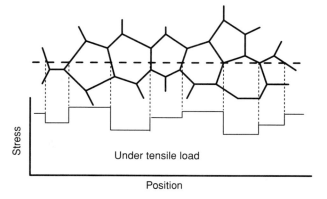

Figure 18.2. When a polycrystal deforms plastically, grains of different orientation are stressed to different levels (*top*). When the load is removed and the grains elastically recover (*bottom*), some grains are left under residual tension and others under residual compression.

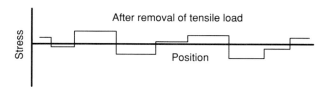

Solution: Under load, the stress in the [100] grain was $2.449/3.067\sigma_{\mathrm{av}} = 0.799\sigma_{\mathrm{av}}$. On unloading, the elastic contraction was $\Delta\varepsilon = \sigma_{\mathrm{av}}/128$ GPa. The decrease of stress in the [100] grain was $\Delta\sigma = 66\,\mathrm{GPa}\,\Delta\varepsilon = (66/128)\sigma_{\mathrm{av}} = 0.516\sigma_{\mathrm{av}}$. The final stress in the [100] grain was $0.799\,\sigma_{\mathrm{av}} - 0.516\sigma_{\mathrm{av}} = +0.28\sigma_{\mathrm{av}}$, so the [100] grain was left in residual tension.

Surface grains are usually left under residual compression after a bar has been stretched plastically in tension and unloaded. The reason is that surface grains are less constrained than interior grains because they have fewer neighbors. With less constraint, fewer slip systems are required, so lower stresses are needed for deformation than in interior grains. After unloading, the surface grains are left in residual compression.

In noncubic crystals, the coefficients of thermal expansion depend on crystallographic direction. Neighboring grains with different orientations respond differently to temperature changes. The elastic deformation required to accommodate the differences in thermal expansion or contraction causes residual stresses.

EXAMPLE PROBLEM 18.2: Estimate the magnitude of the residual stresses in zinc after cooling from an annealing treatment. Consider cooling of a bicrystal consisting of crystal A with [0001] parallel to an external direction, x, and B, with [0001] perpendicular to x as sketched in Figure 18.3. Assume that the cross-sections of A and B normal to x are equal.

For zinc, $\alpha_{\mathrm{B}} = 15 \times 10^{-6}/^\circ\mathrm{C}$, $\alpha_{\mathrm{A}} = 61.5 \times 10^{-6}/^\circ\mathrm{C}$, $E_{\mathrm{A}} = 119\,\mathrm{GPa}$ and $E_{\mathrm{B}} = 34.8\,\mathrm{MPa}/^\circ\mathrm{C}$.

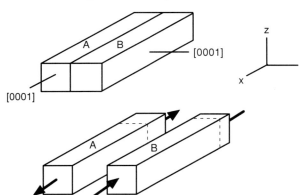

Figure 18.3. Cooling of a bicrystal consisting of two grains with different thermal expansion coefficients. If the grains were not attached, A would contract more than B. Because the contraction must be the same in both grains, grain A is left under residual tension, and B under residual compression.

Solution: An x-direction force balance requires that $\sigma_{\mathrm{xA}} = -\sigma_{\mathrm{xB}}$. If A and B were not attached, A would contract more on cooling than B. They are bonded, however, on a surface parallel to x, so $e_{\mathrm{xA}} = e_{\mathrm{xB}}$. Therefore, $(\sigma_{\mathrm{xA}}/E_{\mathrm{A}}) + \alpha_{\mathrm{A}}\Delta T = (\sigma_{\mathrm{xB}}/E_{\mathrm{B}}) + \alpha_{\mathrm{B}}\Delta T$. Substituting $\sigma_{\mathrm{xB}} = -\sigma_{\mathrm{xA}}$, $\sigma_{\mathrm{xA}} = (\alpha_{\mathrm{B}} - \alpha_{\mathrm{A}})\Delta T/(1/E_{\mathrm{A}} + 1/E_{\mathrm{B}})$.

Thus,

$$\sigma_{\mathrm{xA}} = (\alpha_{\mathrm{A}} - \alpha_{\mathrm{B}})\Delta T/(1/E_{\mathrm{A}} + 1/E_{\mathrm{B}})$$
$$= (61.5 - 15) \times 10^{-6}/^\circ/(1/119 - 1/34.8)\mathrm{GPa}^{-1} = 1.2\,\mathrm{MPa}/^\circ\mathrm{C}!$$

Thus, a temperature change of a few degrees increases the stress enough to cause yielding in some grains. After cooling from an annealing treatment, the residual tensile stress in some grains equals the yield strength of the grain. Any additional stress will cause yielding of that grain. This explains why there is no linear elastic region in tension tests of polycrystalline zinc. Some grains start to deform plastically as soon as any external stress is applied. It is impossible to produce stress-free polycrystalline zinc by annealing because of the intergranular stresses created during cooling from any appreciable annealing temperature.

In multiphase materials, each phase has a different yield strength, different elastic modulus and different thermal expansion coefficient. These differences cause residual stresses to develop after plastic deformation or after temperature changes.

Bauschinger Effect

If a material is loaded first in tension and then compression, the absolute value of the new yield strength in compression is usually lower than the last value of the tensile flow stress. This is called the *Bauschinger effect*. Similar effects occur whenever there is a significant change in loading path. Figure 18.4 shows the effect in torsion when the direction of torsion is reversed. There are several possible causes for this effect. One is dislocation pile-ups at obstacles. See Figure 9.7. In reverse flow, the dislocations can easily move away from the obstacle, even reversing the direction of operation of a Frank-Read source.

Different yield strengths of different regions, A and B, will cause a Bauschinger effect. Figure 18.5 shows the stress–strain curves of two neighboring regions of a material. They must undergo the same strains under tension and unloading. During unloading, the same elastic contraction leaves the weaker grain, A, under residual compression. During subsequent loading in compression, grain A will yield at a stress well below the level of the average tensile stress before unloading. The composite curve will therefore exhibit a low yield stress in compression. The effect can be large enough so that the weaker grain yields in compression during the unloading.

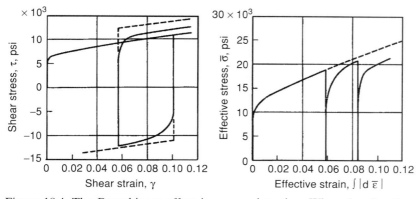

Figure 18.4. The Bauschinger effect in reversed torsion. When the direction of torsion is reversed, the new yield strength is below the last flow stress. From F. McClintock and A. Argon, *Mechanical Behavior of Materials*, Addison-Wesley (1966). From unpublished data of J. A. Meyer, 1957.

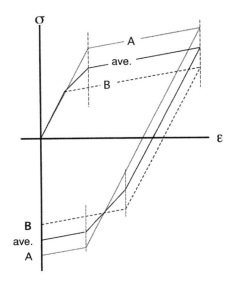

Figure 18.5. Different stress–strain curves of two adjacent grains will result in a Bauschinger effect. Both grains are constrained to undergo the same strains. After unloading, grain A will be left under residual compression when the average stress is zero. On loading in compression, it will yield while the compressive stress in grain B is very small. The average stress–strain curve will show a Bauschinger effect.

The Bauschinger effect can be interpreted as a shift in the yield locus. See the discussion of kinematic hardening in Chapter 5.

Nonuniform Cooling

Larger-scale residual stress patterns are of engineering significance. One source of these is the different cooling rates in different parts of a body after heat treatment. On cooling, the surface always cools faster. The temperature difference between surface and interior temperatures, ΔT, causes a difference in thermal contraction, $\alpha \Delta T$. This difference in thermal contraction is often large enough that it must be accommodated by either a plastic or a viscoelastic contraction of the interior or by expansion of the surface itself. In either case, subsequent cooling and thermal contraction of the interior will leave the surface under residual compression and the center under residual tension. It is often stated that during quenching, compatibility between the cooler surface and interior is maintained by plastic or viscous flow of the "hotter, softer" interior. However, this is not possible in three dimensions. Consider cooling of a sphere. The interior cannot deform to accommodate the surface contraction. That would require its volume to decrease. The same is true of a long thin rod. The interior cannot accommodate both the lengthwise and the circumferential contractions. Axial flow of the interior will be suppressed by the surface. The accommodation requires that the cooler surface deform plastically. Only with large sheets or plates does the flow of the interior accommodate the surface contraction. In this case, the interior is free to thicken.

The situation is more complex if a phase change occurs during the cooling, as in the transformation of austenite to martensite during quenching of steel. The austenite-to-martensite transformation involves a volume expansion. The potential expansion of the surface resulting from the transformation must be accommodated by the surface itself because the volume of the interior cannot change. The volume expansion of the surface must be accommodated by a strain normal to the surface.

Later, transformation and expansion of the interior leaves the surface under residual tension. This tendency toward residual tension at the surface is partially compensated by the normal thermal contraction after the transformation. The net effect depends on the rate of cooling and on the M_s and M_f temperatures.

Nonuniform Material

In a composite material, systematic patterns of residual stresses may form because of different thermal expansion coefficients. Consider a simplified one-dimensional analysis of a composite composed of two materials, A and B. The same dimensional change must occur in both materials when the temperature changes, $\Delta L_A = \Delta L_B$, so

$$\alpha_A \Delta T + \sigma_A / E_A = \alpha_B \Delta T + \sigma_B + E_B. \tag{18.3}$$

A balance of forces requires that

$$\sigma_A A_A = \sigma_B A_B = 0. \tag{18.4}$$

Combining these,

$$\sigma_A = (\alpha_B - \alpha_A) \Delta T / [1/E_A + (A_A/A_B)/E_B]. \tag{18.5}$$

If the problem is two dimensional, as in the case of the composites being arranged as parallel plates or sheets, with no stress normal to the sheets, the compatibility equation for the x-direction is

$$\alpha_A \Delta T + (1/E_A)\sigma_{Ax} - (\upsilon_A/E_A)\sigma_{Ay} = \alpha_B \Delta T + (1/E_B)\sigma_{Bx} - (\upsilon_B/E_B)\sigma_{By}. \tag{18.6}$$

Realizing that symmetry requires that $\sigma_{Ax} = \sigma_{Ay}$ and $\sigma_{Bx} = \sigma_{By}$, the subscripts x and y may be dropped so,

$$\alpha_A \Delta T + (1/E_A)\sigma_A - (\upsilon_A/E_A)\sigma_A = \alpha_B \Delta T + (1/E_B)\sigma_B - (\upsilon_B/E_B)\sigma_B. \tag{18.7}$$

Substituting the force balance,

$$\sigma_A = (\alpha_B - \alpha_A) \Delta T / [(1 - \upsilon_A)/E_A + (A_A/A_B)(1 - \upsilon_B)/E_B]. \tag{18.8}$$

If A is very thin compared to B, $(A_A \ll A_B)$, equation (18.8) simplifies to

$$\sigma_A = [E_A/(1 - \upsilon_A)](\alpha_B - \alpha_A)\Delta T. \tag{18.9}$$

Stresses from Welding

During welding, the region near the weld is hot, whereas the regions remote from the weld are still cool. As the weld region cools, its tendency to undergo a thermal contraction is resisted by the material outside the weld. This results in the weld metal being left in a state of residual tension parallel to the weld. There must be residual compression outside the weld. This is indicated in Figure 18.6. There may also be stresses perpendicular to the weld if it is restrained in the lateral direction or if bending caused by the temperature gradient through the thickness is prevented.

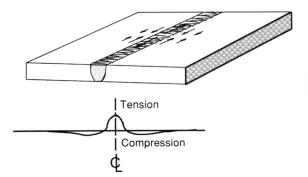

Figure 18.6. Pattern of residual stresses in a welded plate. The weld is left under residual tension.

Stresses from Mechanical Working

Extremely light reductions by rolling or drawing can produce a state of residual compression in the surface. Shot peening causes the same effect. When plastic deformation is limited to the region near the surface and does not penetrate through a piece, the plastic expansion of the surface must be accommodated by an elastic compression of the surface together with an elastic stretching of the interior. This is shown in Figure 18.7.

Except for extremely small reductions, however, the stress at the surface of the work piece is almost always tensile in the direction of prior extension. The magnitudes of the stresses depend on the geometry of the deformation zone. The stresses

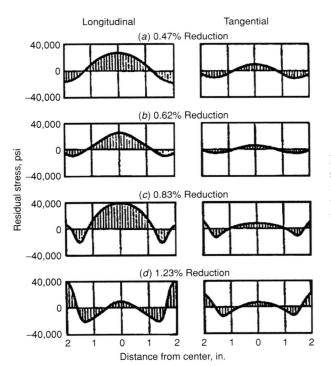

Figure 18.7. Residual stresses in drawn steel rods. Note that the surface is left under residual compression only for very light reductions. From W. M. Baldwin, *Proc. ASTM*, v. 49 (1949).

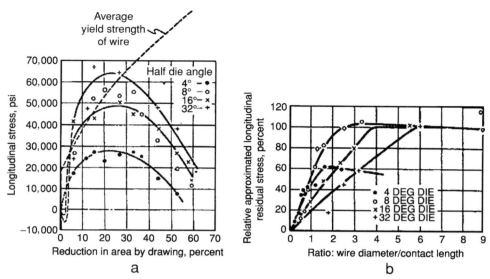

Figure 18.8. (a) Residual stresses in brass wire. (b) Same data replotted as function of Δ (ratio of wire diameter to contact length). From W. M. Baldwin, *Proc. ASTM*, v. 49 (1949).

depend on the ratio, Δ, of the mean thickness, *H*, of the material in the work zone to the contact length between tools and work piece, *L*,

$$\Delta = H/L. \tag{18.10}$$

If Δ ≤ 1, the residual stresses are very small but if Δ ≥ 1 they increase with Δ. Figure 18.8 shows how the longitudinal stress in cold drawn brass wire varies with reduction and the diameter/contact length ratio, Δ. Residual stresses are greatest for low reductions per pass (Figure 18.8a.) They increase with the ratio of wire diameter to contact length, Δ (Figure 18.8b).

When a sheet or plate is bent plastically, the stresses on the outside of the bend are tensile, and those on the inside compressive. On removal of the forces causing the bending, the spring-back reverses the stress pattern, leaving residual compression on the outside of the bend and residual compression on the inside, as shown in Figure 18.9. Tension applied during bending reduces the magnitude of the residual stresses.

EXAMPLE PROBLEM 18.3: A sheet of an ideally plastic material is bent with a pure bending moment and released. Find the pattern of residual stresses.

Solution: Let *t* be the sheet thickness, and let z be the distance from the neutral plane toward the tensile side of the bend. While the bending moment is applied, the stress pattern will be $\sigma = Y$ for $0 \le z \le t/2$ and $\sigma = -Y$ for $0 \ge z \ge t/2$. Unloading of the sheet will be elastic, so the change in stress, $\Delta\sigma$, is proportional to the change in strain, which is in turn proportional to z. Therefore, $\Delta\sigma = Cz$, where C is a constant. For equilibrium after unbending, the internal bending moment must be zero. Therefore, for both the tensile side ($z > 0$) and the compressive side ($z < 0$), $M = \int \sigma z \mathrm{d}z = 0$. Expressing $\sigma = Y - \Delta\sigma = Y - Cz$,

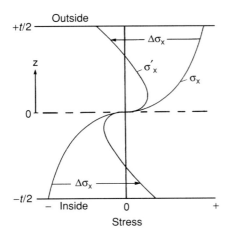

Figure 18.9. Stress distribution under bending and residual stresses after unloading. From W. F. Hosford and R. M. Caddell, *Metal Forming: Mechanics and Metallurgy*, 2nd ed., Prentice Hall (1993).

for the tensile side, $\int (Y - Cz)z\,dz = 0$. Integrating from the centerline to $z = t/2$, $Y(t/2)^2/2 - C(t/2)^3/3 = 0$, so $C = (3/2)Y/(t/2)$. Now substituting, $\sigma = Y - Cz = Y[1 - (3/2)z/(t/2)]$. At the outer surface, $z = t/2$, so $\sigma = -Y/2$ (compression). Similarly for $z < 0$, $\sigma = -Y - Cz$ and at the inner surface where $z = -t/2$, $\sigma = +Y/2$ (tension). See Figure 18.10.

Consequences of Residual Stresses

Among the consequences of residual stresses are the lack of sharp yielding in subsequent tension and changed susceptibilities to fatigue, stress corrosion, and brittle fracture. Figure 18.11 shows a beam that spontaneously split in half as a result of residual stresses from rolling. Fatigue failures, stress corrosion cracks, and brittle fractures all initiate at the surface of a part and tend to occur under tension rather than compression. Therefore, a state of residual compression in the surface is beneficial. Greater external tensile forces can be tolerated for the same level of tensile in the surface layers. Residual tension in the surface is detrimental. Residual stresses cause warping during machining. As soon as a stressed region is removed, the part will deform in such a way as to maintain internal force and moment balances. For example, removal of a surface region that is under tension will cause the part to bend enough to increase the tensile stresses on that side and restore a moment balance.

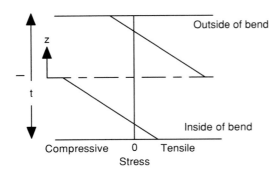

Figure 18.10. Residual stress pattern in a sheet after bending.

Figure 18.11. An I-beam in which residual stresses caused a spontaneous fracture with no external load. At the time of the fracture, the beam was lying flat and the temperature was normal. From E. R. Parker, *Brittle Behavior of Engineering Structures*, Wiley (1957).

Measurement of Residual Stresses

There are two common methods of measuring residual stresses. One is the use of x-ray diffraction to measure changes in lattice spacing, d, between crystallographic planes. The spacing change reflects an elastic strain of the lattice, $e = \Delta d / d$. X-ray measurements of d can be used to find the normal strains in three directions. Then Hooke's law can be used to calculate the state of stress. However, only residual stresses at or near the surface can be determined because x-rays do not penetrate deeply.

The other method relies on measuring dimensional changes when part of the piece is removed by machining, sawing, or etching. The state of stress in the region that was removed can be found from the dimensional changes of the remaining material, using force and moment balances together with Hooke's law. Several examples follow.

If a thin surface layer of thickness, Δt, is removed from a flat plate of thickness, t, the remaining plate will bend with two principal radii of curvature, ρ_x and ρ_y. Here, ρ_x is the radius of curvature about an axis parallel to y, which causes bending a strain parallel to x, and ρ_y is the radius of curvature about an axis parallel to x, as shown in Figure 18.12. Let σ_x and σ_y be the stress in the layer that has been removed. The removal of this layer will cause a change in bending moment per length,

$$\Delta M_{xL} = (t/2)\Delta t \sigma_y. \qquad (18.11)$$

This change in moment per length must be accommodated by an elastic bending of the sheet about the x axis such that

$$\Delta M_x = 2\int_0^{t/2} \Delta \sigma_y z \, dz \qquad (18.12)$$

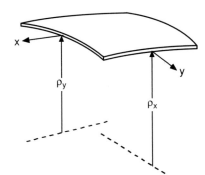

Figure 18.12. Bent sheet with two radii of curvature, ρ_x and ρ_y.

where

$$\Delta\sigma_y = E(\Delta\varepsilon_y + v\Delta\varepsilon_x)/(1 - v^2).$$
(18.13)

Because $\Delta\varepsilon_y$ and $\Delta\varepsilon_x$ must vary linearly from center to surface,

$$\Delta\varepsilon_y = z/\rho_y \quad \text{and} \quad \Delta\varepsilon_x = z/\rho_x.$$
(18.14)

and

$$\Delta\varepsilon_x = z/\rho_x.$$
(18.15)

$\Delta M_x = [2/(1 - v^2)] \int_0^{t/2} E(1/\rho_y + 1/\rho_x)z^2 dz.$ Integrating,

$$\Delta M_x = [E/(1 - v^2)](1/\rho_y + v/\rho_x)[(t/2)^3/3]$$
$$= (1/12)[E/(1 - v^2)](1/\rho_y + v/\rho_x)t^3.$$
(18.16)

Combining equations (18.16) and (18.11),

$$(t/2)\Delta t\sigma_y = (1/12)[E/(1 - v^2)](1/\rho_y + v/\rho_x)t^3,$$
(18.17)

or

$$\sigma_y = [E/(1 - v^2)](1/\rho_y + v/\rho_x)t^2/(6\Delta t).$$
(18.18)

Similarly,

$$\sigma_x = [E/(1 - v^2)](1/\rho_x + v/\rho_y)t^2(6\Delta t).$$
(18.19)

If one of the radii is infinite,

$$\sigma = [E/(1 - v^2)]t^2(6\rho\Delta t),$$
(18.20)

whereas, if the two radii are equal,

$$\sigma_x = [E/(1 - v^2)](1 + v)t^2/(6\rho\Delta t).$$
(18.21)

A second layer can be removed to find the stress in that layer. Now, however, equations (18.18) and (18.19) will yield the stress in second layer after the first layer was removed. The change of stresses induced by the removal of the second layer is

$$\Delta\sigma_x = [E/(1 - v^2)][\Delta(1/\rho_x) + v\Delta(1/\rho_y)]t^2(6\Delta t) \text{ and}$$

$$\Delta\sigma_y = [E/(1 - v^2)][\Delta(1/\rho_y) + v\Delta(1/\rho_x)]t^2(6\Delta t).$$
(18.22)

The residual stress profile in a rod or tube can be found by drilling successively larger holes and noting the changes in length and diameter.

EXAMPLE PROBLEM 18.4: A thin layer (25 μm) was removed from a thin sheet (1.0 mm) of a plastic. After the layer was removed, the sheet became dish shaped with equal radii of curvature, $\rho = 50$ cm. Assume that the elastic modulus of the plastic is 4 GPa and $\upsilon = 0.4$. Calculate the stress that existed in the layer that was removed.

Solution: Using equation (18.21), $\sigma_x = (4 \text{ GPa})(1.4)(10^{-3} \text{ m})^2 / [(6)(0.5 \text{ m})(25 \times 10^{-6} \text{ m})(1 - 0.4^2)] = 75$ MPa.

Relief of Residual Stresses

To remove residual stresses in a material, elastic strains must be converted to plastic strains. During a stress relief anneal, the elastic strains in the material are converted to plastic strains by creep. Higher temperatures accelerate the process. As the magnitude of the residual stresses diminishes, the driving force for additional creep also decreases, and the process of stress relief slows. For this reason, complete stress relief is not possible during conventional stress relief anneals. Stress relief can be modeled as a spring and a dashpot in series under a fixed total strain (Figure 18.13). As the dashpot operates, the spring relaxes lowering the stress. This is analogous to the series spring–dashpot model in Chapter 16, except that the dashpot should be modeled by a power law creep (e.g., equation (16.6)), instead of Newtonian viscosity.

Figure 18.14 shows the effects of temperature and time on stress relief of a titanium alloy.

EXAMPLE PROBLEM 18.5: At a specific location in a material, the level of residual stress is σ_o. The material is given a stress relief anneal. At the annealing temperature, the material has a plastic creep rate of $\dot{e}_p = -C\sigma^5$. Assuming the model of stress relief in Figure 18.13 for which the dimensions remain constant, write an equation that describes how the level of residual stress, σ, decreases with time. Let the elastic modulus be E, and for simplicity, assume a uniaxial stress.

Solution: The elastic strain is $\varepsilon_e = \sigma/E$. Because the dimensions remain constant, $\dot{e}_p = -\dot{e}_e$, so $(1/E)d\sigma/dt = -C\sigma^5$ or $\sigma^5 d\sigma = -CEdt$. Integrating from $\sigma = \sigma_o$ at $t = 0$ $(\sigma^4 - \sigma_o^{-4})/4 = CEt$, $\sigma(\sigma_o^{-4} + 4CEt)^{-0.25}$.

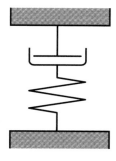

Figure 18.13. Model for stress relief by creep. The dashpot obeys a creep law, $\dot{e} = C\sigma^n$. This model predicts that the rate of stress relief will decrease with time.

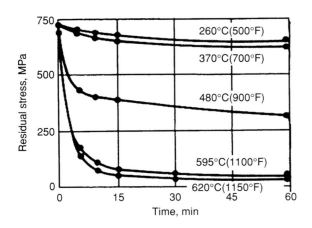

Figure 18.14. Effect of temperature and time on the relief of residual stresses in titanium alloy Ti-6Al-4V. Note that the rate of stress relief increases with temperature but becomes very slow at long times. From *Metals Handbook*, 9th ed. v. 4., ASM (1981).

The use of ultrasonic vibrations has been promoted as an alternative method of relieving residual stresses. The vibrations cause local stresses to cycle about their at-rest values. Wherever the stress rises above that required for local plastic deformation, some of the elastic strain is converted to plastic deformation. This reduces the residual stress in a manner similar to annealing. Complete relief of stresses is not possible without plastically deforming the entire part.

An almost complete stress relief can be achieved in rods by applying enough tension to cause plastic elongation. The amount of deformation need only be great enough to cause the entire cross-section to yield in tension. Unloading will be entirely elastic, so the differences in stress after unloading will be the same as the differences under load. These are small and depend on the slope of the initial portion of the stress–strain curve.

REFERENCES

F. A. McClintock and A. S. Argon, *Mechanical Behavior of Materials*, Addison-Wesley (1966).

G. E. Dieter, *Mechanical Metallurgy*, chapter 15, McGraw-Hill (1961). (Only the first edition treats residual stresses.)

W. F. Hosford and R. M. Caddell, *Metal Forming: Mechanics and Metallurgy*, 3rd ed. Cambridge (2007).

Notes

Tempered safety glass is manufactured by cooling of hot glass sheets by forced air. This leaves the surface under residual compression, making the glass much more resistant to fracture under bending. More important, if a crack starts and penetrates through to the interior, which is under residual tension, it rapidly spreads. The glass breaks into many small pieces that are much less dangerous than the large shards formed when ordinary glass fractures.

Cartridge brass (65 Cu/35 Zn) is susceptible to stress corrosion cracking in atmospheres containing ammonia. Cracks run along grain boundaries under tension. Stress relieving of brass was not common before the 1920s. During World War I, the U.S. Army commandeered barns in France to use as ammunition depots. Unfortunately, the ammonia produced by decomposition of cow urine caused brass cartridge cases to split open. Today, the army requires stress relief of all brass cartridge cases.

Johann Bauschinger (1833–1893) was the director of the Polytechnical Institute of Munich. In this position, he installed the largest testing machine of the period and instrumented it with a mirror extensometer so that he could make very accurate strain measurements. In addition to the discovery that prior straining in tension lowered the compressive yield strength and vice versa (an effect named after him), Bauschinger did the original research on the yielding and strain aging of mild steel.

Problems

1. Consider a piece of polycrystalline iron that has been plastically deformed in tension under a stress of 220 MPa and then unloaded. Because of the orientation dependence of the Taylor factor, it is reasonable to assume that the stress before unloading was 20% higher in grains oriented with <111> parallel to the tensile axis than in the average stress. Young's modulus for polycrystalline iron is listed as 208 GPa but for crystals oriented in a <111> direction, it is 283 GPa. Determine the level of residual stress in the <111>-oriented grains.

2. The residual stresses adjacent to a long butt weld between two deck plates of a ship were found by x-rays. The average values of the lattice parameter, a, were $a = 2.8619$ parallel to the weld bead and $a = 2.86106$ perpendicular to the weld bead. For unstrained material, $a = 2.8610$. Find the values of the residual stresses parallel and perpendicular to the weld. Assume $E = 30 \times 10^6$ psi, $\upsilon = 0.29$ and that the stress normal to the plate is zero.

3. The stresses in the walls of a deep-drawn stainless steel cup are suddenly released when the walls split by stress corrosion cracking as shown in Figure 18.15. After splitting, the segments of the wall curved to a radius of curvature of 10 in. Assume the stresses drop to zero. The wall thickness is 0.030 in., and Young's modulus is 30×10^6 psi. What were the residual stresses in the wall before stress corrosion cracking? Note that if a narrow strip is bent elastically, the change of stress is given by $\Delta \sigma = E \Delta \varepsilon$ and $\Delta \varepsilon = z/\rho$, where z is the distance from the neutral axis and ρ is the radius of curvature. Therefore, $\Delta \sigma$ varies linearly through the thickness and is a maximum at the surfaces.

Figure 18.15. Stress corrosion cracks in a drawn cup. From W. F. Hosford and R. M. Caddell, *Metal Forming: Mechanics and Metallurgy*, 3rd ed., Cambridge (2007).

4. Consider a material composed of equal volumes of having different yield strengths, Y_A, and $Y_B = 1.5 Y_A$, but the same elastic moduli ($E_A = E_B$). Assume that during straining both regions undergo the same strains ($\varepsilon_A = \varepsilon_B$) and that there is no strain hardening in either region.

A. Let the material be subjected to a tensile strain $\varepsilon_A = \varepsilon_B = 2Y_B/E$. Sketch the individual and overall stress–strain curves, σ_A, σ_B and $\sigma_{av} = (\sigma_A + \sigma_B)/2$. What is the overall yield strength of the material (i.e., value of σ_{av} corresponding to the first deviation from linearity)?

B. Now consider the behavior on unloading to $\sigma_{av} = 0$. Add plots of σ_A, σ_B, and σ_{av} to your sketch. What is the level of residual stress in each region?

C. After loading in tension and unloading, consider the behavior on loading under compression until the entire material yields. Assume that the tensile and compressive yield strength of each region is the same. ($|Y_{Acomp}| = Y_{Atens}$) and ($|Y_{Bcomp}| = Y_{Btens}$). Add the compressive behaviors to your plot. What is the new overall yield strength in compression, and how does this compare with what the overall yield strength would be if it had not been first loaded in tension?

5. A 1-cm-diameter steel ball is cooled after austenitization. At one point during the cooling, a 0.5-mm-thick layer at the surface transforms to martensite while the center is still austenite. For simplicity, assume that the interior is at an average temperature of 200°C and the surface is at 20°C, and that there are no stresses in the material at this point. When the center cools to 20°C, it transforms to martensite. For this steel, the austenite-to-martensite transformation is accompanied by a 1.2% volume expansion. Assume this expansion occurs equally in all directions. For steel, the coefficient of linear thermal expansion is $\alpha = 6 \times 10^{-6}/°C$, and Young's modulus is 30×10^6 psi and $\upsilon = 0.29$. Assume these apply to both martensite and austenite. Find the stress state at the surface.

6. If a metal sheet is bent plastically under a bending moment without applied tension, there will be spring-back after the moment is released, and there will be residual stresses in the sheet as shown in Figure 18.9. If the surface layers of such a sheet are removed by etching or corrosion, how will the bend change? Will the radius of the bend increase, decrease, or remain unchanged? Explain.

7. A surface layer, 0.001 in. thick was removed from an aluminum sheet, 0.015 in. thick. On removal of the layer, the sheet curled to form a dish with a radius of curvature of 30 in. What was the stress in the surface layer before it was removed. For aluminum, $E = 10 \times 10^6$ psi and $\upsilon = 0.30$.

8. For polycrystalline magnesium the coefficient of thermal expansion is $25.2 \times 10^{-6}/°C$ and Young's modulus is 68.4 GPa. Parallel to the c axis of a single crystal, the coefficient of thermal expansion is $24.3 \times 10^{-6}/°$ and Young's modulus is 80.9 GPa. Estimate the stress parallel to the c-axes of grains in a randomly oriented polycrystal when the temperature is changed by 100°C. Assume that the dimensional change parallel to the c-axis of the grain is the same as that in the polycrystal. For simplicity, assume also that the stress in the polycrystal is negligible.

9. A typical residual stress pattern in an extruded bar is shown in Figure 18.16. To find the residual stresses in an extruded bar of brass, 1.000 in. diameter and 10.000 in. long, the bar was put in a lathe and machined to a diameter of 0.900 in. After machining, the length was found to be 10.004 in. What was the average residual stress in the layer that was machined away? For the brass, $E = 110$ GPa. For simplicity, neglect the Poisson effects.

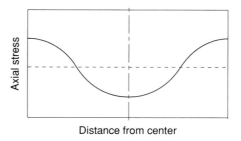

Figure 18.16. Residual stress pattern in an extruded bar.

10. During a stress relief anneal, creep converts elastic strains (and therefore stresses) into plastic strains. The decrease in stress should be $\Delta\sigma = -Ee$, or $d\sigma/dt = -Ede/dt = -E\dot{e}$. Assume that the stress relief anneal is done in a temperature range where the basic creep rate is given by $de/dt = \dot{e} = C\sigma^m$ and that the residual stress before annealing was σ_o.

A. If $m = 5$, as is typical of creep at the low temperatures used for stress relieving, $\sigma_\mathrm{o} = 5{,}000$ psi and $E = 10 \times 10^6$ psi, and $C = 6 \times 10^{-22}$ (psi)$^{-5}$/hr at the annealing temperature, how long would it take for the stress to drop to 2,500 psi?

B. For the same temperature, how long would it take the stress to drop to 1,000 psi?

19 Ceramics and Glasses

Introduction

Ceramics are compounds of metals and nonmetals. Most ceramics are hard and have limited ductility at room temperature. Their tensile strengths are limited by brittle fracture, but their compressive strengths are much higher. Ceramics tend to retain high hardnesses at elevated temperatures; thus, they are useful as refractories such as furnace linings and as tools for high-speed machining of metals. Most refractories are oxides, so unlike refractory metals, oxidation at high temperature is not a problem. The high hardness of ceramics at room temperature leads to their use as abrasives, either as loose powder or bonded into grinding tools. The low ductility of ceramics limits the structural use of ceramics mainly to applications in which the loading is primarily compressive. This chapter also covers glasses, both ceramic and metallic.

Elastic Properties

Ceramics, particularly those with strong bonding, tend to have high elastic moduli. For ionically bonded binary compounds, the modulus is proportional to the derivative of the bonding energy,

$$dU/dr = a|z_A||z_B|r_o^2, \tag{19.1}$$

where z_A and z_B are the ionic valences and r_o is the distance between ions (sum of the ionic radii). The *Madelung constant*, a, is characteristic of the crystal structure. Table 19.1 lists the moduli of some ceramics.

EXAMPLE PROBLEM 19.1: Predict the ratio of the Young's moduli for NaCl and MgO, both of which have the rock salt structure. The radii of the ions Na^+ and Cl^- are 0.097 and 0.181 nm, respectively, and the radii of Mg^{+2} and O^{-2} are 0.066 and 0.140 nm, respectively.

Solution: Because both have the same structure, α is the same for both so

$$E_{NaCl}/E_{Mgo} = [|(+1)(-1)|/(0.097 + 0.181)^2]/[|(+2)(-2)|/(0.66 + 0.140^2]$$
$$= 0.138.$$

The experimental values of the moduli are 36 GPa for NaCl and 305 GPa for MgO, so the ratio is $36/305 = 0.12$, which is reasonably close to the predictions.

Table 19.1. *Young's moduli for some ceramics*

Material	Young's modulus	
	GPa	(10^6 psi)
Al_2O_3	400	58.0
B_4C	445	65.0
CaF_2	110	16.0
CsCl	25	3.6
Diamond	1,000	145.0
Fe_3O_4	200	29.0
LiF	103	15.0
MgO	305	44.0
NaCl	36	5.2
SiC	450	65.0
Soda-lime glass	75	11.0
SiO_2	96	14.0
TiC	437	63.0
ZnS	86	12.5
ZrO_2	200	29.0

Slip Systems

Ceramics tend to be brittle because the number of independent slip systems is limited. In an ionically bonded crystal, the slip plane and direction must be such that during slip there is not close contact between ions of like sign. This is illustrated in Figure 19.1. For some crystal structures, there are more than one type of slip system, each requiring different stresses to operate.

Table 19.2 lists the slip systems for a number of structures that can be easily be activated. As discussed in Chapter 7, five independent slip systems are required to accommodate an arbitrary shape change are required for deformation of polycrystals. Few of the ceramics have five, except at elevated temperatures.

Other slip systems can operate at high temperatures. Table 19.3 gives the critical resolved shear stress for several slip systems in Al_2O_3 at high temperatures.

Hardness

Both hardness and modulus increase with bond strength, so a ceramics with a high modulus of elasticity tends to have a high hardness. Table 19.4 lists the hardnesses of several of ceramics.

Figure 19.1. Schematic illustration of the principle that the slip systems are such that, on translation, like-sign ions do not come in contact. In (a), the mutual repulsion between ions of like sign during slip would cause fracture. In (b), there is no close contact between ions of like sign during slip.

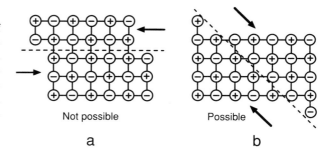

Not possible   Possible

a   b

Table 19.2. *Slip systems of several ceramic structures*

Structure			Slip systems	No. of independent slip systems
Rock salt	NaCl(LiF, MgO)	(RT)	$<110>\{110\}$	2
		(high temp)	$\&<110>\{001\}$	3
Rock salt	AgCl	(RT)	$<110>\{001\}, <110>\{001\}$	
			$\&<110>\{111\}$	5
Cubic cesium Chloride	CsCl(CsBr)	(RT)	$<100>\{001\}$	3
Cubic fluorite	CaF$_2$(BaF$_2$)	(RT)	$<110>\{001\}$	3
	" &UO$_2$	(high temp)	$\&<110>\{110\}$	3
Rutile	TiO$_2$	(high temp)	$<110>\{1\bar{1}0\}\&<110>\{001\}$	3
Hexagonal	graphite,			
	Al$_2$O$_3$, BeO	(high temp)	$<11\bar{2}0>(0001)$	2

Table 19.3. *Critical resolved shear stresses for basal and prism slip in Al$_2$O$_3$ of Al$_2$O$_3$ at 25°C and 1,000°C*

Slip system	Critical resolved shear stress, τ_c, 25°C	(MPa) 1,000°C
$(0001)<2\bar{1}\bar{1}0>$	30,000 (extrapolated)	130
$(0\bar{1}10)<2\bar{1}\bar{1}0>$	5,000	330

Table 19.4. *Hardness of several ceramics*

Material	Vickers hardness (kg/mm^2)
Soda-lime glass	600
Fused silica	650
Si$_3$N$_4$	1,700
Al$_2$O$_3$	1,750
SiC	2,600
B$_4$C	3,200

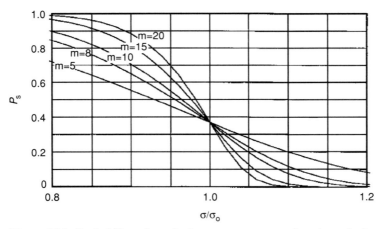

Figure 19.2. Probability of survival at a stress, s, as a function of σ/σ_o. The relative scatter depends on the "modulus," m.

Weibull Analysis

Data on fracture strength typically have a large amount of scatter. In Chapter 14, it is shown that the fracture strength, σ_f, of a brittle material is proportional to $K_{Ic}/\sqrt{(\pi a)}$, where a is the length of a preexisting crack. The scatter of fracture strength data is caused mainly by statistical variations in the length of preexisting cracks in specimens. For engineering use, it is important to determine not only the average strength, but also the amount of its scatter. If the scatter is small, one can safely apply a stress only slightly below the average strength. On the other hand, if the scatter is large, the stress in service must be kept far below the average strength. Weibull* suggested that in a large number of samples, the fracture data could be described by

$$P_\sigma = \exp[-(\sigma/\sigma_o)^m], \qquad (19.2)$$

where P_σ is the probability that a given sample will survive a stress of σ without failing. The terms σ_o and m are constants characteristic of the material. The constant σ_o is the stress level at which the survival probability is $1/e = 0.368$ or 36.8%. A large value of m in equation (19.2) indicates very little scatter, and conversely, a low value of m corresponds to a large amount of scatter. This is shown in Figure 19.2.

Because

$$\ln(-\ln P_\sigma) = m\ln(\sigma/\sigma_o), \qquad (19.3)$$

a plot of P_σ on a log(log) scale versus σ/σ_o on a log scale is a straight line with the slope of m, as shown in Figure 19.3.

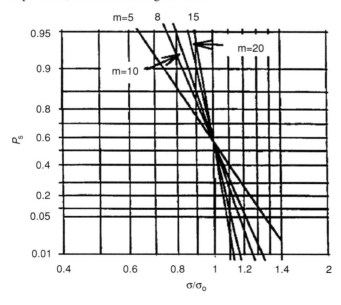

Figure 19.3. Probability of survival at a stress, σ, as a function of σ/σ_o. Note that this is the same as Figure 19.2, except here P_s is plotted on a log(log) scale and σ/σ_o on a log scale. The slopes of the curves are the values of m.

* W. Weibull, *J. Appl. Mech.*, v. 18 (1951), p. 293, and *J. Mech. Phys. Solids*, v. 8 (1960).

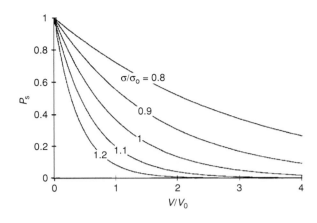

Figure 19.4. Expected variation of strength with volume according to equation (19.5) with m = 5. The survival rate decreases with increasing stress level, σ, and increasing sample volume, V.

EXAMPLE PROBLEM 19.2: Analysis of 100 tests on a ceramic indicated that 11 specimens failed at 85 MPa or less and that 53 failed at a stress of 100 MPa or less. Estimate the stress that would cause 1 failure in 10,000.

Solution: $P_s = \exp[-(\sigma/\sigma_o)^m]$, $\ln P_s = -(\sigma/\sigma_o)^m$, $\ln P_{s2}/\ln P_{s1} = (P_2/\sigma_1)^m$, $m = \ln[\ln P_{s2}/\ln P_{s1}]/\ln(\sigma_2/\sigma_1)$. Substituting $P_{s2} = 1 - 0.53 = 0.47$ at $\sigma_2 = 100$ MPa and $P_{s1} = 1 - 0.11 = 0.89$ at $\sigma_1 = 85$ MPa, $m = \ln[\ln(0.47)/\ln(0.89)]/\ln(100/85) = 11.50$. $(\sigma_3/\sigma_1) = (\ln P_{s3}/\ln P_{s1})^{1/m} = [\ln(0.9999)/\ln(0.47)]^{1/11.5} = 0.46$. $\sigma_3 = (46)(85) = 39$ MPa.

For brittle materials, P_s depends on the volume of the specimen as well as on the stress. This is because the probability of a large flaw is greater in larger specimens. Let the survival rate for samples of volume, V_o, be P_{so}. If the sample volume is increased to V, the survival probability, P_s, decreases to

$$P_s = (P_{so})^{V/V_o}. \tag{19.4}$$

Substituting equation (19.2),

$$P_s = \exp[-(V/V_o)(\sigma/\sigma_o)^m]. \tag{19.5}$$

Figure 19.4 shows the decrease of survival rate with sample volume for several levels of σ/σ_o.

Testing

Because of brittleness, special care is needed in measuring the strength of ceramics. Tension tests are seldom used because of the difficulty in making specimens with reduced sections. Also, any slight misalignment of the specimen would cause bending stresses in addition to the applied tension. (This is not a problem with materials that deform plastically because the bending can be accommodated with a very small amount of plastic deformation.) Instead, most testing is done with three- or four-point bending tests. See Chapter 3.

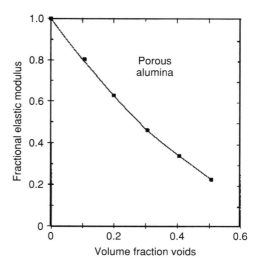

Figure 19.5. Decrease of the elastic modulus with porosity. The fractional modulus is ratio of the modulus of porous alumina to that of 100% dense alumina. Data from R. L. Coble and W. D. Kingery, *J. Amer. Cer. Soc.*, v. 29 (1956).

Porosity

Most ceramic objects are made from powder by pressing and sintering, so most contain some porosity. Porosity is desired in materials used for insulation and filters, but for most applications, porosity is undesirable because it adversely affects the mechanical properties. The elastic modulus decreases with porosity. This is shown in Figure 19.5 for the case of alumina. It can be seen that a 10% porosity causes a decrease of the modulus of about 20%. The effects of porosity on strength and on creep rate at elevated temperatures are even more pronounced, as illustrated in Figures 19.6 and 19.7. The solid line in Figure 19.6 is the prediction of the equation

$$\sigma = \sigma_0 \exp(-bP), \tag{19.6}$$

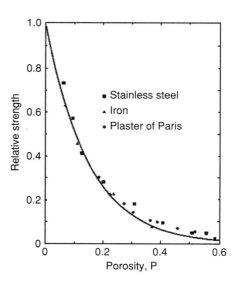

Figure 19.6. Decrease of fracture strength of ceramics and metals with porosity. The solid line is equation (20.6) with b = 6. From R. L. Coble and W. D. Kingery, *J. Amer. Cer. Soc.*, v. 29 (1956).

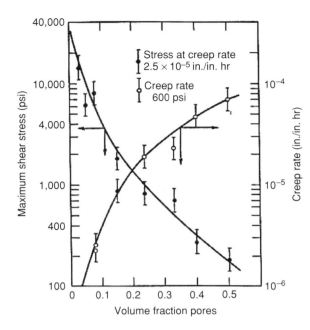

Figure 19.7. Effect of porosity on the strength and creep rate of alumina at 1,000°C. From R. L. Coble and W. D. Kingery, *J. Amer. Cer. Soc.*, v. 29 (1956).

where P is the volume fraction porosity and the constant, b, is about 7. Fracture toughness also falls precipitously with porosity.

EXAMPLE PROBLEM 19.3: When the porosity of a certain ceramic is reduced from 2.3% to 0.5% the fracture stress drops by 12%. Assuming equation (19.8), how much would the fracture stress increase above the level for 1/2% porosity if all porosity were removed?

Solution: $\sigma_2/\sigma_1 = \exp(-bP_2)/\exp(-bP_1) = \exp[(P_1 - P_2)]$, $b = \ln(\sigma_2/\sigma_1)/(P_1 - P_2)$. Substituting $\sigma_2/\sigma_1 = 1/0.88$ and $P_1 - P_2 = 0.023 - 0.005 = 0.018$, $b = \ln(1/0.88)/0.018 = 7.1$. $\sigma_3/\sigma_2 = \exp[-b(P_3 - P_2)] = \exp[-7.1(0 - 0.005)] = 1.036$, or a 3.6% increase.

High-Temperature Behavior

Yield strengths drop with increasing temperature. Creep becomes significant at temperatures above about half of the melting point. Figure 19.8 shows the effect of temperature on the strength of Al_2O_3 single crystals.

Figure 19.9 is an Arrhenius plot of creep rate in mullite versus reciprocal temperature.

The mechanisms of creep are the same as those found in metals, except that there is the additional possibility of viscous deformation of a glassy phase in the grain boundaries. Figure 19.10 summarizes the deformation mechanisms in MgO of 10 mm grain size.

Fracture Toughness

Although all ceramics are brittle, there are significant differences in toughness among ceramics. Typical values of K_{Ic} given in Table 19.5 range from less than 1 to about 7 MPa$\sqrt{m}$. If K_{Ic} is less than about 2, extreme care must be exercised in

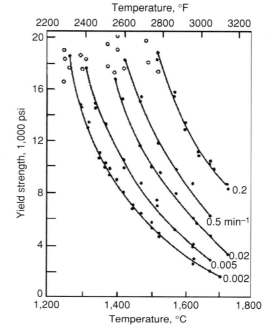

Figure 19.8. Dependence of yield strength of single crystal Al_2O_3 on temperature and strain rate at 1,275°C. From M. L. Kronberg, *J. Amer. Cer. Soc.*, v. 29 (1956).

handling the ceramics. They will break if they fall on the floor. This limits severely the use of ceramics under tensile loading. Ceramics having K_{Ic} greater than about 4 are quite robust. For example, partially stabilized zirconia is used for metal-working tools.

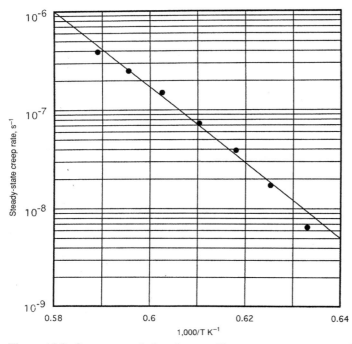

Figure 19.9. Creep rate of glass-free mullite at a constant stress of 100 MPa as a function of temperature. Data from *Engineered Materials Handbook, Vol. 4, Ceramics and Glass,* ASM (1991).

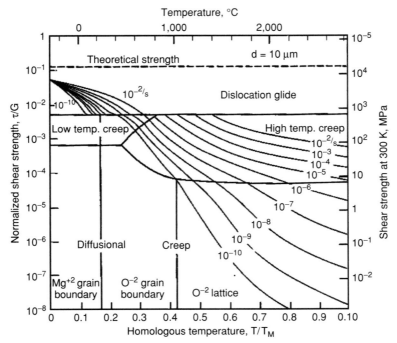

Figure 19.10. Deformation mechanism map for MgO with a grain size of 10 mm. From M. Ashby, *Acta Met.*, v. 20 (1972).

Anstis et al.[*] proposed that the toughness, K_c, of a brittle material can be found by examining the radial cracks around a hardness indentation and applying the following relation

$$K_c = A(E/H)^{1/2}\left(P/c^{3/2}\right),\qquad(19.7)$$

where A is a material-independent constant $= 0.016 \pm 0.004$; E, H, and P are the Young's modulus, hardness, and indenting force; and c is the crack length. See Figure 19.11. Good agreement was found with conventional testing measurements, as shown in Figure 19.12.

Toughening of Ceramics

Ceramics can be toughened by energy-absorbing mechanisms. There are three basic ways: by deflection of cracks, by bridging, and by phase transformation.

Polycrystalline ceramics are usually tougher than single crystals of the same material because grain boundaries deflect cracks. In a polycrystal, the plane of a cleavage crack must change from grain to grain. Therefore, the crack is generally not normal to the tensile stress. The stress intensity factor, K, at the tip of a crack that is not normal to the tensile stress is lower than that at the tip of a crack that is. This effect is evident in Table 19.5 by comparing the toughnesses of single-crystal Al_2O_3 with polycrystalline Al_2O_3 and the toughnesses of typical glasses with crystallized glass *(glass ceramic)*.

---

[*] G. R. Anstis, P. Chantikul, B. R. Lawn, and D. B. Marshall, *J. Amer. Cer. Soc.*, v. 60 (1981).

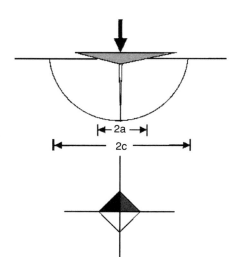

Figure 19.11. Schematic drawing of cracks around a hardness indenter.

Fibers, whiskers, or elongated grains of a second phase may form ligaments that bridge across an open crack and continue to carry some load (Figure 19.13). In this case, shearing of the bonds between the matrix and these ligaments will absorb additional energy. This principle was used in ancient times to toughen clay bricks

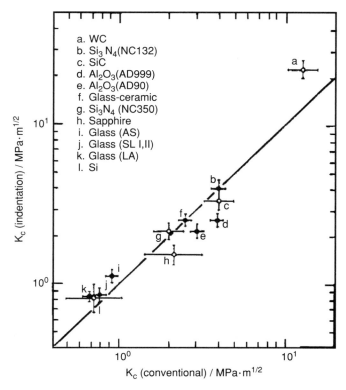

a. WC
b. Si_3N_4(NC132)
c. SiC
d. Al_2O_3(AD999)
e. Al_2O_3(AD90)
f. Glass-ceramic
g. Si_3N_4 (NC350)
h. Sapphire
i. Glass (AS)
j. Glass (SL I,II)
k. Glass (LA)
l. Si

Figure 19.12. Correlation of the values of fracture toughness measured from hardness indentations with the values measured by conventional means. Vertical bars are one standard deviation of $P/c_o^{3/2}$. Horizontal bars are the error in conventional measurements. From Anstis et al. *Ibid.*

Table 19.5. *Fracture toughness of some ceramics*

Material	K_{Ic} (MPa$\sqrt{m}$)
Al_2O_3 (single crystal)	2.20
Al_2O_3 (polycrystal)	4.00
Mullite (fully dense)	2.00–4.00
ZrO_2 (cubic)	3.00–3.60
ZrO_2 (partially stabilized)	3.00–15.00
MgO	2.50
SiC (hot pressed)	3.00–6.00
TiC	3.00–6.00
WC	6.00–20.00
Silica (fused)	0.80
Soda-lime glass	0.82
Glass ceramics	2.50

with straw or horsehair. The toughening effect can be increased by increasing the volume fraction of the reinforcing phase, weakening the bonds between the fibers and the matrix, or by using fibers of lower elastic modulus. Figure 19.14 shows the toughening effect of SiC fibers in several ceramics.

A martensitic transformation in zirconia is often used to increases its toughness. With the addition of 7% CaO (or an appropriate amount of MgO), the high temperature structure of ZrO_2 consists of two solid solutions: one cubic and one tetragonal. At low temperatures, the tetragonal zirconia should transform to a monoclinic structure, but the transformation is normally suppressed. A sufficient stress at room temperature, however, will cause the transformation to occur martensitically, with a volume expansion of about 4% and a shear strain of about 14%. The high stresses at the tip of an advancing crack trigger this reaction. Some of the energy applied externally is absorbed by the transforming particles, so less energy is available for crack propagation. The volume expansion of the particles causes compressive stresses in the matrix, which "shields" the crack tip from the full stress intensity. This is illustrated schematically in Figure 19.15.

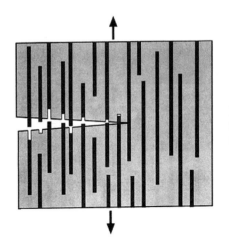

Figure 19.13. Schematic illustration of toughening by reinforcing ligaments. Shearing between the fibers and matrix during pull-out absorbs additional energy.

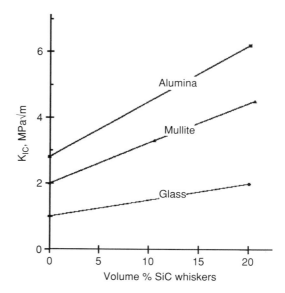

Figure 19.14. Toughening of several ceramics by addition of SiC fibers. Data from P. Becher *J. Amer. Cer. Soc.*, v. 74 (1991).

Fatigue

Fatigue is caused by plastic deformation under cyclic loading. Most ceramics do not deform plastically at room temperature, so room temperature fatigue is not a problem in ceramic materials. However, fatigue is possible at temperatures high enough to allow creep.

Silicate Glasses

The structure of pure silica, SiO_2, consists of tetrahedra with a Si^{+4} ion in the center and an O^{-2} ion at each corner. Thus each Si^{+4} ion is covalently bonded to four O^{-2} ions, and each O^{-2} ion is covalently bonded to two Si^{+4} ions, as shown schematically in Figure 19.16. This entirely covalently bonded structure has a very high melting point and is too viscous to shape at temperatures that can be reached economically.

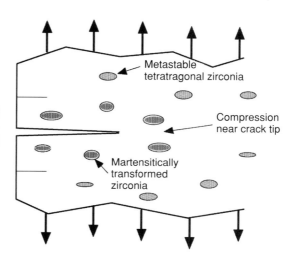

Figure 19.15. Stress-induced transformation of zirconia from tetragonal to monoclinic structures, creating a region of compression around the crack tip.

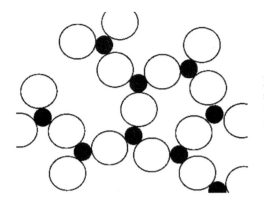

Figure 19.16. Schematic illustration of the structure of silica. Each O^{-2} ion bonds two tetrahedra together. In three dimensions, the Si^{+4} ions are surrounded by four O^{-2} ions and the packing is tighter.

Additions of soda (Na_2O), lime (CaO), and other alkali and alkaline earth oxides modify the network and decrease the working temperature. The added O^{-2} ions replace some of covalent Si-O-Si covalent bonds with weaker $(Si-O)^{-}$-Ca^{+2}-$(O-Si)^{-}$ ionic bonds, as illustrated in Figure 19.17. This lowers the viscosity and allows shaping at temperatures that can be reached easily.

Figure 19.18 is a plot of the viscosities of several glass compositions as a function of temperature. Note that above the glass transition temperatures, the temperature dependence of the viscosity can be described by an Arrhenius equation,

$$\eta = A \exp[-Q/(RT)]. \tag{19.8}$$

Below their glass transition temperatures, however, glasses should be regarded as solids rather than super cooled liquids. The viscosity increases much more rapidly with decreasing temperature than indicated by equation (19.8).

The "working range" (10^3–10^7 Pa-s) is the temperature range in which glass can be shaped economically. Stress relief occurs in the annealing range ($10^{11.5}$–$10^{12.5}$ Pa-s). Increased contents of alkali and alkaline earth oxides increase coefficients of thermal expansion, while lowering viscosities. Table 19.6 shows coefficients of thermal expansion of several glass compositions, and Figure 19.19 shows the relationship between the coefficient of thermal expansion and viscosity.

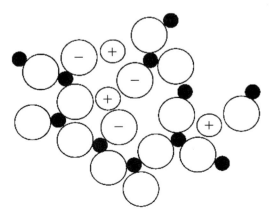

Figure 19.17. Schematic illustration of the structure of a glass-containing alkali or alkaline earth ions. Some covalent O-Si bonds are replaced by weaker ionic bonds between oxygen and Na or K ions. In three dimensions, the Si^{+4} ions are surrounded by four O^{-2} ions.

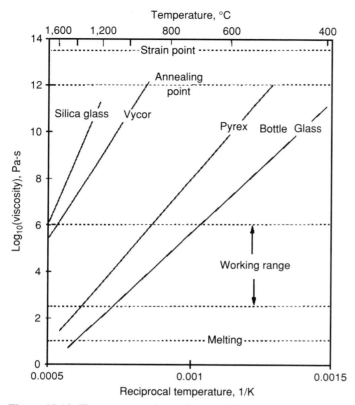

Figure 19.18. Temperature dependence of viscosity for several glasses. The "working range" is the temperature range in which glasses can be economically shaped. The straight lines on the semilog plot do not extend below the glass transition temperature.

Glasses are brittle at room temperature. Their fracture resistances can be improved by inducing a pattern of residual stress with the surface under compression. The surface compression must, of course, be balanced by a residual tension in the center. Residual compression in the surface increases fracture resistance because fractures almost always start at the surface because with any applied bending or torsion, the stresses are highest there. Furthermore, defects are much more likely to be present at the surface. With residual compression at the surface, greater external loads can be tolerated before without fracture.

Table 19.6. *Compositions and coefficients of thermal expansion of several glasses*

Glass	Composition, wt. %*	$\alpha \times 10^{-6}$
Silica	$100SiO_2$	0.5
Vycor	$4B_2O_5$**	0.6
Pyrex	$12B_2O_5, 4Na_2O, 4Al_2O_3$	2.7
Plate	$13CaO, 13Na_2O, 2MgO, 1Al_2O_3$	9.0

* In each case, balance is SiO_2.
** The composition of finished Vycor. The composition before forming has more of other oxides.

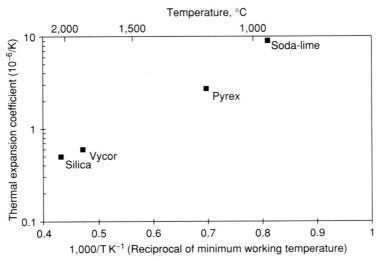

Figure 19.19. Relation between thermal expansion coefficient and the temperature at which the viscosity is 10^7 Pa-s. Compositions that promote lower working temperatures have higher coefficients of thermal expansion.

Compressive residual stresses can be induced in the surface by either cooling rapidly from high temperature or by chemical treatment. In the former process, called *tempering*, the glass is cooled with jets of air. During the cooling, the surface undergoes a thermal contraction before the interior. Compatibility is maintained by viscous flow. When the interior finally cools, compatibility is maintained by elastic contraction (compression) of the surface. Glass may also be *chemically tempered* by ion exchange. Glass is immersed in a molten salt bath containing potassium ions. Some of these ions diffuse into the glass, replacing sodium ions. Because the K^+ ions are larger than the Na^+ ions they replace, the region affected is left in compression. One important difference between the two processes is the depth of the compressive layer. In the chemically tempered glass, the depth of the region under compression is much less than in the thermally tempered material. Not only is tempered glass more resistant to fracture than untempered glass, but it also breaks into much smaller pieces. These are less dangerous than the large shards produced in fracture of untempered glass. Tempered glass is used for the side and rear windows of automobiles for these reasons. The windshields of automobile are made from safety glass produced by laminating two pieces of glass with a polymeric material that keeps the broken shards from causing injury. Figure 19.20 shows fracture patterns typical of untempered glass, laminated safety glass, and tempered glass.

Strength of Glasses

The tensile strengths of glass and other brittle materials are much more sensitive to surface flaws than more ductile materials because stress concentrations cannot be relieved by deformation. Therefore, strengths of glass are quite variable. Typical

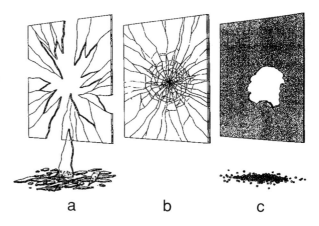

Figure 19.20. Typical fracture patterns of three grades of glass: (a) annealed, (b) laminated and (c) tempered. From *Engineering Materials Handbook 4, Ceramics and Glasses,* ASM (1991).

a   b   c

strengths of various grades of glass are listed in Table 19.7, but there is considerable scatter from these values.

Thermally Induced Stresses

Glass, like most ceramics, is susceptible to fracturing under stresses caused by temperature gradients. Internal stresses in a material arise when there are different temperature changes in adjacent regions. In the absence of stress, a temperature change causes a fractional dimensional change, $\Delta L/L = \varepsilon = \alpha \Delta T$. Under stress, the total strain is

$$\varepsilon_x = \alpha \Delta T + (1/E)[\sigma_x - \upsilon(\sigma_y - \sigma_z)]. \tag{19.9}$$

When two regions, A and B, are in intimate contact they must undergo the same strains, $(\varepsilon_{xA} = \varepsilon_{xB})$. If there is a temperature difference, $\Delta T = T_A - T_B$, between the two regions,

$$\alpha \Delta T + (1/E)[\sigma_{xA} - \sigma_{xB} + \upsilon(\sigma_{yA} + \sigma_{zA} - \sigma_{yB} - \sigma_{zB})] = 0. \tag{19.10}$$

EXAMPLE PROBLEM 19.4: The temperature of the inside wall of a tube is 200°C, and the outside wall temperature is 40°C. Calculate the stresses at the outside of the wall if the tube is made from a glass having a coefficient of thermal expansion of $\alpha = 8 \times 10^{-6}/°C$, an elastic modulus of 10×10^6 psi, and a Poisson's ratio of 0.3.

Table 19.7. *Typical fracture strengths of several glasses*

Type of glass	Strength (MPa)
Pristine fibers	7,000
Typical fibers (E-glass)	3,500
Float glass (windows)	70
Tempered window glass	150
Container glass	70
Fused silica	60

Solution: Let x, y, and z be the axial, hoop, and radial directions. The stress normal to the tube wall, $\sigma_z = 0$, and symmetry requires that $\sigma_y = \sigma_x$. Let the reference position be the midwall, where $T = 120°C$. ΔT at the outside is $40° - 120° = -80°C$. The strains ε_x and ε_y must be zero relative to the midwall. Substituting in equation (19.9), $0 = \alpha \Delta T + (1/E)\sigma_x + \upsilon(\sigma_y + \sigma_z)$, $\alpha \Delta T + (1 - \upsilon)\sigma_x/E = 0$, so $\sigma_x = \alpha E \Delta T/(1 - \upsilon)$. $\sigma_x = (8 \times 10^{-6}/°C)(80°C) \times (10 \times 10^6 \, \text{psi})/0.7 = 9,140 \, \text{psi}$.

In general, the stresses reached will be proportional to α, $E/(1 - \upsilon)$, and ΔT. The parameter,

$$R_1 = \sigma_f(1 - \upsilon)/(E\alpha), \tag{19.11}$$

describes the relative susceptibility to thermal shock. A different thermal shock parameter,

$$R_2 = K_{Ic}/(E\alpha), \tag{19.12}$$

is based on the fracture toughness. If the length of preexisting cracks is constant, these are equivalent because σ_f is proportional to K_{Ic}. Thermal conductivity has some influence on susceptibility to thermal shock because it influences the term ΔT. The higher the thermal conductivity is, the lower is the value of ΔT. It should be noted that these parameters apply to only materials that are brittle. In materials that flow plastically, the stresses never rise to the levels predicted from equation (19.10). Instead, the right-hand side of equation (19.9) should be modified by the addition of a plastic strain term.

Because E does not differ greatly among the various grades of glass, differences of thermal shock resistance are related to differences in thermal expansion. The compositions and thermal expansion coefficients of several glasses are listed in Table 19.6. The thermal shock resistance of silica glass and Vycor is much better than that of plate glass.

Delayed Fracture

Glass that has been under stress for a period of time may fracture suddenly. Such delayed fracture is not common in metals (except in cases of hydrogen embrittlement of steels), but sometimes does occur in polymers. It is often called *static fatigue*. The phenomenon is sensitive to temperature (Figure 19.21). It depends on the level of stress and the amount of prior abrasion of the surface (Figure 19.22). Most important, it is very sensitive to environment. Cracking is much more rapid with exposure to water than if the glass is kept dry (Figure 19.23) because water breaks the Si-O-Si bonds by the reaction -Si-O-Si- + $H_2O \rightarrow$ Si-OH + HO-Si.

Glassy Metals

Special compositions of alloys have very low liquidus temperatures. Under rapid cooling, these may freeze to an amorphous glass rather than a crystal. For many compositions, the required cooling rates are so high that glass formation is limited to very thin sections. However several alloys have been developed that form

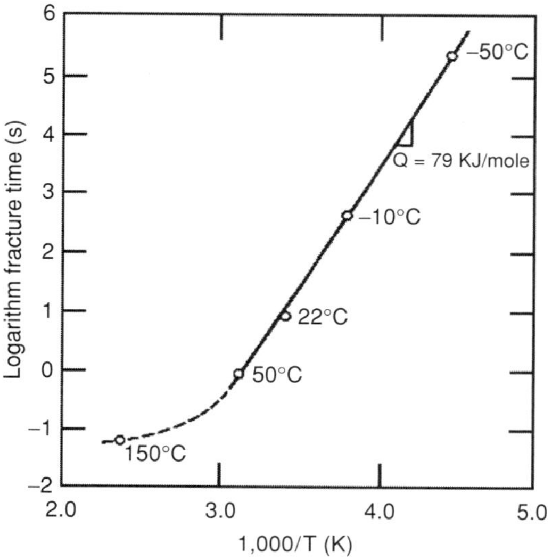

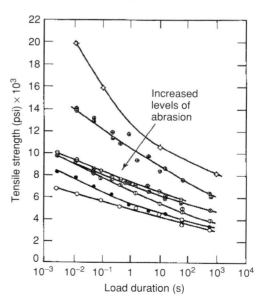

Figure 19.21. Temperature dependence of the time to failure for soda-lime glass rods in bending under a stress of 10,000 psi. From R. J. Charles, *Progress in Ceramic Science*, v. 1 (1961).

glasses at moderate cooling rates. Relatively large samples of these can be frozen to a glass. One of these, $Zr_{41.2}Ti_{13.8}Cu_{12.5}Ni_{10}Be_{22.5}$, has a very low solidus temperature (670°C) relative to the melting points of the constituents, Zr (MP = 1,852°C), Ti (MP = 1,670°C), Be (MP = 1,277°C), Cu (MP = 1,083°C) and Ni (MP = 1,852°C). It will solidify to a glass at cooling rates that can be achieved in large pieces.

Metal glasses have extremely high elastic limits. The elastic strains at yielding may be as high as 2.5%. This permits storage of a very large amount of elastic energy because the elastic energy per volume, E_v, is proportional to the square of the strain.

$$E_v = 1/2\sigma\varepsilon = 1/2\varepsilon^2/E. \qquad (19.13)$$

These metal glasses do not work harden. Rather when shear occurs on a plane, the atoms along the plane become less densely packed. Because of this, further shear

Figure 19.22. Static fatigue of glass specimens After varying amounts of abrasion. Note that for a given material, the life increases as the stress is lowered. From R. E. Mould and D. D. Southwick, *J. Amer. Cer. Soc.*, v. 42 (1959).

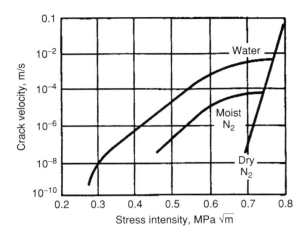

Figure 19.23. The effect of environment on crack velocity in a silicate glass under load. From *Engineering Materials Handbook, IV, Ceramics and Glasses*, ASM (1991).

on the plane becomes easier. This work softening and the heating along the plane of shear cause a rapid localization of the strain onto a single plane of shear. This results in a shear failure that occurs almost immediately after yielding. Plastic elongations in a tension test are less than 1%. The tensile strength and yield strength are the same, even though failure occurs by ductile shearing. The fracture toughnesses are low (35–50 MPa$\sqrt{m}$) because very little energy is absorbed.

Another limitation of glassy metals is the very low fatigue resistance. There seems to be no endurance limit. The stress for failure at a large number of cycles may drop below 5% of the yield strength. A significant improvement in tensile elongation, fatigue strength, and toughness has been achieved by modifying the composition so that the glass has small crystalline regions of β-titanium. This causes some strain hardening, thus eliminating the localization of flow.

REFERENCES

Y.-M. Chiang, D. Birney, and W. D. Kingery, *Physical Ceramics*, 2nd ed., Wiley (1997).

W. D. Kingery, K. Bowen, and D. R. Uhlmann, *Introduction to Ceramics*, 2nd ed., Wiley (1960).

M. W. Barsoum, *Fundamentals of Ceramics*, McGraw-Hill (1997).

Engineered Materials Handbook, vol. 4, Ceramics and Glasses, ASM International (1991).

Notes

Does glass really creep at room temperature? In many high school science classes, it is taught that glass is just a very cool liquid with a very high viscosity at room temperature. An observation used to support this claim is that the panes of stained glass windows in the very old European cathedrals are usually thicker at the bottom that at the top. It is said that the glass has crept over the centuries under its self-weight. There are two difficulties with this story. Observable thickening of the bottoms by creep under self-weight would make the windows shorter. A 10% thickening at the bottom would be accompanied by approximately a 5% shortening, which would be much more obvious to the observer than the thickening. The other problem with the story is that any creep would be much too small to observe the deformation that is attributed to creep. The strain rate to produce 10% deformation over 750 years

Figure 19.24. Before 1800, plate glass was made by spinning a glob of hot glass on the end of a rod, allowing centrifugal force to form a disc about 1.2 m in diameter. Panes were cut from the disc. Courtesy of Broadfield House Glass Museum, Kingsford, UK.

is $\dot{\varepsilon} = 0.1/(750 \times 365 \times 24 \times 3{,}600\,\text{s}) = 5 \times 10^{-12}\,\text{s}^{-1}$. The stress due to self-weight at the bottom of a pane 1/4 m high is $\rho g h = (3 \times 10^3\,\text{kg/m}^3)(10\,\text{m/s}^2)(1/4\,\text{m}) = 7.5\,\text{kPa}$. This combination of stress and strain rate corresponds to a viscosity of $\eta = 3\,\text{s}/\dot{\varepsilon} = 3 \times 7.5 \times 10^3/5 \times 10^{-12} = 4.5 \times 10^{15}\,\text{Pa/s}$. This amount of deformation would require that the viscosity be 4.5×10^{15} Pa/s or less. This should be compared with the actual viscosity. If the plot of viscosity for the softest glass (Figure 19.19) is extrapolated to 20°C ($1/T = 0.0034/\text{K}$), the viscosity would be about $\eta = 10^{47}$ Pa/s.

The true explanation is that the glass varied in thickness when it was installed. Until the nineteenth century, sheet glass was made by spinning a hot viscous glob on a rod. Centrifugal force caused the glob to form into a disc, as shown in Figure 19.24. The disc was thicker near the center than at the edges, so panes cut from it had a thickness variation. A good artisan would naturally install a pane with the thicker section at the bottom.

One interesting application of metal glass is as plates in the heads of golf clubs. The energy of impact of the club and the ball is absorbed elastically in both the ball and the head of the club. Because there are large losses in the ball, it is advantageous to have as much of the energy as possible absorbed elastically by the head of the club. Glassy metal plates are used to absorb energy as they bend. A large amount of energy can be stored because the metal glass can undergo high elastic strains. There is very little loss of this energy, being almost completely imparted to the ball as it leaves the face of the club.

Problems

1. Estimate the ratio of the elastic moduli of cubic AlSb and cubic ZnSe. Both have the sphalerite structure. The lattice parameters are 0.6136 nm. and 0.5669 nm, respectively.

2. Find the number of independent slip systems in rutile (TiO) for which slip occurs on the $<110>\{1\bar{1}0\}$ and $<110>\{001\}$ systems.

3. From a large number of tests on a certain material, it has been learned that 50% of them will break when loaded in tension at stress equal to or less than 520 MPa and that 30% will break at stress equal to or less than 500 MPa. Assume that the fracture statistics follow a Weibull distribution.

  A. What are the values of σ_o and m in the equation $P_s = \exp[-(\sigma/\sigma_o)^m]$?
  B. What is the maximum permissible stress if the probability of failure is to be kept less than 0.001% (i.e., 1 failure in 100,000)?

4. Twenty ceramic specimens were tested to fracture. The measured fracture loads in N were 279, 195, 246, 302, 255, 262, 164, 242, 197, 224, 255, 269, 213, 172, 179, 143, 206, 233, 246, and 295.

  A. Determine the Weibull modulus, m.
  B. Find the load for which the probability of survival is 99%.

5. Equations (19.6) and (19.7) relate the surface stress, σ, in bending to the load and dimensions of the bending specimens, assuming elastic behavior. The corresponding strain on the surface also depends on the elastic properties of the material and the ratio of t/b. Express the strain, e, in terms of the load and dimensions for

  A. $t/b \ll 1$,
  B. $t/b \gg 1$.

6. What percent reduction of the elastic modulus of alumina would be caused by 1% porosity? See Figure 19.6.

7. Pottery is generally fired at a high temperature, then it is cooled and a glaze is applied. On reheating, the glaze melts and spreads over the surface. On cooling again, residual stresses may develop in the glaze. To ensure that these are compressive, what relation is necessary between the properties of the body of the ceramic and of the glaze?

8. Why is there so little concern about thermal shock in metals and polymers, whereas there is much concern with ceramics? Why aren't the parameters $R_1 = \sigma_f(1 - 2\upsilon)/(E\alpha)$ and $R_2 = K_{Ic}/(E\alpha)$ useful in predicting fracture in metals?

9. Rapid heating and cooling can cause cracks to start at the surface of a ceramic material. Sometimes the cracks meet the surface at 90 degrees and sometimes at 45 degrees to the surface. By noting the orientation of a crack, how can one tell whether it initiated on the heating or the cooling portion of the thermal cycle?

10. A glass retort is to be used under conditions that the temperature of the inside of the wall is 100°C and the outside is 0°C. What stress will develop on the outside? The properties of the glass are $\alpha = 1.5 \times 10^{-6}/°C$, $E = 70\,GPa$, $\upsilon = 0.30$.

Hint: Assume a linear temperature gradient, and for simplicity, take the midpoint strain as zero.

20 Polymers

Introduction

A separate chapter is devoted to polymers because of their engineering importance and because their mechanical behavior is so different from that of metals and ceramics. The mechanical response of polymers is far more time dependent than that of crystalline materials. Viscoelastic effects (Chapter 15) are much more important in polymers than in metals or ceramics. The properties of polymers are also much more sensitive to temperature than those of other materials. Changes of molecular orientation with deformation cause large changes in properties and a much greater degree of anisotropy than is observed in metals or ceramics. The phenomena of crazing and rubber elasticity have no analogs in crystalline materials. Some polymers exhibit very large tensile elongations. Although a few alloys exhibit shape-memory behavior, the effect is much greater in polymers, more common, and of greater technological importance.

Elastic Behavior

Elastic strains in metals and ceramics occur by stretching of primary metallic, covalent, or ionic bonds. The elastic modulus of most crystals varies with direction by less than a factor of 3. The effects of alloying, thermal, and mechanical treatments on the elastic moduli of crystals are relatively small. As the temperature is increased from absolute zero to the melting point, Young's modulus usually decreases by a factor of no more than 5. For polymers, however, a temperature change of 30°C may change the elastic modulus by a factor of 1,000. Elastic deformation of polymeric involves stretching of the weak van der Waals bonds between neighboring molecular chains and rotation of covalent bonds. This accounts for the fact that the elastic moduli of random linear polymers are often at least two orders of magnitude lower than those of metals and ceramics. However, highly oriented polymers may have Young's moduli that are higher than the stiffest metals when they are tested parallel to the direction of chain alignment. Figure 20.1 is a schematic plot showing how Young's modulus increases with the fraction of covalent bonds aligned with the loading direction.

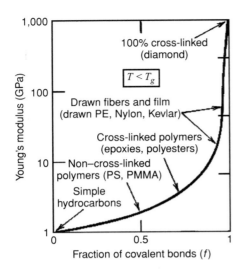

Figure 20.1. The elastic moduli of polymeric materials increase with the number of covalent bond aligned with the loading axis. A typical metal E is about 100 GPa. From M. F. Ashby and D. R. H. Jones, *Engineering Materials 2*, 2nd ed., Butterworth Heinemann (1998).

Figure 20.2 illustrates the temperature dependence of the elastic moduli of several types of polymers. The temperature dependence is greatest near the glass transition temperature and near the melting point. The decrease of E for the "crystalline" polymer at the glass transition suggests that it is not 100% crystalline and that the drop is attributable to its amorphous portion. Polymer C has a higher melting point than polymer A because of its higher molecular weight. The cross-linked polymer cannot melt without breaking the covalent bonds in the cross-links. Actual data are given for polystyrene in Figure 20.3. The stiffness of a polymer at room temperature depends on whether its glass transition temperature is above or below that temperature. Below the glass transition temperature, the elastic moduli are much higher

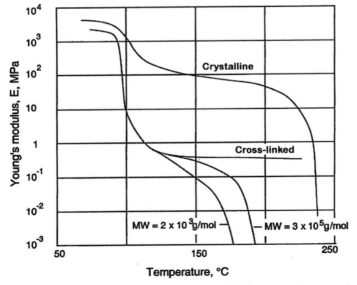

Figure 20.2. Temperature dependence of E for several types of polymers. Crystallization, cross-linking, and increased molecular weight increase stiffness at higher temperatures. Data from A. V. Tobolsky, *J. Polymer Sci.*, Part C, Polymer Symposia No. 9 (1965).

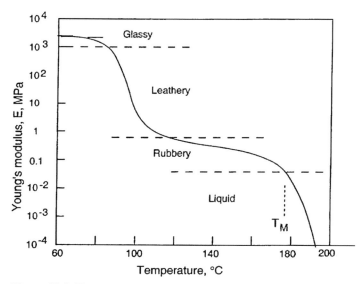

Figure 20.3. Temperature dependence of Young's modulus of polystyrene. As the temperature is increased from 80°C to 120°C through the glass transition, the modulus drops by more than a factor of 3. Data from A. V. Tobolsky, *J. Polymer Sci., Part C*, Polymer Symposium No. 9 (1965).

than above it. Figure 20.3 indicates that the modulus of polystyrene changes by a factor of more than 10^3 between 85°C and 115°C. The elastic moduli of thermoplastics increase with the degree of crystallinity, as shown in Figures 20.4 and 20.5.

A strong time (and, therefore, strain rate) dependence of the elastic modulus accompanies the large temperature dependence, as shown in Figure 20.6. This is because the polymers undergo time-dependent viscoelastic deformation when stressed. If a stress is suddenly applied, there is an immediate elastic response. More deformation occurs with increasing time. The effect is so large near the glass transition temperature that it is customary to define the modulus in terms of the time of loading.

Figure 20.7 illustrates the time dependence of Young's modulus. Under constant load, the strain increases with time, so $E = s/e$ decreases with time.

One might expect that the time and temperature dependence of the modulus could be interrelated by an Arrhenius rate law in a manner similar to the

Figure 20.4. Variation of the Young's modulus of polyethylene at room temperature with degree of crystallinity. Data from Wang, *J. Appl. Phys.*, v. 44 (1973).

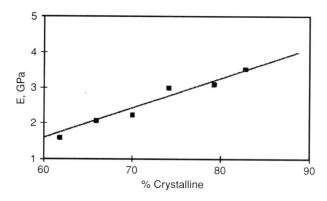

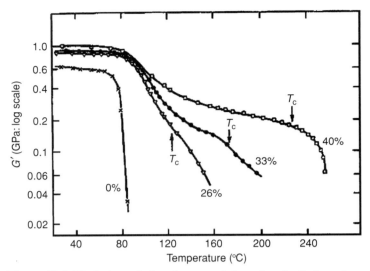

Figure 20.5. Variation of the shear modulus of polyethylene terephthalate with varying amounts of crystallinity. The specimens were initially quenched below T_g (60°C). Three of them were subsequently reheated to allow crystallization at the temperatures indicated by T_c. From N. G. McCrum, C. P. Buckley, and C. B. Bucknall, *Principles of Polymer Engineering*, Oxford (1988).

Zener-Hollomon treatment of flow stress in metals. If straining were a rate process that could be described by an Arrhenius rate law, the time to reach a given strain (and value of E) would be expressed as

$$t = A \exp(+Q/RT), \qquad (20.1)$$

where A is a function of E. The value of Q could be computed from two combinations of time, t, and temperature, T, that give the same value of E. In that case, $t_1/t_2 = \exp[(Q/R)(1/T_1 - 1/T_2)]$, so

$$Q = R \ln(t_1/t_2)/(1/T_1 - 1/T_2). \qquad (20.2)$$

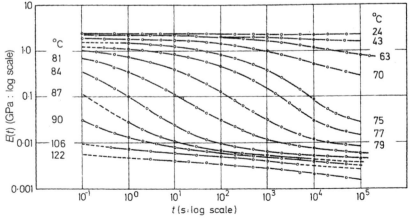

Figure 20.6. Time dependence of Young's modulus of PVC between 24°C and 122°C. From N. G. McCrum, C. P. Buckley, and C. B. Bucknall, *Ibid.*

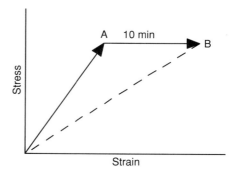

Figure 20.7. Schematic drawing showing why E is time dependent. After loading, the specimen undergoes an instantaneous elastic strain to point A. After 10 minutes under constant stress, the specimen undergoes additional viscoelastic (anelastic) strain to point B. For the same stress, the strain is greater, so the modulus is lower.

However, this type of calculation results in activation energies that depend on temperature near the glass transition. Such a change of Q with temperature indicates that the dominant relaxation mechanism changes. Example problem 20.1 and Figure 20.9 illustrates this.

EXAMPLE PROBLEM 20.1: Use the data in Figure 20.6 to calculate the activation energy, Q, for $E(t)$ at different temperatures, and plot the calculated Q as a function of temperature.

Solution: Pairs of adjacent time–temperature combinations that result in the same value of $E(t)$ were noted and substituted into equation (20.2). The calculated values of activation energy, Q, were plotted in Figure 20.8 at the average temperature of the interval. Obviously, there is an increase of the apparent value of Q near the glass transition temperature. It can be concluded, therefore, that a Zener-Hollomon type analysis cannot be applied to polymers near the glass transition temperature.

Examination of the curves for $E(t)$ versus $\log(t)$ at different temperatures in Figure 20.6 indicates that they have similar shapes. The curve for one temperature

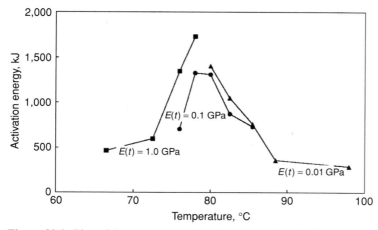

Figure 20.8. Plot of the apparent activation energy for the time–temperature dependence of $E(t)$ for PVC. The plot was constructed using the data in Figure 20.6 with equation (20.2). Adjacent points were used for three different levels of $E(t)$. The peak in the apparent activation energy occurs near 80°C. This plot illustrates the danger of assuming that relaxation is controlled by a single activation energy.

can be generated from that at another temperature by simply shifting it in time by a constant shift factor, $\log(a) = \log(t_2) - \log(t_1) = \log(t_2/t_1)$. Williams, Landel, and Ferry* proposed that the shift factor needed to bring the curve for temperature, T, into coincidence with that for a reference temperature, T_s, could be expressed as

$$\log(a) = C_1(T - T_s)/[C_2 + (T - T_s)], \tag{20.3}$$

where C_1 and C_2 are constants. This empirical equation works very well near the glass transition temperature. If the reference temperature is taken as the glass transition temperature, $T_s = T_g$,

$$\log(a) = C_{1g}(T - T_g)/[C_{2g} + (T - T_g)]. \tag{20.4}$$

It was suggested that values of about $C_{1g} = 17.4$ and $C_{2g} = 51.6$ K are appropriate for many polymers.

> **EXAMPLE PROBLEM 20.2:** The glass transition temperature, T_g, for an ethylene-propylene copolymer is $-59°$C. Find the temperature at which the toughness in a Charpy test is the same as in slow bending of a Charpy bar at $-59°$C. Assume the deformation in the Charpy test occurs in 1 ms, whereas the slow bend test takes 10 s.
>
> **Solution:** Assume the WLF relation (equation 20.4), and substitute $\log(a) = \log(10/10^{-3}) = 4$, $4 = 17.4 \Delta T/(51.6 + \Delta T)$ where $\Delta T = (T - T_g)$. $\Delta T(17.4 - 4) = 206.4$. $\Delta T = 206.4/13.4 = 15.4$ $T = 15.0 + (-50) = -35°$C.

Rubber Elasticity

The elastic behavior of rubber is very different from that of crystalline materials. Rubber is a flexible polymer in which the molecular chains are cross-linked. The number of cross-links is controlled by the amount of cross-linking agent compounded with the rubber. Elastic extension occurs by straightening of the chain segments between the cross-links. With more cross-linking, the chain segments are shorter and have less freedom of motion so the rubber is stiffer. Originally, only sulfur was used for cross-linking, but today other cross-linking agents are used. Under a tensile stress, the end-to-end lengths of the segments increase, but thermal vibrations of the free segments tend to pull the ends together, just as the tension in violin strings is increased by their vibration. A mathematical model that describes the elastic response under uniaxial tension or compression predicts:

$$\sigma = G(\lambda - 1/\lambda^2), \tag{20.5}$$

where λ is the extension ratio, $\lambda = L/L_o = 1 + e$, where e is the engineering strain. The shear modulus, G, depends on the temperature and on the length of the free segments,

$$G = NkT, \tag{20.6}$$

where N is the number of chain segments/volume and k is Boltzmann's constant. Figure 20.9 shows the theoretical predictions of equation with σ normalized by G.

---

* M. L. Williams, R. F. Landel, and J. D. Ferry, *J. Amer. Chem. Soc.*, v. 77 (1995).

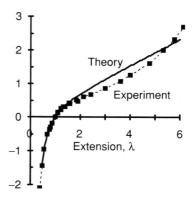

Figure 20.9. Stress–extension (λ) curve of vulcanized natural rubber with $G = 0.39$ MPa. The continuous curve is the prediction of equation (20.5). The dashed curve is from data of L. R. G. Treloar, *The Physics of Rubber Elasticity*, 3rd ed., Oxford (1975). Theory and experiment agree closely for $\lambda < 1.5$.

Agreement with experiment is very good for $\lambda < 1.5$, but is poorer at higher values of λ.

EXAMPLE PROBLEM 20.3: Use equation (20.5) to find an expression for Young's modulus and Poisson's ratio at low strains.

Solution: Differentiating equation (20.5), $d\sigma/d\lambda = G(1 + 2\lambda^{-3})$. $E = d\sigma/d\varepsilon = d\sigma/d\lambda = G(1 + 3\lambda^3)$. For small strains ($\lambda \approx 1$), $E = 3G = 3NkT$.

Substitution of $E = 3G$ into equations (2.6b) and (2.6d) shows that Poisson's ratio, $\upsilon = 1/2$ and the bulk modulus, $B = E/[3(1 - 2\upsilon)] = \infty$ (rubber is incompressible).

Note that because $G = NkT$, the elastic modulus *increases* with temperature in the rubbery range above the glass transition temperature (Figure 20.10). The increase of modulus with increasing sulfur content is attributable to the greater number of cross-links, N.

EXAMPLE PROBLEM 20.4: Use equations (20.5) and (20.6) to find an expression for the coefficient of thermal expansion, α, of rubber as a function of λ.

Solution: Combining equations (20.5) and (20.6), $\sigma = NkT(\lambda - \lambda^{-2})$. Now differentiating at constant stress, $0 = Nk(\lambda - \lambda^{-2})\,dT + NkT(d\lambda + 2\lambda^{-3}d\lambda)$;

$$\alpha = d\lambda/dT = -(1/T)(\lambda - \lambda^{-2})/(1 + 2\lambda^{-3}). \qquad (20.7)$$

This implies that α is zero if the rubber is not strained and that it is negative for $\lambda > 1$. As the rubber is stretched, λ becomes increasingly negative. Figure 20.11 is a plot of equation (20.7).

The negative thermal expansion coefficient under tension can be demonstrated by stretching a rubber band and hanging a fixed weight (constant σ) on it so $\lambda > 1$, and then heating it with a hair dryer. The weight will rise as the rubber band contracts.

Damping

The viscoelastic behavior that is responsible for the time and temperature dependence of the modulus also causes damping. The two phenomena are interrelated.

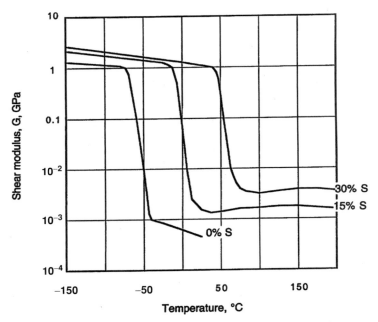

Figure 20.10. The shear modulus of natural rubber above T_g increases with a greater degree of vulcanization. Note also that the modulus increases with temperature above T_g as predicted by equation (20.6). Data from N. G. McCrum, C. P. Buckley, and C. B. Bucknall, *Ibid*, attributed to Wolf.

Figure 20.12 shows schematically that the damping is greatest in the temperature range in which the decrease of modulus with temperature, $-dE/dT$, is greatest. The experimental data in Figure 20.13 shows the correlation of damping and the temperature dependence of the shear modulus, G, in polymethyl methacrylate. The damping in low- and high-density polyethylene is shown in Figure 20.14.

Mechanical hysteresis is the energy loss per volume per cycle. The area under the loading curve represents the mechanical energy per volume put into the material during loading, whereas the area under the unloading curve represents the mechanical energy recovered during unloading. The area between the two curves is the magnitude of the hysteresis and represents the energy that remains in the material, which is almost entirely converted to heat. The mathematics of damping was treated in Chapter 15.

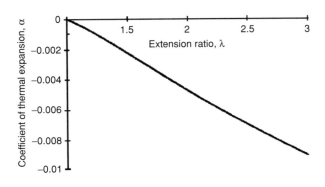

Figure 20.11. Coefficient of thermal expansion for rubber becomes increasingly negative as the material is stretched.

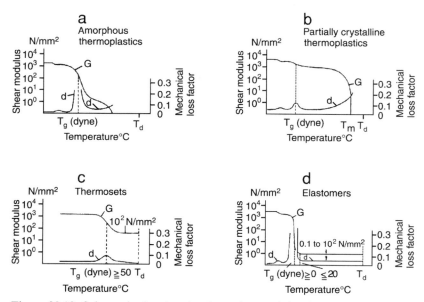

Figure 20.12. Schematic showing the dependence of the shear modulus, G, and the damping mechanical loss factor, Λ, on temperature for amorphous thermoplastics, partially crystalline thermoplastics, thermosets, and elastomers. From L. Engel et al., *An Atlas of Polymer Damage*, Prentice Hall (1978).

Yielding

Figure 20.15 shows typical tensile stress–strain curves for a thermoplastic. The lower strengths at higher temperatures are obvious. Increased strain rates have the same effect as decreased temperatures. At low temperatures, PMMA is brittle. At higher temperatures, the initial elastic region is followed by a drop in load that accompanies yielding. A strained region or neck forms and propagates the length of the tensile specimen (Figures 20.16 and 20.17). Only after the whole gauge section is strained, does the stress again rise. Superficially, this is similar to the upper and lower yield points and propagation of Lüder's bands in low-carbon steel after strain aging. However, there are several notable differences. The engineering strain in the

Figure 20.13. Variation of G and the damping parameter, Λ, for polymethyl methacrylate with temperature. The relaxation peak corresponds to movement of main backbone segments and the β peak to the rotation of side groups. From N. G. McCrum, C. P. Buckley, and C. B. Bucknall, *Principles of Polymer Engineering*, Oxford (1988).

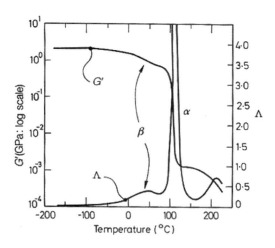

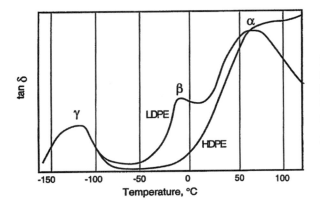

Figure 20.14. Variation of tan with temperature in high- and low-density polyethylene. The extra β peak for the LDPE must be caused by the side branches. Adapted from I. M. Ward and D. W. Hadley, *Mechanical Properties of Solid Polymers*, Wiley (1993).

necked region of a polymer is several hundred percent in contrast to 1% or 2% in a Lüder's band in steel. Also, the mechanism is different. In polymers, the deformation is associated with the reorientation of the polymer chains in the deformed material, so that after the deformation the chains are aligned with the extension axis (Figure 20.18.) This causes a very large increase in the elastic modulus. Continued stretching, after the necked region has propagated the length of the gauge section, can be achieved only by continued elastic deformation. This now involves opening of the bond angle of the C-C bonds. The stress rises rapidly until the specimen fractures. If the specimen is unloaded before fracture and a stress is applied perpendicular to the axis of prior extension, the material will fail at very low loads because only van der Waals bonds need be broken.

Polymers with low molecular weights may fail before necking. Low molecular weights may result from the polymerization process or from environmental degradation. Although high molecular weights tend to result in stronger and more ductile polymers, they also raise the viscosity of the molten polymer, which is often undesirable in injection molding.

Effect of Strain Rate

The effect of temperature and strain rate on flow stress can be treated by the Eyring equation in a manner similar to the treatment in Chapter 6,

$$\dot{\varepsilon} = \dot{\varepsilon}_o \exp[-Q/(RT)]\sinh[V\sigma/(RT)], \tag{20.8}$$

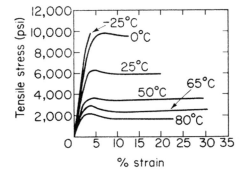

Figure 20.15. Tensile stress–strain curves for polymethyl methacrylate (PMMA) at several temperatures. Note that the strength decreases and the elongation increases with higher temperatures. From Carswell and H. K. Nason, *ASTM Symposium on Plastics*, Philadelphia (1944).

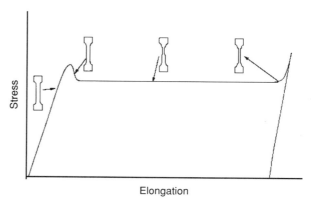

Figure 20.16. Stress–strain curve for polyethylene. Note the formation and propagation of a necked region along the gauge
section.

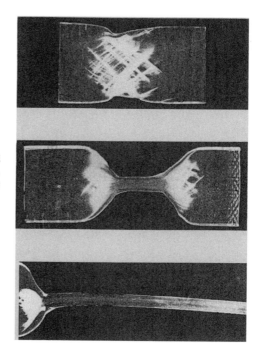

Figure 20.17. Stages of neck formation and drawing
in HDPE. From F. W. McClintock and A. S. Argon,
Mechanical Behavior of Materials, Addison-Wesley
(1966).

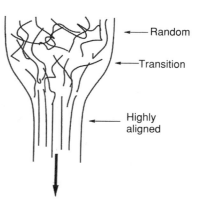

Figure 20.18. Schematic illustration of alignment of molecules
during neck formation.

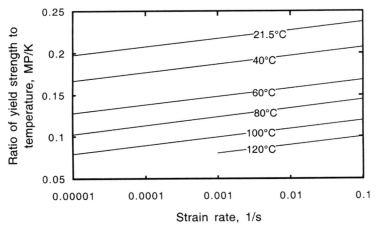

Figure 20.19. Variation of ratio of yield strength to temperature with strain rate of polycarbonate. Data from Bauwens-Cowet, Bauwens, and Homës, *J. Polymer Sci.*, v. A2 (1969).

where V is the activation volume, Q is an activation energy, and $\dot{\varepsilon}_o$ is a constant. If $V\sigma/(RT) \gg 1$, $\sinh[V\sigma/(RT)] \approx (1/2)\exp[V\sigma/(RT)]$, so $\dot{\varepsilon} = (1/2\,\dot{\varepsilon}_o)\exp[-Q/(RT + V\sigma/(RT)]$. Solving for stress,

$$\sigma/T = (R/V)[Q/(RT) + \ln(2\dot{\varepsilon}/\dot{\varepsilon}_o)]. \tag{20.9}$$

A plot of σ/T versus strain rate on a log scale for polycarbonate (Figure 20.19) is linear as predicted by equation (20.9). However, if the range of strain rates is too large, such a plot may have a break. That would indicate another mechanism with a different activation energy is involved.

Effect of Pressure

For polymers, the stress–strain curves in compression and tension can be quite different. Figures 20.20 and 20.21 are stress–strain curves for epoxy and PMMA in tension and compression. Figure 20.22 compares the yield strengths of polycarbonate as tested in tension, shear, and compression. The effect of pressure on the yield strength of PMMA is plotted in Figure 20.23.

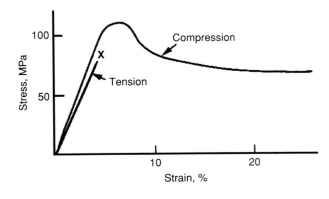

Figure 20.20. Stress–strain curves of an epoxy measured in tension and compression tests. Data from P. A. Young and R. J. Lovell, *Introduction to Polymers*, 2nd ed., Chapman and Hall (1991).

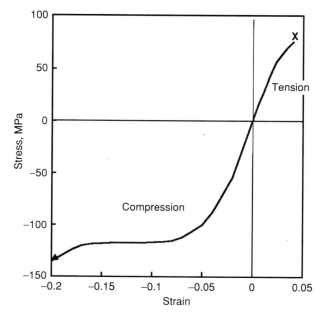

Figure 20.21. Stress–strain curves for PMMA (plexiglas) in tension and compression. Note that the strength in compression is higher than that in tension. Data from C. W. Richards, *Engineering Materials Science*, Wadsworth (1961).

The difference between the yield strengths in tension and compression indicates that yielding is sensitive to the level of hydrostatic pressure. This dependence is evident in the shapes of the yield loci of several randomly oriented (isotropic) thermoplastics (Figure 20.24). The loci are not centered on the origin. To account for this, a pressure-dependent modification of the von Mises criterion has been suggested:

$$(\sigma_2 - \sigma_3)^2 + (\sigma_3 - \sigma_1)^2 + (\sigma_1 - \sigma_2)^2 + 2(C - T)(\sigma_1 + \sigma_2 + \sigma_3) = 2CT, \qquad (20.10)$$

where T is the yield strength in tension and C is the absolute value of the yield strength in compression. Another possible pressure-dependent modification of the

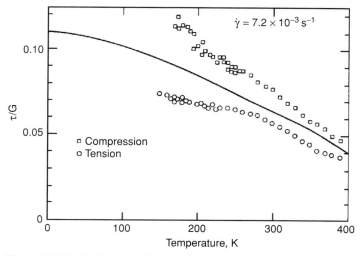

Figure 20.22. Yield strengths of polycarbonate in tension, shear, and compression as functions of temperature. From A. S. Argon, *Phil. Mag.*, v. 28 (1973).

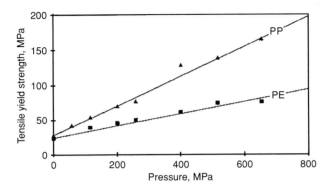

Figure 20.23. The effect of pressure on the yield stresses of polyethylene and polypropylene. Data from Mears, Pae and Sauer, *J. Appl. Phys.* v. 40 (1969).

von Mises yield criterion is

$$(\sigma_2 - \sigma_3)^2 + (\sigma_3 - \sigma_1)^2 + (\sigma_1 - \sigma_2)^2 + K_1(\sigma_1 + \sigma_2 + \sigma_3)|\sigma_1 + \sigma_2 + \sigma_3| = K_2.$$
(20.11)

EXAMPLE PROBLEM 20.5: Express K_1 and K_2 in equation (20.11) in terms of the tensile yield strength, T, and the absolute magnitude of the compressive yield strength, C.

Solution: Consider a uniaxial tensile test in the 1-direction. At yielding, $\sigma_1 = T$ and $\sigma_2 = \sigma_3 = 0$. Substituting into equation (20.11), $2T^2 + K_1T^2 = K_2$. Now consider a uniaxial compression test in the 3-direction. At yielding, $\sigma_3 = -C$ and $\sigma_1 = \sigma_2 = 0$. Substituting into equation (20.11), $2C^2 - K_1C^2 = K_2$. Combining, $2C^2 - K_1C = 2T^2 + K_1T^2$, $K_1(T^2 + C^2) = 2(C^2 - T^2)$. $K_1 = 2(C^2 - T^2)/(C^2 + T^2)$. $K_2 = 2T^2 + K_1T^2 = 2T^2 + T^2[2(C^2 - T^2)/(C^2 + T^2)] = (2T^4 + 2T^2C^2 + 2T^2C^2 - 2T^4)/(T^2 + C^2) = 4T^2C^2/(T^2 + C^2)$.

Note, that if $T = C$, $K_1 = 0$ and $K_2 = 2T^2 = 2Y^2$, so equation (20.9) reduces to the von Mises criterion.

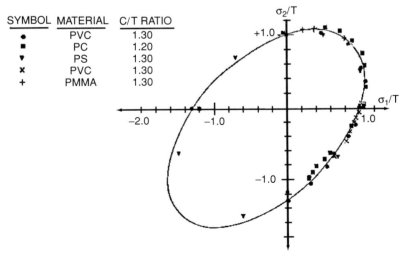

SYMBOL	MATERIAL	C/T RATIO
●	PVC	1.30
■	PC	1.20
▼	PS	1.30
✕	PVC	1.30
+	PMMA	1.30

Figure 20.24. Pressure-dependent yield loci. The solid line represents the predictions of equation (20.10), whereas the dashed line represents the von Mises criterion. From R. S. Ragava, PhD thesis, University of Michigan (1972).

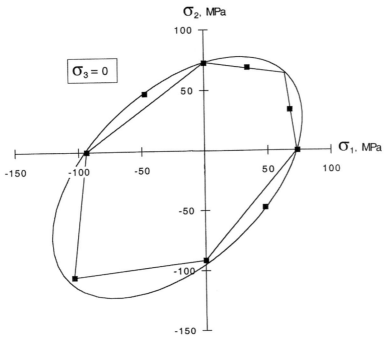

Figure 20.25. Pressure-dependent modifications of the von Mises (equation 20.10) and Tresca (equation 20.12) yield criteria for polystyrene. Data from Whitney and Andrews, *J. Polymer Sci.*, c-16 (1967).

EXAMPLE PROBLEM 20.6: The compressive yield strength of a polymer is 15% greater than its tensile yield strength. Calculate the ratio of the yield strength in balance biaxial tension, σ_b, to the yield strength in uniaxial tension according to equation (20.10). Repeat using equation (20.11).

Solution: Substituting $\sigma_1 = \sigma_2 = \sigma_b$, $\sigma_3 = 0$, $K_1 = (C^2 - T^2)/(C^2 + T^2)$ and $K_2 = 4T^2C^2/(C^2 + T^2)$ into equation (20.10), $2\sigma_b^2 + 2(0.15)(2\sigma_b) = 2(1.15)T$, $\sigma_b^2/T^2 + 2(1.15)(\sigma_b/T) - 1.15 - 0$, $\sigma_b/T = 0.9328$. Now substituting $C = 1015T$, $2\sigma_b^2 + 0.2555\,\sigma_b^2 = 2.2777T^2$), $(\sigma_b/T)^2 = 0.88915$, $\sigma_b/T = 0.944$.

Note that the predictions of equations (20.10) and (20.11) are very similar.

The Coloumb-Mohr criterion is a pressure-modified Tresca criterion,

$$\sigma_1 - \sigma_3 + m(\sigma_1 + \sigma_2 + \sigma_3) = 2\tau_u, \qquad (20.12)$$

where $2\tau_u$ is the yield strength in pure shear. The yield locus corresponding to this is plotted in Figure 20.25.

EXAMPLE PROBLEM 20.7: Evaluate m and $2\tau_u$ in equation (20.12) in terms of C and T. Then calculate the ratio of the yield strength in biaxial tension to that in uniaxial tension for a polymer for which the compressive yield strength is 15% higher than the tensile yield strength.

Solution: Consider a uniaxial tensile test in the 1-direction. At yielding, $\sigma_1 = T$ and $\sigma_2 = \sigma_3 = 0$ so $T + mT = 2\tau_u$. At yielding in a compression test, $\sigma_3 = -C$ and $\sigma_1 = \sigma_2 = 0$, so $+C - mC = 2\tau_u$. Combining, $T + mY = +C - mC$, so $m = (C - T)/(C + T)$. $T + mT = 2\tau_u$. At yielding in a compression test, σ_3

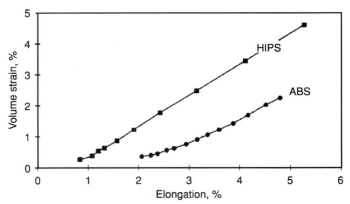

Figure 20.26. Volume change accompanying elongation. Data from Bucknall, Partridge, and Ward, *J. Mat. Sci.*, v. 19 (1984).

$= -C$ and $\sigma_1 = \sigma_2 = 0$, so $+C - Cm - 2\tau_u$. Combining these expressions, $T + mY = +C - mC$ so $m = (C - T)/(C + T). 2\tau_u = T(1 + m) = T[1 + (C-T)/(C + T) = 2CT/(C + T) = 2(1.15)T^2/(2.15T) = 1.07Tm = (C - T)/(C + T) = (C/T - 1)/(C/T + 1) = 0.15/2.15 = 0.679$. In biaxia tension, $\sigma_1 = \sigma_2 = \sigma_b$ and $\sigma_3 = 0$ so $\sigma_b(1 + 2m) = 2\tau_u = 1.07T$. $\sigma_b/T = 1.07(1 + 2\text{x}0.697) = 0.939$.

In biaxial tension, $\sigma_1 = \sigma_2 = \sigma_b$, $\sigma_3 = 0$, $\sigma_b(1 + 2m) = 2t_u = 1.070T$, $\sigma_b/T = 1.070/(1 + 2 \times 0.0697) = 0.939$.

The fundamental flow rule, $d\varepsilon_{ij} = d\lambda(\partial f/\partial \sigma_{ij})$ (equation 6.14), predicts that that if the yield criterion is sensitive to the level of hydrostatic pressure, $-(\sigma_1 + \sigma_2 + \sigma_3)/3$, yielding must be accompanied by a volume change. If $C > T$, the yielding should cause a volume increase. Figure 20.26 is a plot of volume strain, $\Delta v/v = (\varepsilon_1 + \varepsilon_2 + \varepsilon_3)$, against elongation for two polymers.

Crazing

Many thermoplastics undergo a phenomenon known as *crazing* when loaded in tension. Sometimes the term, *craze yielding*, is used to distinguish this from the usual *shear yielding*. A craze is an opening resembling a crack. However, a craze is not a crack in the usual sense. Voids form and elongate in the direction of extension. As a craze advances, fibers span the opening, linking the two halves. Figure 20.27 is a schematic drawing of a craze, and Figure 20.28 shows crazes in polystyrene.

Crazing occurs under tension. Bucknall[*] has proposed a modified Griffith criterion (equation 14.17) for crazing, where the tensile stress necessary to initiate crazing, σ_1, must equal or exceed

$$\sigma_1 \geq \sqrt{\{2EG_{Ic}/[(1 - v^2)\pi a]\}}. \tag{20.13}$$

Here, G_{Ic} is the critical energy release rate.

Figure 20.29 shows how the competition between crazing and fracture depends on the degree of chain entanglement, defined as M_v/M_e, where M_v and M_e are the viscosity average and entanglement average molecular weights, respectively.

Crazing predominates if tension is applied perpendicular to the fiber alignment as shown by Figure 20.30.

[*] C. B. Bucknall, *Polymers* v. 48 (2007).

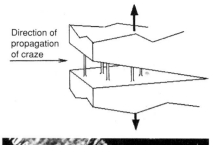

Figure 20.27. Schematic drawing of a craze. Note the fibrils connecting both sides of the craze.

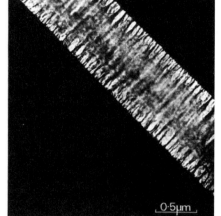

Figure 20.28. Typical microstructure of a thin craze in polystyrene. Note the fibrils that span the crack. From P. Behan, M. Bevis, and D. Hull, *Phil. Mag.*, v. 24 (1971).

Figure 20.29. Dependence of the competition between crazing and fracture on the number of entanglements per molecule. Open circles are data for polystyrene and filled circles are for PMMA. From C. B. Bucknall, *Polymer*, v. 48 (2007).

Figure 20.30. Effect of testing direction on the stresses necessary for yielding and for crazing at 20°C in polystyrene that has been oriented by stretching 260%. Data from L.Cramwell and D. Hull, *Int. Symp. on Macromolecules* (1977).

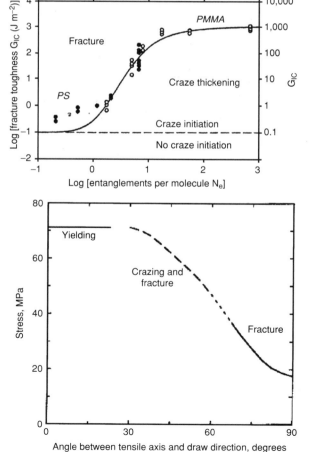

Table 20.1. *Compressive strengths of organic fibers*

Fiber	Tensile strength (GPa)	Compressive strength (GPa)
Spectra 1000¥	3.0	0.2
Kevlar 149†	3.4	0.4
PBZR*	4.0	0.4
PBO°	5.0	0.3

¥ Polyethylene.
† Poly(paraphenylene terephthalic acid).
* Poly(paraphenylene benzobistriazole).
° Poly(paraphenylene benzobisoxazole), now called Zylon.

Yielding of Fibers in Compression

The compressive yield strengths of fibers with highly aligned molecules are considerably lower than their tensile strengths. The reason is that under compression they can yield by buckling of bundles of fibrils or molecules. A semischematic stress–strain curve for PBZT is shown in Figure 20.31. Table 20.1 compares tensile and compressive properties of several fibers. The low compressive strength is of particular importance when the loading involves bending, with buckling occurring on the inside of the bend as shown in Figure 20.32. Figure 20.33 is a schematic yield locus.

Fracture

Polymers like most metals are embrittled at low temperatures. Network polymers tend to be brittle at all temperatures, and thermoplastics become brittle at temperatures below the glass transition. Figure 20.34, shows the Charpy impact behavior of PVC. There is a sharp ductile-brittle transition that depends on the notch radius.

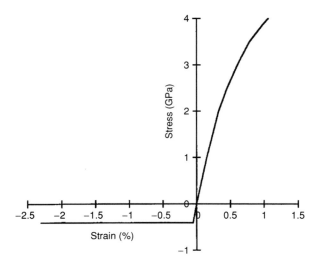

Figure 20.31. Schematic stress strain curve for oriented PBZT. The tensile and compressive yield strengths and the modulus correspond to actual data. The curvature is schematic. Courtesy of David Martin.

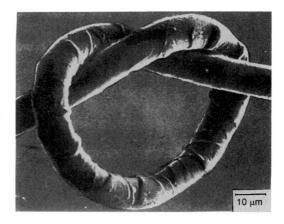

Figure 20.32. Bent fiber of Kevlar 49. Note the kinks on the compression side of the bend. From K. K. Chawla, *Composite Materials; Science and Engineering*, 2nd ed., Springer (1998).

   The ductile-brittle transition is associated with the glass transition temperature. Considerable effort has gone into developing polymers that are both strong and tough. Polycarbonate is useful in this respect. Rubber particles in polymers such as polystyrene add greatly to the toughness with some decrease of strength. Figures 20.35 and 20.36 show that the ductility of PMMA and the toughness of epoxy are increased by additions of rubber. The microstructure of commercial HIPS (high-impact polystyrene) is shown in Figure 20.37.

Deformation Mechanism Maps

The predominant deformation and fracture mechanisms change with stress and temperature. Deformation mechanism maps, such as Figure 20.38, are a convenient way to summarize these changes.

Shape-Memory Effect

Many linear polymers show an interesting and useful shape-memory behavior. If films are stretched at $T_g + 50°C$, they become biaxially oriented. In the case of PE and PP, the orientation is sufficient to cause a substantial increase in the degree of crystallization. If cooled to room temperature while stretched, the biaxial orientation and/or crystallization is frozen in. However, if the films are reheated,

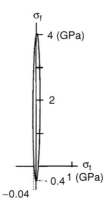

Figure 20.33. Schematic yield locus for an oriented PBZT fiber. The stresses parallel and perpendicular to the fiber axis are, σ_f and σ_t. The tensile yield stress along the fiber axis and both compressive yield stresses are based on data supplied by David Martin. The tensile yield strength perpendicular to the fiber axis is conjectural.

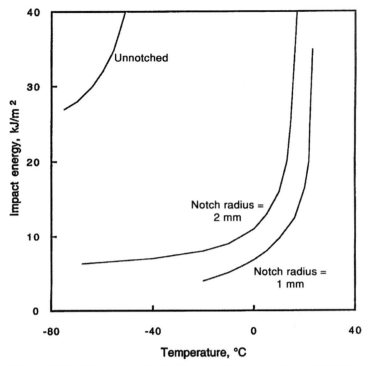

Figure 20.34. Charpy ductile-brittle transition behavior of PVC. Note that the ductile-brittle transition temperature increases as the notch radius becomes sharper. Data from P. I. Vincent, *Impact Tests and Service Performance of Thermoplastics*, Plastics Institute (1971).

without application of stress, they will undergo a biaxial shrinkage of up to 50%. This phenomenon finds extensive commercial use in shrink-wrap packaging of consumer products. It is also used for temporary storm windows. The customer can staple a prestretched film on a frame and then heat it with a hot air dryer to remove any wrinkles. Still another application is for electrical insulation. Preexpanded plastic tubes can be slipped over wire joints and similarly heated with hot air to provide a tight insulating sleeve.

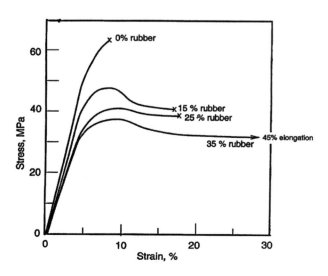

Figure 20.35. Stress–strain curves for rubber-toughened PMMA. Note that the elongation increases with the weight-fraction rubber. Data from P. A. Young and R. J. Lovell, *Introduction to Polymers*, 2nd ed., Chapman and Hall (1991), attributed to D. E. J. Saunders.

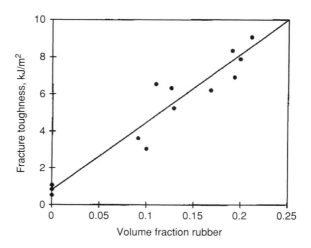

Figure 20.36. Increase of toughness of epoxy by the addition of rubber. Data from Bucknall and Yoshi, *Brit. Polymer J.*, v. 10 (1978).

Figure 20.37. Transmission electron micrograph of deformed HIPS. The dark networks are the rubber particles, and the lighter region is polystyrene. The dark spots are caused by the staining technique. The tensile axis is vertical. Note that the crazes emanate from the rubber particles. From P. A. Young and R. J. Lovell, *Introduction to Polymers*, 2nd ed., Chapman and Hall (1991). Courtesy Hu Xioa.

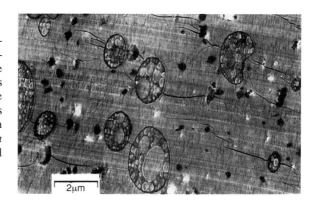

Figure 20.38. Deformation mechanism map for PMMA. From M. F. Ashby and D. R. H. Jones, *Engineering Materials 2*, 2nd ed., Butterworth Heinemann (1998).

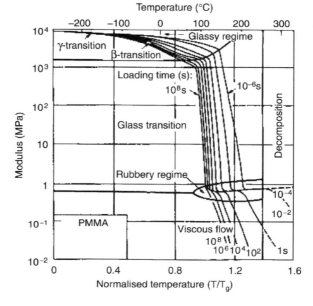

REFERENCES

N. G. McCrum, C. P. Buckley, and C. B. Bucknall, *Principles of Polymer Engineering*, Oxford (1988).

R. J. Young and P. A. Lovell, *Introduction to Polymers*, 2nd ed., Chapman and Hall (1991).

I. M. Ward and D. W. Hadley, *Mechanical Properties of Solid Polymers*, Wiley (1993).

Engineered Materials Handbook, Vol. 2, Engineering Plastics, ASM International (1988).

S. B. Warner, *Fiber Science*, Prentice-Hall (1995).

Notes

Before the discovery of vulcanization, natural rubber obtained from the sap of tropical trees became a sticky mass in the heat of summer and hard and brittle in the cold of winter. It was the dream of Charles Goodyear to find a way to make rubber more useful. Without any training in chemistry, he started his experiments in a debtor's prison. In a kitchen of a cottage on the prison grounds, he blended rubber with anything he could find. He found that nitric acid made the rubber less sticky, but when he treated mailbags for the U.S. Post Office and left them for a time in a hot room they became a sticky mess. In 1839, he accidentally dropped a lump of rubber mixed with sulfur on a hot stove. After scraping the hard mess from the stove, he found that it remained flexible when it cooled. This process of vulcanization was improved and patented a few years later. However, his patent was pirated, and he landed in debtor's prison again.

Celluloid was the first commercial plastic. It was made by reacting cellulose, extracted from wood pulp, with nitric acid to form nitrocellulose that was then mixed with camphor. The process was developed by John Wesley Hyatt in the 1860s for billiard balls as a substitute for ivory, which was becoming scarce. It was quickly adopted for dominoes, cuffs, and collars. Among its few remaining uses are ping-pong balls and nail polish.

The first synthetic polymer Bakelite was invented by Leo Baekelund in 1907. Baekelund had immigrated from Belgium a few years earlier, with all his worldly possessions in four locked suitcases. He found a way to react phenol and formaldehyde to form a solid polymer. The key was learning to control the temperature of the reaction. This new material, called Bakelite, found extensive use as handles for knife and pot handles, as electrical insulators, in telephone handsets, and in many automotive applications. Unlike Goodyear, he was able to control his process, and he retired as a multimillionaire.

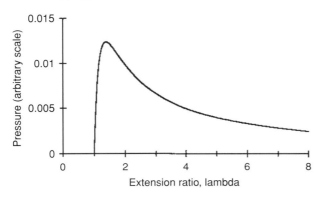

Figure 20.39. Variation of pressure in a balloon with the stretch ratio, $\lambda = D/D_0$.

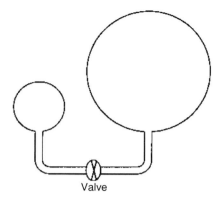

Figure 20.40. Two identical balloons, one inflated more than the other. The pressure in the smaller balloon is higher, so when the valve is opened air will flow from the smaller balloon into the larger one.

Valve

The number of plastics increased dramatically in the 1920s and 1930s. B. F. Goodrich introduced polyvinyl chloride (PVC) in 1926. In 1930, I. G. Farben made polystyrene commercially available. Clear polymethyl merthacrylate (Plexiglas) was introduced by Rohm and Haas in 1935. In the late 1930s, W. H. Carothers, a du Pont chemist, helped develop neoprene, the first synthetic rubber. However, his major contribution was the synthesizing of nylon in 1938.

Stretch a rubber band and hold it to your lips. Do you sense an increase or decrease in temperature? If it warms, that indicates that coefficient of thermal expansion of stretched rubber is negative! You can test this by hanging a weight on a rubber band and then heating the rubber band. Does it shrink or contract as it is heated under tension?

The shuttle disaster was caused by loosening of the rubber O-rings caused by the low temperature at launch. As noted in this chapter, rubber under tension has a negative coefficient of thermal expansion. This means potential expansion as the temperature is lowered, lowering the tensile stress.

For rubber stretched under biaxial tension, $\sigma_x = \sigma_y = \sigma$, the stress is given by $\sigma = NkT(\lambda^2 - 1/\lambda^4)$, where l is the stretch ratio, $L_x/L_{xo} = L_y/L_{yo}$. It is interesting to consider what this equation predicts about how the pressure in a spherical rubber balloon varies during the inflation. A force balance gives $\pi Dt\sigma = \pi D^2 P/4$, so $P = (4t/D)s$, where D is the diameter and t the thickness. The diameter increases with λ so $D = D_o\lambda$, where D_o is the initial radius. Rubber is almost incompressible, so its volume remains constant. The volume of the rubber in the balloon is $\pi D^2 t = 4\pi D_o^2 t_o$, so $t = t_o(D_o/D)^2 = t_o/\lambda^2$. Combining these expressions, $P = (2t_o/r_o)(1/\lambda - 1/\lambda^7)$. A plot of P versus λ (Figure 20.39) shows that after a reaches about 1.4, the pressure decreases as the radius (diameter) increases. This prediction leads itself to an interesting demonstration. Inflate two balloons to different diameters and then connect them in such a way as to allow the pressures to come to equilibrium. The diameter of the larger diameter balloon will grow, and the diameter of the smaller balloon will shrink (Figure 20.40).

Problems

1. What stress would be required to stretch a piece of PVC by 1% at 75°C. See Figure 20.6. If it were held in the stretched position, what would the stress be after a minute? After an hour? After a day?

2. Examine Figure 20.11, which relates the shear modulus of rubber to temperature. At 100°C, G increases with increasing percent sulfur.

    A. Find the slope of a plot of G versus %S at 100°C.

    B. Because $G = NkT$, this slope must be related to N. Write an expression relating N to %S at 100°C.

3. Evaluate the coefficient of thermal expansion, α, for rubber under an extension of 100% at 20°C. A rubber band was stretched from 6 in. to 12 in. While being held under a constant stress, it was heated from 20°C to 40°C. What change in length, ΔL, would the heating cause?

4. The stress–strain curve in Figure 20.17 indicates that the true strain associated with necking of polyethylene stretched is about 2.8. Neglecting any volume change, calculate the ratio of the cross-sectional areas before and after necking. Compare your answer with Figure 20.18.

5.   A. Use the general statement of the flow rule, $d\varepsilon_{ij} = d\lambda(\partial f/\partial\sigma_{ij})$, with the yield criteria, equation (20.10), to derive the expression predicting the relative volume change, $(dv/v)d\varepsilon_1$, in tension as a function of C/T.

    B. Use these expressions and the data in Figure 20.27 to predict C/T for HIPS.

6. An isotropic pressure-dependent yield criterion for polymers of the form, $f = A(\sigma_1 + \sigma_2 + \sigma_3) + B[(\sigma_2 - \sigma_3)^2 + (\sigma_3 - \sigma_1)^2 + (\sigma_1 - \sigma_2)^2]^{1/2} = 1$ has been proposed.

    A. Evaluate A and B in terms of T and C. Here, T is the yield strength in tension, and C is the absolute magnitude of the compressive strength. Consider a one-direction tension test in which $\sigma_1 = T$, $\sigma_2 = \sigma_3 = 0$ at yielding and a one-direction compression test in which $\sigma_3 = C$, $\sigma_2 = \sigma_3 = 0$ at yielding.

    B. Use the flow rule, $d\varepsilon_{ij} = d\lambda(\partial f/\partial\sigma_{ij})$, to evaluate the relative volume change on yielding. Express $d\delta/d\varepsilon_1$ as a function of T/C. Here, $d\delta = dv/v = d\varepsilon_1 + d\varepsilon_2 + d\varepsilon_3$ is the volume strain.

7. Cut a strip from a commercial food wrap. Using your fingers pull it in tension. Then pull it in tension 90 degrees to the original direction of stretching. Compare your observations with Figure 20.31. Explain why crazing is more likely when the loading is normal to the direction of prior extension.

8. Consider equation (20.13), which relates crazing to stress state.

    A. Write an expression for the value of σ_1 to cause crazing in terms of the stress ratio, $\alpha = \sigma_2/\sigma_1$, and A and B. Assume plane stress, $\sigma_3 = 0$.

    B. Compare the values of σ_1 for $\alpha = 0$ and $\alpha = 1$.

9. According to equations (20.10) and (20.11), yielding is possible under a state of pure hydrostatic tension, σ_H. Find that value of σ_H according to both equations if the tensile and compressive yield strengths are 80 and 100 MPa, respectively.

10. The compressive strength of Kevlar is about one-eighth of its tensile strength. If it is bent to small radius of curvature, it will kink as shown Figure 20.32. Estimate the smallest diameter rod on which Kevlar 49 fiber of 12-μm-diameter fibers can be wound without kinking. The tensile strength is 2.8 GPa, Young's modulus is 125 GPa, Poisson's ratio is 1 : 3, and the tensile strain to fracture is 2.3%.

Composites

Introduction

Throughout history, mankind has used composite materials to achieve combinations of properties that could not be achieved with individual materials. The Bible describes mixing of straw with clay to make tougher bricks. Concrete is a composite of cement paste, sand, and gravel. Today, poured concrete is almost always reinforced with steel rods. Other examples of composites include steel-belted tires, asphalt blended with gravel for roads, plywood with alternating directions of fibers, and fiberglass-reinforced polyester used for furniture, boats, and sporting goods. Composite materials offer combinations of properties otherwise unavailable.

The reinforcing material may be in the form of particles, fibers, or sheet laminates.

Fiber-Reinforced Composites

Fiber composites may also be classified according to the nature of the matrix and the fiber. Examples of a number of possibilities are listed in Table 21.1.

Various geometric arrangements of the fibers are possible. In two-dimensional products, the fibers may be unidirectionally aligned, at 90 degrees to one another in a woven fabric or cross-ply, or randomly oriented (Figure 21.1). The fibers may be very long or chopped into short segments. In thick objects, short fibers may be random in three dimensions. The most common use of fiber reinforcement is to impart stiffness (increased modulus) or strength to the matrix. Toughness may also be of concern.

Elastic Properties of Fiber-Reinforced Composites

The simplest arrangement is long parallel fibers. The strain parallel to fibers must be the same in both the matrix and the fiber, $\varepsilon_f = \varepsilon_m = \varepsilon$. For loading parallel to the fibers, the total load, F, is the sum of the forces on the fibers, F_f, and the matrix, F_m. In terms of the stresses, $F_f = |\sigma_f A_f$ and $F_m = \sigma_m A_m$, where σ_f and σ_m are the stresses in the fiber and matrix and where A_f and A_m are the cross-sectional areas of

Table 21.1. *Various combinations of fibers and matrices*

Fiber	Metal matrix	Ceramic matrix	Polymer matrix
Metal	W/Al	Concrete/steel	Rubber/steel (tires)
Ceramic	B/Al	C/glass	Fiberglass/polyester
	C/Al	SiC/Si-Al-O-N	Fiberglass/epoxy
	Al$_2$O$_3$/Al		
	SiC/Al		
Polymer		Straw/clay	Kevlar/epoxy

the fiber and matrix. The total force, F, is the sum of $F_f + F_m$,

$$\sigma A = \sigma_f A_f + \sigma_m A_m, \tag{21.1}$$

where A is the overall area, $A = A_f + A_m$, and σ is the stress, F/A. For elastic loading, $\varepsilon = E\varepsilon$, $\sigma_f = E_f\varepsilon_f$, and $\sigma_m = E_m\varepsilon_m$, so $E\varepsilon A = E_f\varepsilon_f A_f + E_m\varepsilon_m A_m$. Realizing that $\varepsilon_f = \varepsilon_m = \varepsilon$, and expressing $A_f/A = V_f$ and $A_m/A = V_m$,

$$E = E_f V_f + E_m V_m. \tag{21.2}$$

This is often called the *rule of mixtures*. It is an upper bound to the elastic modulus of a composite.

Consider the behavior of the same composite under tension perpendicular to the fibers. It is no longer reasonable to assume that $\varepsilon_f = \varepsilon_m = \varepsilon$. An alternative, although extreme, assumption is that the stresses in the matrix and the fibers are the same, $\sigma_f = \sigma_m = \sigma$. In this case, $\varepsilon = \sigma/E$, $\varepsilon_f = \sigma_f/E_f$, and $\varepsilon_m = \sigma_m/E_m$, and the overall (average) strain is $\varepsilon = \varepsilon_f V_f + \varepsilon_m V_m$. Combining, $\sigma/E = V_f\sigma_f/E_f + V_m\sigma_m E_m$. Finally realizing that $\sigma_f = \sigma_m = \sigma$,

$$1/E = V_f/E_f + V_m/E_m. \tag{21.3}$$

Equation (21.3) is a lower bound for the modulus. Figure 21.2 shows the predictions of equations (21.2) and (21.3). The actual behavior for loading perpendicular to the fibers is between these two extremes.

Now consider the orientation dependence of the elastic modulus of a composite with unidirectionally aligned fibers. Let 1 be the axis parallel to the fibers, and let the 2 and 3 axes be perpendicular to the fibers. If a uniaxial stress applied along a

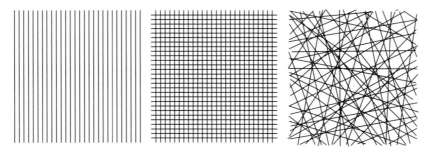

Figure 21.1. Several geometric arrangements of fiber reinforcements.

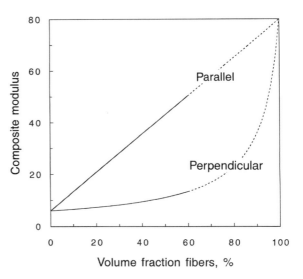

Figure 21.2. Upper and lower bounds to Young's modulus for composites. The upper bound is appropriate for loading parallel to the fibers. Loading perpendicular to the fibers lies between these two extremes. The lines are dashed above $V_f = 60\%$ because this is a practical upper limit to the volume fraction of fibers.

direction, x, at an angle, θ, from 1-axis and $90 - \theta$ from 2-axis, the stresses on the 1, 2, 3 axis system may be expressed as

$$\sigma_1 = \cos^2 \theta \sigma_x$$
$$\sigma_2 = \sigma_3 \sin^2 \theta \sigma_x,$$
$$\tau_{12} = \sin \theta \cos \theta \sigma_x.$$
$$\sigma_3 = \tau_{23} = \tau_{31} = 0. \tag{21.4}$$

Hooke's laws give the strains along the 1- and 2-axes,

$$e_1 = (1/E_1)[\sigma_1 - \upsilon_{12}\sigma_2],$$
$$e_2 = (1/E_2)[\sigma_2 - \upsilon_{12}\sigma_1], \quad \text{and}$$
$$\gamma_{12} = (\tau_{12}/G_{12}). \tag{21.5}$$

The strain in the x-direction can be written as

$$\begin{aligned} e_x &= e_1 \cos^2 \theta + e_2 \sin^2 \theta + 2\gamma_{12} \cos \theta \sin \theta \\ &= (\sigma_x/E_1)[\cos^4 \theta - \upsilon_{12} \cos^2 \theta \sin^2 \theta] + (\sigma_x/E_2)[\sin^4 \theta - \upsilon_{12} \cos^2 \theta \sin^2 \theta] \\ &\quad + (2\sigma_x/G_{12}) \cos^2 \theta \sin^2 \theta. \end{aligned} \tag{21.6}$$

The elastic modulus in the x-direction, E_x, can be found from equation (21.6).

$$E_x = \sigma_x/e_x. \tag{21.7}$$

Figure 21.3 shows that the modulus drops rapidly for off-axis loading. The average for all orientations in this figure is about 18% of E_1.

A crude estimate of the effect of a cross-ply can be made from an average of the stiffnesses due to fibers at 0 degrees and 90 degrees, as shown in Figure 21.4. Although the cross-ply stiffens the composite for loading near 90 degrees, it has no effect on the 45-degree stiffness. Figure 21.5 illustrates why this is so. For loading at 45 degrees, extension can be accommodated by rotation of the fibers without any extension.

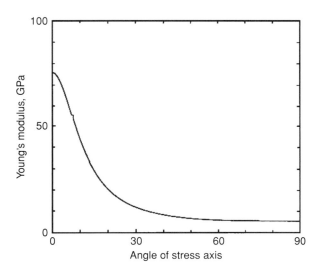

Figure 21.3. Orientation dependence of the elastic modulus in a composite with unidirectionally aligned fibers. Here, it is assumed that $E_1 = 75$ MPa, $E_2 = 5$ MPa, $G = 10$ MPa, and $\upsilon_{12} = 0.3$. Note that most of the stiffening is lost if the loading axis is degrees misoriented from the fiber axis by as little as 15 degrees.

The simple averaging used to calculate Figure 21.4 underestimates the stiffness for most angles of loading, θ. It assumes equal strains in both plies in the loading direction but neglects the fact that the strains perpendicular to the loading direction in both plies must also be equal. When this constraint is accounted for, a somewhat higher modulus is predicted. This effect, however, disappears at $\theta = 45°$ because composites with both sets of fibers have the same lateral strain.

With randomly oriented fibers, the orientation dependence disappears. One might expect that the modulus would be the average of the moduli for all directions of uniaxially aligned fibers. However, this again would be an underestimate because it neglects the fact that the lateral strains must be the same for all fiber alignments. A useful engineering approximation for randomly aligned fibers is

$$E \approx (3/8) E_{\text{para}} + (5/8) E_{\text{perp}}, \tag{21.8}$$

where E_{para} and E_{perp} are the E moduli parallel and perpendicular to uniaxially aligned fibers.

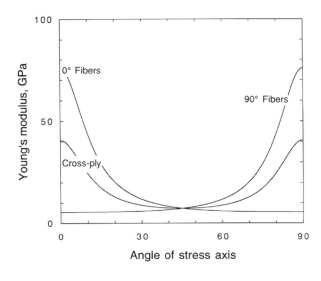

Figure 21.4. Calculated orientation dependence of Young's modulus in composites with singly and biaxially oriented fibers. For this calculation, it was assumed that $E_{\text{para}} = 75$ GPa and $E_{\text{perp}} = 75$ GPa. Note that even with biaxially oriented fibers, the modulus at 45 degrees is very low.

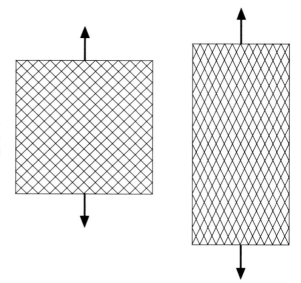

Figure 21.5. Rotation of fibers in a woven cloth. Stress in at 45 degrees to the fibers allows deformation with little or no stretching of the fibers.

Strength of Fiber-Reinforced Composites

The rule of mixtures cannot be used to predict the strengths of composites with uniaxially aligned fibers. The reason can be appreciated by considering the stress–strain behavior of both materials, as show in Figure 21.6. The strains in the matrix and fibers are equal, so the fibers reach their breaking strengths long before the matrix reaches its tensile strength. Thus, the strength of the composite, $UTS < V_m UTS_m + V_f UTS_f$.

If the load carried by the fibers is greater than the breaking load of the matrix, both will fail when the fibers break.

$$UTS = V_m \sigma_m + V_f (UTS)_f, \qquad (21.9)$$

where $\sigma_m = (E_m/E_f)(UTS)_f$ is the stress carried by the matrix when the fiber fractures.

For composites with low-volume fraction fibers, the fibers may break at a load less than the failure load of the matrix. In this case, after the fibers break, the whole

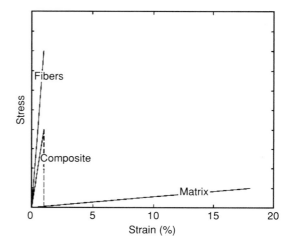

Figure 21.6. Stress–strain curves for the matrix, the fibers, and the composite.

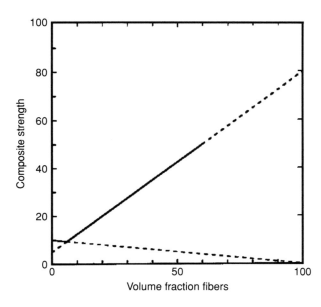

Figure 21.7. Dependence of strength on the volume fraction fibers. It is assumed that the fiber strength, $UTS_f = 80$, $UTS_m = 10$, and $\sigma_m = 5$. Note that the fibers lower the composite strength less than the tensile strength of the matrix for very low-volume fraction fibers.

load must be carried by the matrix so the predicted strength is

$$UTS = V_m(UTS)_m. \tag{21.10}$$

Figure 21.7 shows how the strength in tension parallel to the fibers varies with volume fraction fibers. Equation (21.10) applies at low-volume fraction fibers, and equation (21.9) applies for higher-volume fraction fibers.

EXAMPLE PROBLEM 21.1: A metal composite composed of metal M reinforced by parallel wires of metal W. The volume fraction W is 50%, and W has a yield stress of 200 MPa and a modulus of 210 GPa. Metal M has a yield stress of 50 MPa and a modulus of 30 GPa. What is the maximum stress that the composite can carry without yielding?

Solution: At yielding, the strains will be the same in the wire and matrix. $\sigma_w/E_w = \sigma_w/E_m$. The wire will yield at a strain of $\sigma_w/E_w = 200\,\mathrm{MPa}/210\,\mathrm{GPa} = 0.001$, and the matrix will yield at a strain of 50 MPa/30 GPa = 0.0017. The composite will yield at the lower strain (i.e., when $\varepsilon = 0.001$). At this point, the stress in the wire would be 200 MPa, and the stress in the matrix would be $0.001 \times 30\,\mathrm{GPa} = 30\,\mathrm{MPa}$. The average stress would be $0.50 \times 200 + 0.50 \times 30 = 115\,\mathrm{MPa}$.

Volume Fraction of Fibers

Although the stiffness and strength of reinforce composites should increase with volume fraction of fibers, there are practical limitations on the volume fraction. Fibers must be separated from one another. Fibers are often precoated to ensure this separation and to control the bonding between fibers and matrix. Techniques of infiltrating fiber arrangements with liquid resins lead to variability in fiber spacing, as shown in Figure 21.8. The maximum possible packing density is greater for

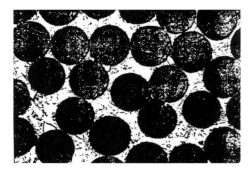

Figure 21.8. Glass fibers in a polyester matrix. Note the variability in fiber spacing. From *Engineered Materials Handbook, v. I, Composites*, ASM International (1987).

0.0127 mm

unidirectionally aligned fibers than for woven or cross-ply reinforcement. A practical upper limit for volume fraction seems to be about 55% to 60%. This is why the calculated lines in Figures 21.2 and 21.6 are dashed above $V_f = 60\%$.

Orientation Dependence of Strength

For unidirectionally aligned fibers, the strength varies with orientation. There are three possible modes of failure. For tension parallel or nearly parallel to the fibers, failure occurs when the stress in the fiber exceeds the fracture strength, S_f, of the fibers. Neglecting the stress in the matrix, the axial stress, σ, at this point is

$$\sigma = S_f / \cos^2 \theta. \tag{21.11}$$

For loading at a greater angle to the fibers, failure can occur by shear in the matrix. The axial stress, σ, is

$$\sigma = \tau_{fm} / \sin \theta \cos \theta, \tag{21.12}$$

where τ_{fm} is the shear strength of the fiber–matrix interface. For loading perpendicular or nearly perpendicular to the fibers, failure is governed by the strength, S_{fm}, of the fiber–matrix interface.

$$\sigma = S_{fm} / \sin^2 \theta. \tag{21.13}$$

These three possibilities are shown in Figure 21.9. The three fracture modes are treated separately. However, the slight increase of failure stress, σ, with θ for low angles is not realized experimentally. This fact suggests an interaction of the longitudinal and shear fracture modes. Also, for angles near 45 degrees, experimental values of fracture strength tend to fall somewhat below the predictions in Figure 21.9, which indicates an interaction between shear and transverse fracture modes.

Fiber Length

Fabrication is much simplified if the reinforcement is in the form of chopped fibers. Chopped fibers can be blown onto a surface to form a mat. Composites with chopped fibers can be fabricated by processes that are impossible with continuous fibers, such as extrusion, injection molding, and transfer molding. The disadvantage of chopped fibers is that some of the reinforcing effect of the fibers is sacrificed

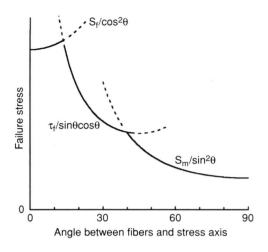

Figure 21.9. Failure strength of a unidirectionally aligned fiber composite as a function of orientation. There are three possible fracture modes: tensile fracture of the fibers, shear failure parallel to the fibers, and tensile failure normal to the fibers.

because the average axial stress carried by fibers is less for short ones than long ones. The reason is that at the end of the fiber, the stress carried by the fiber is vanishingly low. Stress is transferred from the matrix to the fibers primarily by shear stresses at their interfaces. The average axial stress in a fiber depends on its aspect ratio, D/L, where D and L are the fiber's diameter and length.

The following development is based on the assumption that the shear stress between the matrix and the fiber, τ, is constant and that no load is transferred across the end of the fiber. Figure 21.10 shows a force balance on a differential length of fiber, dx, which results in $(\sigma_x + d\sigma_x)A = \sigma_x A + \tau(\pi D\, dx)$ or $(D^2/4)\, d\sigma_x = \tau(D\, dx)$. Integrating,

$$\sigma_x = 4\tau(x/D). \tag{21.14}$$

This solution is valid only for $\sigma_x \leq \sigma_\infty$, where σ_∞ is the stress that would be carried by an infinitely long fiber. Figure 21.11 illustrates three possible conditions, depending on x^*/L, where $x^* = (D/4)(\sigma_\infty/\tau)$ is the distance at which $\sigma_x = \sigma_\infty$. The length, $L^* = 2x^*$ is called the *critical length*.

If $L > L^*$ (Figure 21.11a),

$$\sigma_{av.} = (1 - x^*/L)\sigma_\infty. \tag{21.15}$$

If $L = L^*$ (Figure 21.11b),

$$\sigma_{av.} = (1/2)\sigma_\infty. \tag{21.16}$$

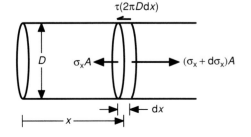

Figure 21.10. Force balance on a differential length of fiber. The difference in the normal forces must be balanced by the shear force.

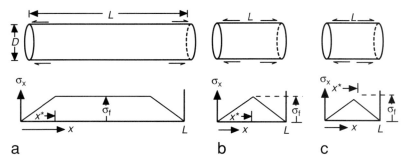

Figure 21.11. Distribution of fiber stress for (a) $L > L^*$, (b) $L = L^*$, and (c) $L < L^*$.

If $L < L^*$ (Figure 21.11c), σ_x never reaches σ_∞, and the average value of axial stress

$$\sigma_{av} = L/(2L^*)\sigma_\infty. \tag{21.17}$$

The composite modulus and strength can be calculated by substituting σ_{av} for σ_f in equation (21.1).

The problem with this analysis is that the shear stress, τ, between matrix and fiber is not constant from $x = 0$ to $x = x^*$ along the length of the fiber. The shear stress, τ, between the fiber and the matrix decreases with x as the stress (and, therefore, the elastic elongation) of the fiber increases.

It is reasonable to assume that t increases as the difference between the strains in the fiber and matrix increases. This difference, in turn, is proportional to $\sigma_\infty - \sigma_x$ because the fiber strain at x is σ_x/E_f and the matrix strain is σ_∞/E_f. If it is assumed that $\tau = C(\sigma_f - \sigma_x)$, where C is a constant into equation (21.10) and integrating,

$$\sigma_x = \sigma_\infty[1 - \exp(-2Cx/D)]. \tag{21.18}$$

This leads to the stress distribution sketched in Figure 21.12 near the end of the fiber. For long fibers ($L > 2x^*$), the average load carried by the fiber is not much different from that calculated with the simpler analysis.

A still more rigorous analysis results in an expression for the shear stress,

$$\tau = E_f(D/4)\exp(\beta)\sinh[\beta(L/2 - x)]/\cosh[\beta(L/2)], \tag{21.19}$$

where β is given by

$$\beta = (1/4)(G_m/E_f)/\ln(\phi/V_f). \tag{21.20}$$

Figure 21.12. More realistic stress distribution near end of a fiber.

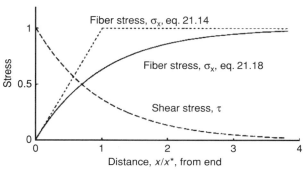

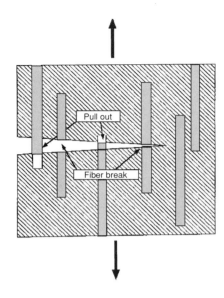

Figure 21.13. Some fibers fracturing at the crack and others pulling out.

Here, ϕ is the ideal packing factor for the arrangements of fibers. For a square array, $\phi = \pi/4 = 0.785$, and for a hexagonal array, $\phi = \pi(2\sqrt{3}) = 0.907$. As the ratio of the matrix shear modulus to the fiber tensile modulus increases, the load is transferred from the matrix to the fibers over a shorter distance.

Failure with Discontinuous Fibers

Failure may occur either by fracture of fibers or by the fibers pulling out of the matrix. Both possibilities are shown in Figure 21.13. If, as a crack in the matrix approaches the fiber, the plane of the crack is near the end of the fiber, pullout will occur. If it is not near the end, the fiber will fracture. Figure 21.14 shows the pullout of boron fibers in an aluminum matrix. Figure 21.14 shows the pullout of boron fibers in an aluminum matrix. To fracture a fiber of diameter, D, the force, F, must be

$$F = \sigma^* \pi D^2/4, \tag{21.21}$$

where σ^* is the fracture strength of the fiber.

Much more energy is absorbed if the fibers pull out. The fiber pullout force is

$$F = \tau^* \pi Dx, \tag{21.22}$$

Figure 21.14. SiC fibers pulling out of a titanium matrix. From T. W. Clyne and P. J. Withers, *An Introduction to Metal Matrix Composites*, Cambridge University Press (1993).

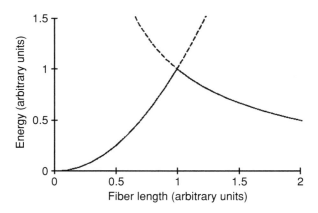

Figure 21.15. Energy expended in fiber pullout increases with fiber length up to a critical length and then decreases with further length increase.

where τ^* is the shear strength of the fiber–matrix interface. Fibers will pull out if $\tau^*\pi Dx$ is less than the force to break the fibers, $\sigma^*\pi D^2/4$. The critical pullout distance, x^*, corresponds to the two forces being equal,

$$x^* = (\sigma^*/\tau^*)D/4. \tag{21.23}$$

Fibers of length L less than $2x^*$ will pull out, and the average pullout distance will be $L/4$. Therefore, the energy, U_{po}, to pull out the fiber is the integral of $\tau^*\pi D\int x\,\mathrm{d}x$ between limits of 0 and $L/4$ is

$$U_{po} = \tau^*\pi DL^2/32. \tag{21.24}$$

However, if the fiber length is greater than $2x^*$, the probability that it will pull out is $2x^*/L$, and the average pullout distance is $x^*/2$, so the average energy expended in pullout is

$$U_{po} = \tau^*\pi D(x^*/2)^2(2x^*/L) = t^*\pi Dx^{*3}/(2L). \tag{21.25}$$

Comparison of equations (21.24) and (21.25) shows that the pullout energy, and hence the toughness, increase with L up to a critical length, $L = 2x^* = (\sigma^*/\tau^*)D/2$. Further increase of L decreases the fracture toughness as shown in Figure 21.15. Because the composite stiffness and strength continue to increase with increasing fiber length, it is often desirable to decrease the fiber–matrix shear strength, τ^*, and thereby increase x^* so longer fibers can be used without decreasing toughness.

Failure under Compression

In compression parallel to aligned fibers, failures can occur by fiber buckling. This involves lateral shearing of the matrix between fibers, and therefore, the compressive strength of the composite depends on the shear moduli. Yielding of composites with polymeric fibers is sensitive to buckling under compression because of the low compressive strengths of the fibers themselves. Figure 21.16 shows in-phase buckling of a Kevlar fiber-reinforced composite in compression. The low strength of some polymer fibers in compression is discussed in Chapter 20.

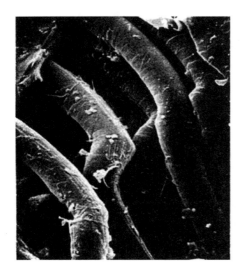

Figure 21.16. Buckling of Kevlar fibers in an epoxy matrix under compression. From *Engineered Materials Handbook, v. 1, Composites*, ASM International (1987).

Typical Properties

Two types of polymer matrixes are common: epoxy and polyester. Most polymers used for matrix materials have moduli of 2 to 3 GPa and tensile strengths in the range of 35 to 70 MPa. Fiber reinforcements include glass, boron, Kevlar, and carbon. Properties of some epoxy matrix composite systems are given in Table 21.2. Properties of some commonly used fibers are shown in Table 21.3.

Other fiber composites include ceramics ceramic reinforced with metal or ceramic fibers. Metals such as aluminum-base alloys may be reinforced with ceramic fibers to increase their stiffness. In some eutectic systems, directional solidification can lead to rods of one phase reinforcing the matrix.

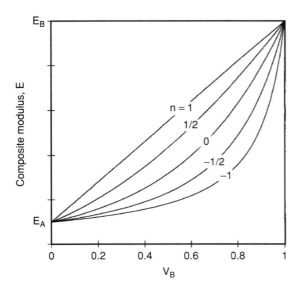

Figure 21.17. Dependence of Young's modulus on volume fraction according to equation 21.26 for several values of n. The subscripts A and B represent the lower and higher modulus phases. The isostrain and isostress models correspond to n = +1 and −1. Here, the ratio of $E_B/E_A = 10$.

Table 21.2. *Properties of epoxy matrix composites*

Fiber	Vol % fiber	Young's modulus (GPa)		Tensile strength (MPa)	
		Longitudinal	Transverse	Longitudinal	Transverse
E-glass unidirectional	60	40	10	780	28
E-glass bidirectional	35	16.5	16.5	280	280
E-glass chopped matte	20	7	7	100	100
Boron unidirectional	60	215	24	1,400	65
Kevlar 29 unidirectional	60	50	5	1,350	–
Kevlar 49 unidirectional	60	76	6	1,350	30
Carbon	62	145		1,850	

Particulate Composites

Composites reinforced by particles rather than long fibers include such diverse materials as concrete (cement–paste matrix with sand and gravel particles), polymers filled with wood flour, and "carbide tools" with a cobalt–base matrix alloy hardened by tungsten carbide particles. Sometimes the purpose is simply economics (e.g., wood flour is cheaper than plastics). Another objective may be increased hardness (e.g., carbide tools). The isostrain and isostress models (equations 21.2 and 21.3) are upper and lower bounds for the dependence of Young's modulus on volume fraction. The behavior of particulate composites is intermediate and can be represented by a generalized rule of mixtures of the form:

$$E^n = V_A E_A^n + V_B E_B^n, \tag{21.26}$$

where A and B refer to the two phases. The exponent, n, lies between the extremes of n = +1 for the isostrain model and n = −1 for the isostress model. Figure 21.17 shows the dependence of E on volume fraction for several values of n. If the modulus

Table 21.3. *Typical fiber properties*

Fiber	Young's modulus (GPa)	Tensile strength (GPa)	Elongation (%)
Carbon (PAN* HS)	250	2.7	1.0
Carbon (PAN* HM)	390	2.2	0.5
SiC	70		
Steel	210	2.5	
E-glass	70	1.75	
B	390	2.0–6.0	
Kevlar 29	65	2.8	4.0
Kevlar 49	125	2.8	2.3
Al_2O_3	379	1.4	
β-SiC	430	3.5	

of the continuous phase is much higher than that of the particles, $n = 1/2$ is a reasonable approximation. For high modulus particles in a low modulus matrix, $n < 0$ is a better approximation.

Brick Wall Model

A simple model for a particulate composite with a large fraction of the harder phase is illustrated in Figure 21.18. The harder phase, A, is a series of cubes. Let the distance between cube centers be 1 and the thickness of the softer phase, B, be t. Then the volume fraction of A is $V_A = (1 - t)^3$. The composite can be considered as a series of columns of alternating A and B loaded in parallel with columns of B.

If the behavior of the columns of alternating A and B is described by the lower bound equation (21.3), the modulus, E', of these columns is

$$1/E' = (1 - t)/E_A + t/E_B. \tag{21.27}$$

and the modulus of the overall composite can be found using equation (22.2) as

$$E_{av} = (1 - t)^2 E' + [1 - (1 - t)^2] E_B. \tag{21.28}$$

The predictions of equation (21.28) are too low because the assumptions allow the soft material in the A-B columns to squeeze out from between the hard cubes. A more realistic assumption is that $\varepsilon_{2A} = \varepsilon_{2B}$. A force balance in the 2-direction gives $\sigma_{2B}t = -\sigma_{2A}(1 - t)$. Applying Hooke's law to find σ_{2B}/σ_1,

$$\sigma_{2B}/\sigma_1 = (v_B/E_B - v_A/E_A)/[(1 - v_B)/E_B + t(1 - t)(1 - v_A)E_A]. \tag{21.29}$$

The 1-direction strains in the A-B columns are

$$\varepsilon_{1B}/\sigma_1 = (1/E_B)[1 - 2v_B(\sigma_{2B}/\sigma_1)] \tag{21.30}$$

$$\text{and} \quad \varepsilon_{1A}/\sigma_1 = (1/E_A)[1 - 2v_A(\sigma_{2A}/\sigma_1)]$$
$$= (1/E_A)[1 - 2v_A(\sigma_{2b}/\sigma_1)t/(1 - t)] \tag{21.31}$$

$$\text{so} \quad E' = 1/[\varepsilon_{1B}/\sigma_1 + \varepsilon_{1A}/\sigma_1] \tag{21.32}$$

The composite modulus, E_{av}, can be found by treating these columns and the material between them with the isostrain model,

$$E_{av} = t(2 - t)E_B + (1 - t)^2 E'. \tag{21.33}$$

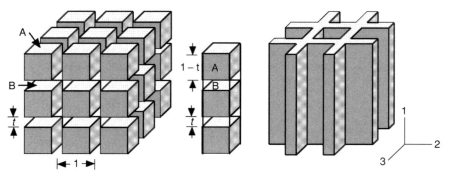

Figure 21.18. Brick wall model of a composite. Hard particles in the shape of cubes are separated by a soft matrix of thickness, t.

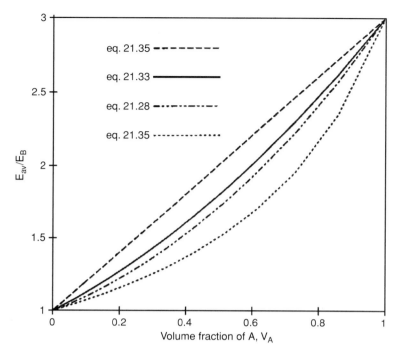

Figure 21.19. Predictions of brick wall model for the elastic modulus of a particulate composite with $E_A = 3E_B$ and $u_A = u_B = 0.3$. Note that the predictions of equations (22.28) and (22.33) are not very different.

This is an upper bound for this geometric arrangement and much better solution than equation (21.28). Figure 21.19 shows the results calculations for $E_A = 3E_B$ and $\upsilon_A = \upsilon_B = 0.3$ using equations (21.28) and (21.33). Also shown for comparison are the upper-bound equation,

$$E_{av} = V_A E_A + (1 - V_A) E_B, \qquad (21.34)$$

and the upper-bound equation,

$$1/E_{av} = V_A/E_A + (1 - V_A)/E_B. \qquad (21.35)$$

Predictions of equation (21.28) with (21.27) are close to those of the empirical equation (21.26) with n = −0.165 and the predictions of equation (21.33) with (21.32) are close to equation (21.26) with n = +0.175.

EXAMPLE PROBLEM 21.2: Estimate the modulus for a composite consisting of 60% WC particles in a matrix of cobalt using the brick wall model. The Young's moduli for WC and Co are 700 GPa and 210 GPa, respectively. WC is the discontinuous phase, A.

Solution: $t = 1 - 0.6^{1/3} = 0.156$. From equation (21.29), $\sigma_{2B}/\sigma_1 = 0.615$
From equation (21.30), $\varepsilon_{1B}/\sigma_1 = 2.81 \times 10^{-3}$ GPa^{-1},
From equation (21.31), $\varepsilon_{1A}/\sigma_1 = 1.32 \times 10^{-3}$ GPa^{-1},
From equation (21.32), $E' = 644$ GPa,
From equation (21.28), $E_{av} = 519$ GPa.

Lamellar Composites

Two or more sheets of materials bonded together can be considered as lamellar composites.

Examples include safety glass, plywood, plated metals, and glazed ceramics. Consider sheets of two materials, A and B, bonded together in the x-y plane and loaded in this plane. The basic equations governing the strain are

$$\varepsilon_{xA} = \varepsilon_{xB} \quad \text{and}$$
$$\varepsilon_{yA} = \varepsilon_{yB}. \tag{21.36}$$

If both materials are isotropic and the loading is elastic, equations (21.36) become

$$\varepsilon_{xA} = (1/E_A)(\sigma_{xA} - \upsilon_A\sigma_{yA}) = \varepsilon_{xB} = (1/E_B)(\sigma_{xB} - \upsilon_B\sigma_{yB})$$
$$\varepsilon_{yA} = (1/E_A)(\sigma_{yA} - \upsilon_A\sigma_{xA}) = \varepsilon_{yB} = (1/E_B)(\sigma_{yB} - \upsilon_B\sigma_{xB}). \tag{21.37}$$

The stresses are

$$\sigma_{xA}t_A + \sigma_{xB}t_B = \sigma_{xav},$$
$$\sigma_{yA}t_A + \sigma_{yB}t_B = \sigma_{yav},$$
$$\sigma_{zA} = \sigma_z = 0, \tag{21.38}$$

where t_A and t_B are the fractional thicknesses of A and B.

Now consider loading under uniaxial tension applied in the x-direction. Substituting $\sigma_{yav} = 0$, $\sigma_{yB} = -(t_A/t_B)\sigma_{yA}$ and $\sigma_{xB} = (\sigma_{xav} - t_A\sigma_{xA})/t_B$ into equations (21.38),

$$\sigma_{xA} - \upsilon_A\sigma_{yA} = (E_A/E_B)[(\sigma_{xav} - t_A\sigma_{xA})/t_B + \upsilon_B(t_A/t_B)\sigma_{yA}] \quad \text{and}$$
$$\sigma_{yA} - \upsilon_A\sigma_{xA} = (E_A/E_B)[(V_A/A_B)\sigma_{yA} + \upsilon_B(\sigma_{xav} - t_A\sigma_{xA})/t_B]. \tag{21.39}$$

Young's modulus according to an upper-bound isostrain model for loading in the plane of the sheet can be expressed as

$$E = N/D, \quad \text{where}$$
$$N = E_B^2 t_B^2(1 - \upsilon_A^2) + 2E_A E_B t_A t_B(1 - \upsilon_A\upsilon_B) + E_A^2 t_A^2(1 - \upsilon_B^2) \quad \text{and}$$
$$D = E_B t_B(1 - \upsilon_A^2) + E_A t_A(1 - \upsilon_B^2). \tag{21.40}$$

For the special case in which $\upsilon_B = \upsilon_A$, these expressions reduce to the upper-bound model,

$$E = E_B t_B + E_A t_A. \tag{21.41}$$

EXAMPLE PROBLEM 21.3: Calculate the composite modulus for a sandwich of two sheets of fiber-reinforced polyester (each 0.5 mm thick) surrounding rubber (4 mm thick). For the fiber-reinforced polyester, $E_B = 7$ GPa and $u_B = 0.3$. For the rubber, $E_A = 0.25$ GPa and $\upsilon_A = 0.5$.

Solution: Substituting $t_A = 0.8$ and $t_B = 0.2$ into equations (21.40) and (21.41),

$$N = (7 \times 0.2)^2(0.75) + 2(7)(0.25)(0.2)(0.8)(0.85) + (0.25)(0.8)^2(0.91) = 2.21$$
$$D = (7 \times 0.2)(0.75) + (0.25 \times 0.8)(0.91) = 1.232$$
$$E = N/D = 2.21/1.232 = 1.79\,\text{GPa}$$

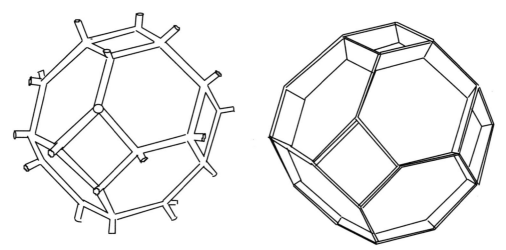

Figure 21.20. Open and closed cell foams modeled by tetrakaidecahedra. From W. F. Hosford, *Materials for Engineers*, Cambridge University Press (2008).

Morphology of Foams

There are two types of foams: closed cell foams and open cell (or reticulated) foams. In open foams, air or other fluids are free to circulate. These are used for filters and as skeletons. They are often made by collapsing the walls of closed cell foams. Closed cell foams are much stiffer and stronger than open cell foams because compression is partially resisted by increased air pressure inside the cells. Figure 21.20 shows that the geometry of open and closed cell foams can be modeled by Kelvin tetrakaidecahedra.

Mechanical Properties of Foams

The elastic stiffness depends on the relative density. In general, the dependence of relative stiffness, E^*/E, where E^* is the elastic modulus of the structure, and E is the modulus of the solid material on relative density is of the form

$$E^*/E = (\rho^*/\rho s)^n, \tag{21.42}$$

where ρ^* and ρ_s are the overall density of the foam and the solid foam from which the foam is formed.

Experimental results shown in Figure 21.21 indicate that for open cells, $n = 2$, so

$$E^*/E = (\rho^*/\rho s)^2. \tag{21.43}$$

For closed cell foams, E^*/E is much higher and $n < 2$. Although deformation under compression of open cell foams is primarily by ligament bending, compression of closed cell wall foams involves gas compression, wall stretching, and wall bending.

Metal Foams

Metal foams are useful in many engineering applications because of their extremely light weight, good energy absorption, high ratios of strength and stiffness to weight, and outstanding damping capability. Figure 21.22 shows the stress–strain curves of

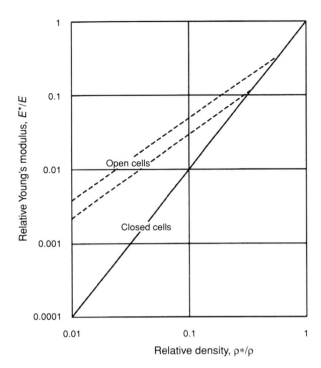

Figure 21.21. Dependence of Young's modulus on density. From W. F. Hosford, *Materials for Engineers*, Cambridge University Press, 2008. Data from Gibson and Ashby, *Cellular Foams*, Cambridge (1999).

Al–Si–Cu–Mg alloy foams different densities. There is an initial nearly linear region where partially reversible cell wall bending occurs. This is followed by a plastic plateau during which cells collapse, buckle, yield, and fracture. Finally, the stress rises rapidly as complete compaction commences.

In most cases, the stress there is an initial peak after which the stress drops significantly. This drop is attributed to the collapse of the lowest local density. The elastic modulus was measured from unloading curves at 1% strain, and plateau stress was taken as the average stress in the range 10% to 50% strain. The densification strain was taken as the strain at the point of intersection between the horizontal plateau stress line and the backward extended densification line.

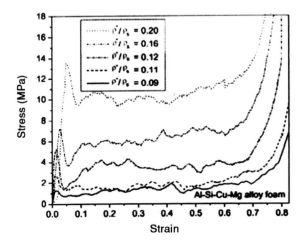

Figure 21.22. Tensile stress–strain curves of Al–Si–Cu–Mg alloy foams of different densities. From L. J. Gibson and M. F. Ashby, *Ibid*.

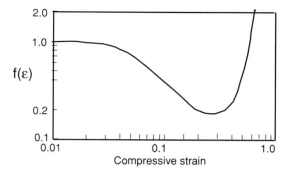

Figure 21.23. Dependence of f(ε) on compressive strain.

Flexible Foams – Open Cell

If an open cell foam is modeled as spheres of diameter, D, each connected by n ligaments of length, l, and cross-sectional area, A, the ratio of the elastic modulus in tension of a foam to that of the material, E_f/E_m, according to Lederman*, is

$$E_f/E_m = (nA/\pi D^2)(D/l^2/[2(1+D/l)]). \qquad (21.44)$$

For a fixed volume fraction polymer, E_f/E_m is a maximum if $nA/\pi D^2 = 0.5$. Rusch[†] found that

$$E_f/E_m = (\phi/12)(2+7\phi+3\phi^2), \qquad (21.45)$$

where ϕ is the volume fraction foam. This expression predicts that $E_f/E_m \rightarrow \phi/6$ as $\phi \rightarrow 0$ and $E_f/E_m \rightarrow 1$ as $\phi \rightarrow 1$.

Rusch proposed that the compressive stress–strain behavior of open cell foams can be approximated by

$$\sigma = E_f\varepsilon f(\varepsilon) \qquad (21.46)$$

where f(ε) is a shape factor that depends on cell geometry and is independent of ϕ. Figure 21.23 shows that initially f(ε) = 1, but with increasing strain f(ε) decreases as the ligaments buckle. Finally, f(ε) rises rapidly as the ligaments collapse on each other.

Flexible Foams – Closed Cell

With closed cells, compression increases the gas pressure in the cells. This raises the stress–strain curve. Adding this effect to equation (21.46),

$$\sigma = E_f\varepsilon f(\varepsilon) + P_o\varepsilon/(1-\phi-\varepsilon) \qquad (21.47)$$

where P_o is the initial pressure.

REFERENCES

K. K. Chawla, *Composite Materials, Science and Engineering*, Springer-Verlag (1978).
P. K. Mallick, *Fiber-Reinforced Composites, Materials, Manufacturing & Design*, Dekker (1988). *Engineered Materials Handbook, Vol. I, Composites*, ASM International (1987).

* J. M. Lederman, *J. Appl. Polym. Sci.*, v. 15 (1971).
† K. C. Rusch, *J. Appl. Polym. Sci.*, v. 13 (1969).

N. C. Hilyard, *Mechanics of Cellular Plastics*, Macmillan (1982).

E. Andrew, W. Sanders, and L. J. Gibson, Compressive and tensile behaviour of aluminium foams, *Mater. Sci. Eng. A*, 270 (1999).

L. J. Gibson and M. F. Ashby, *Cellular Foams*, Cambridge (1999).

A. Kim, M. A. Hasan, S. H. Nahm, and S. S. Cho, *Composite Structures*, v. 71 (2005).

Notes

The engineering use of composites dates back to antiquity. The Bible, in Exodus, cites the use of straw in clay bricks (presumably to prevent cracking during fast drying under the hot sun in Egypt). The Mayans and Incas incorporated vegetable fibers into their pottery. English builders used sticks in the "wattle and daub" construction of timber frame buildings. In colonial days, hair was used to strengthen plaster, the hair content being limited to 2% or 3% to ensure workability.

Saracens made composite bows composed of a central core of wood with animal tendons glued on the tension side and horn on the compression side. These animal products are able to store more elastic energy in tension and compression, respectively, than wood.

Shortly after Leo Baekelund (1863–1944) developed phenol-formaldehyde, "Bakelite," in 1906, he found that adding fibers to the resin greatly increased its toughness. The first use of this molding compound was for the gearshift knob of the 1916 Rolls Royce.

Widespread commercial use of glass-reinforced polymers began soon after World War II. Freshly formed glass fibers of 0.005- to 0.008-mm-diameters have strengths up to 4 GPa, but the strength is greatly reduced if the fibers contact one another. For that reason, they are coated with an organic compound before being bundled together.

Problems

1. Calculate the volume fraction fiber in the several composites described:

A. Maximum possible fiber fraction for unidirectionally aligned cylindrical fibers with negligible spacing between. (Assume a hexagonal array.)

B. Maximum possible fiber fraction for unidirectionally aligned cylindrical fibers of 100 µm diameter coated with 10-µm-thick coating. (Assume a hexagonal array.)

C. Maximum possible fiber fraction for alternating layers of unidirectionally aligned fibers, as shown in Figure 21.24.

D. Maximum possible fiber fraction alternating layers of unidirectionally aligned fibers of 100 µm diameter coated with 10-µm-thick coating, as shown in Figure 21.24.

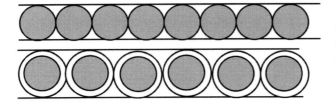

Figure 21.24. A ply of unidirectionally aligned fibers (*top*) and a ply of unidirectionally aligned coated fibers (*bottom*).

2. Calculate the elastic modulus of a composite of 40 volume % continuous aligned boron fibers in an aluminum matrix ($E = 70$ GPa),

    A. Parallel to the boron fibers,
    B. Perpendicular to the boron fibers.

3. In all useful fiber-reinforced composites, the elastic moduli of the fibers are higher than those of the matrix. Explain why.

4. Consider the matrix of elastic constants for a composite consisting of an elastically soft matrix reinforced by a 90-degree cross-ply of stiff fibers. The general form of the matrix of elastic constants is

s_{11}	s_{12}	s_{13}	0	0	0						
s_{12}	s_{11}	s_{13}	0	0	0						
s_{13}	s_{13}	s_{33}	0	0	0	0	0	0	s_{44}	0	0
0	0	0	0	s_{44}	0	0	0	0	0	0	s_{66}

Young's modulus for the composite when loaded parallel to one of the sets of fibers is 100 GPa, so $s_{11} = 10 \times 10^{-12}$ Pa^{-1}. Of the values listed, which is most likely for s_{12}? for s_{66}?

    A. 10×10^{-10} Pa^{-1};
    B. 30×10^{-12} Pa^{-1};
    C. 10×10^{-12} Pa^{-1};
    D. 3×10^{-12} Pa^{-1};
    E. 100×10^{-12} Pa^{-1};
    F. 10×10^{-10} Pa^{-1};
    G. -30×10^{-12} Pa^{-1};
    H. 10×10^{-12} Pa^{-1};
    I. -3×10^{-12} Pa^{-1};
    J. -100×10^{-12} Pa^{-1}.

5. What would be the critical length, L^*, for maximum load in a 10-μm-diameter fiber with a fracture strength of 2 GPa embedded in a matrix such that the shear strength of the matrix–fiber interface is 100 MPa.

6. Estimate the greatest value of the elastic modulus that can be obtained by long randomly oriented fibers of E-glass embedded in an epoxy resin if the volume fraction is 40%. Assume the modulus of the epoxy is 5 GPa.

7. Carbide cutting tools are composites of hard tungsten carbide particles in a cobalt matrix. The elastic moduli of tungsten carbide and cobalt are 102×10^6 psi and 30×10^6 psi, respectively. It was experimentally found that the elastic modulus of a composite containing 52 volume % carbide was 60×10^6 psi. What value of the exponent, n, in equation (21.26) would this measurement suggest? A trial-and-error solution is necessary to solve this. (Note that n = 0 is a trivial solution.)

8. A steel wire (1.0 mm diameter) is coated with aluminum, 0.20 mm thick.

    A. Will the steel or the aluminum yield first as tension is applied to the wire?
    B. What tensile load can the wire withstand without yielding
    C. What is the composite elastic modulus?
    D. Calculate the composite thermal expansion coefficient.

	Young's modulus	Yield strength	Poisson's ratio	Linear coef. of thermal	(GPa)(MPa)	Expans (K^{-1})
Aluminum	70	65	0.3	24×10^{-6}		
Steel	210	280	0.3	12×10^{-6}		

9. Consider a carbon-reinforced epoxy composite containing 45 volume % unidirectionally aligned carbon fibers.

A. Calculate the composite modulus.
B. Calculate the composite tensile strength. Assume both the epoxy and carbon are elastic to fracture.

	Young's modulus	Tensile strength
Epoxy	3 GPa	55 MPa
Carbon	250 GPa	2.5 GPa

10. Derive the term, $P_o \varepsilon / (1 - \phi - \varepsilon)$, for the effect of gas pressure in equation (21.47).

22　Mechanical Working

Introduction

The shapes of most metallic products are achieved by mechanical working. The exceptions are those produced by casting and by powder processing. Mechanical shaping processes are conveniently divided into two groups, bulk forming and sheet forming. Bulk-forming processes include rolling, extrusion, rod and wire drawing, and forging. In these processes, the stresses that deform the material are largely compressive. One engineering concern is to ensure that the forming forces are not excessive. Another is ensuring that the deformation is as uniform as possible so as to minimize internal and residual stresses. Forming limits of the material are set by the ductility of the work piece and by the imposed stress state.

Products as diverse as cartridge cases, beverage cans, automobile bodies, and canoe hulls are formed from flat sheet by drawing or stamping. In sheet forming, the stresses are usually tensile, and the forming limits usually correspond to local necking of the material. If the stresses become compressive, buckling or wrinkling will limit the process.

Bulk-Forming Energy Balance

An energy balance is a simple way of estimating the forces required in many bulk-forming processes. As a rod or wire is drawn through a die, the total work, W_t, equals the drawing force, F_d, times the length of wire drawn, ΔL, $W_t = F_d \Delta L$. Expressing the drawing force as $F_d = \sigma_d A$, where A is the area of the drawn wire and σ_d is the stress on the drawn wire, $W_t = \sigma_d A \Delta L$ (Figure 22.1). Because $A \Delta L$ is the drawn volume, the actual *work per volume*, w_t, is

$$w_t = \sigma_d. \tag{22.1}$$

The total work per volume can also be expressed as the sum of the individual work terms,

$$w_t = w_i + w_f + w_r. \tag{22.2}$$

The *ideal work*, w_i, is the work that would be required by an *ideal process* to create the same shape change as the real process. The ideal process is an imaginary

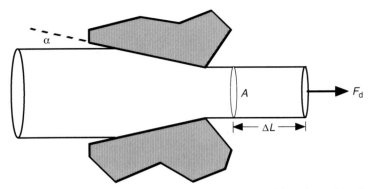

Figure 22.1. Drawing of a wire. The force, F_d, working through a distance of ΔL deforms a volume $A\Delta L$.

process in which there is no friction and no redundancy. It does not matter whether such a process is possible. For example, the ideal work to draw a wire from 10 mm to 7 mm in diameter can be found by considering uniform stretching in tension, even though in reality the wire would neck. In this case, the ideal work/volume,

$$w_i = \int \sigma\,d\varepsilon. \tag{22.3}$$

For a material that does not work harden, this reduces to

$$w_i = \sigma\varepsilon. \tag{22.4}$$

The work of friction between the wire drawing die and the wire can also be expressed on a per-volume basis, w_f. In general, the total frictional work, w_f, is roughly proportional to the flow stress, σ, of the wire and to the area of contact, A_c, between the wire and the die. This is because the normal force between the wire and die is $F_n = \sigma_n A_c$, where the contact pressure between tool and die, σ_n, depends on the flow stress, σ. The frictional force is $F_f = \mu F_n$, where μ is the coefficient of friction. For a constant reduction, the contact area, A_c, increases as the die angle, α, decreases so w_f increases as the die angle, α, decreases. A simple approximation for low die angles is

$$w_f = w_i(1 + \varepsilon/2)\mu\cot\alpha. \tag{22.5}$$

This predicts that the frictional work increases with increasing reduction and decreasing die angle.

The redundant work is the energy expended in plastic straining that is not required by the ideal process. During drawing, streamlines are bent as they enter the die and again as they leave the die. This involves plastic deformation that is not required in the ideal process of pure stretching. It also causes shearing of the surface relative to the interior as schematically illustrated in Figure 22.2. The redundant work per volume, w_r, increases with die angle, α. A simple approximation for relatively low die angles is

$$w_r = (2/3)\sigma\tan\alpha. \tag{22.6}$$

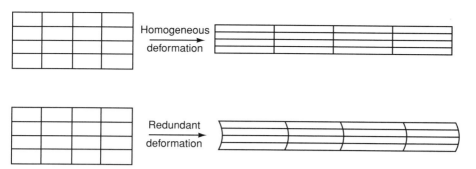

Figure 22.2. Redundant deformation involves shearing of surface relative to the interior.

This equation predicts that w_r increases with increasing die angle but does not depend on the total strain, ε. Therefore, the ratio, w_r/w_i decreases with increasing reduction.

A mechanical efficiency, η, can be defined such that

$$\eta = w_i/w_t. \tag{22.7}$$

Note that η is always less than 1. For wire drawing, typically drawing efficiencies are in the range of 50% to 65%.

Figure 22.3 shows how each of the work terms and the efficiency depend on die angle for a fixed reduction. There is an optimum die angle, α^*, for which the efficiency is a maximum. Figure 22.4 shows that, in general, with increased reduction the efficiency increases and the optimum die angle also increases. The reduction is defined as $r = (A_o - A)/A_o$.

For wire drawing, there is a maximum reduction that can be made in a single drawing pass. If a greater reduction is attempted, the stress on the drawn section will exceed its tensile strength, and the wire will break instead of drawing through the die. Therefore, multiple passes are required to make wire. After the first few dies, little additional work hardening occurs so the flow stress, σ, becomes the tensile strength. The limit then can be expressed as $\sigma = \sigma_d = w_t$. Substituting

Figure 22.3. Dependence of w_a, w_i, w_r, and w_f on die angle, α. Note that there is an optimum die angle, α^*, for which the total work per volume is a minimum.

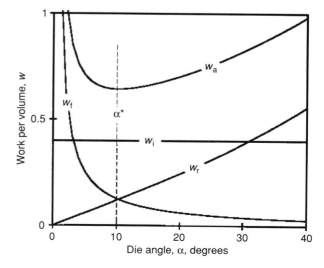

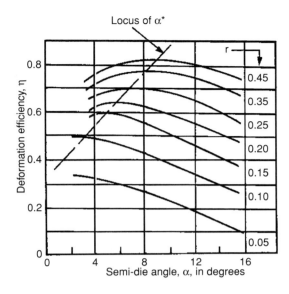

Figure 22.4. Variation of efficiency with die angle and reduction. Note that the efficiency and the optimum die angle, α^*, increase with reduction. From W. F. Hosford and R. M. Caddell, *Metal Forming: Mechanics and Metallurgy*, 3rd ed., Cambridge University Press (2007). Adapted from data from of J. Wistreich, *Metals Rev.*, v. 3 (1958).

equation (22.7) ($w_a = w_i/h$) and equation (22.4) ($w_i = \sigma\varepsilon$) the drawing limit corresponds to

$$\varepsilon = \eta. \qquad (22.8)$$

The diameter reduction, $(D_o - D_f)/D_o = 1 - D_f/D_o$ can be expressed as $\Delta D/D_o = 1 - \exp(-\varepsilon/e)$. The maximum diameter reduction for an efficiency of 50% is then $\Delta D/D_o = 1 - \exp(0.5/2) = 22\%$, and for an efficiency of 65%, it is 28%. Because the reductions per pass are low in wire drawing, the optimum die angles are also very low as suggested by Figure 22.4.

A work balance for extrusion is very similar, except now $W_t = F_{ext}\Delta L_o$, where F_{ext} is the extrusion force, and A_o and ΔL_o are the cross-sectional area and length of billet extruded (Figure 22.5). The actual work per volume $w_t = W_t/(A_o\Delta L_o) = F_{ext}/A_o$. This can be expressed as

$$w_t = P_{ext}, \qquad (22.9)$$

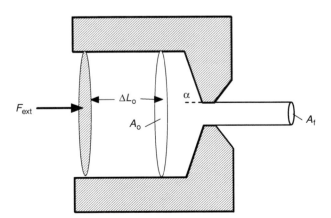

Figure 22.5. Direct extrusion. The extrusion force, F_{ext}, working through a distance, ΔL, deforms a volume, $A_o\,\Delta L$.

where P_{ext} is the extrusion pressure. Because material is being pushed through the die instead of being pulled, there is no inherent limit to the possible reduction per pass as there is in drawing. Therefore, extrusions are made in a single operation with extrusion ratios, A_o/A_f, as high as 16 or more. Because of the high reductions, high die angles are common. Often dies with 90-degree die angles are used because they permit more of the billet to be extruded before the die is opened. Similar work balances can be made for rolling and forging.

EXAMPLE PROBLEM 22.1: Consider the 16:1 extrusion of a round aluminum billet from a diameter of 24 cm to a diameter of 6 cm. Assume that the flow stress is 15 MPa at the appropriate temperature and extrusion rate. Estimate the extrusion force, assuming an efficiency of 0.50. Calculate the lateral pressure on the side of the extrusion chamber.

Solution: $P_{ext} = (1/\eta)\sigma\varepsilon$. Substituting $\eta = 0.50$, $\sigma = 15$ MPa, and $\varepsilon = \ln(16) = 2.77$, $P_{ext} = (1/0.5)(15)(2.77) = 83.2$ MPa. $F_{ext} = (\pi D^2/4)P_{ext} = 3.76$ MN.

For axially symmetric flow, the radial stress on the billet, σ_r, and the hoop stress, σ_c, must be equal, and they must be large enough so the material in the chamber is at its yield stress. Using von Mises, $(\sigma_2 - \sigma_3)^2 + (\sigma_3 - \sigma_1)^2 + (\sigma_1 - \sigma_2)^2 = 2Y^2$. With $\sigma_2 = \sigma_3 = \sigma_r$, $\sigma_r - \sigma_1 = -Y$, $\sigma_r = -Y + \sigma_1$. Substituting $\sigma_1 = -P_{ext} = -83.2$ MN, and $Y = 15$ MPa, $\sigma_r = -98.2$ MPa.

EXAMPLE PROBLEM 22.2: Find the horsepower that would be required to cold roll a 48-in.-wide sheet from a thickness of 0.030 in. to 0.025 in. if the exit speed is 80 ft/s and the flow stress is 10,000 psi. Assume a deformation efficiency of 80% and neglect work hardening.

Solution: The rate of doing work = (work/volume)(volume rolled/time) = $[(1/\eta)\sigma\varepsilon](vwt) = (1/0.80)(10 \times 10^3 \text{ lb/in.}^2)[\ln(0.030/0.025)](80 \times 12 \text{ in./s})(48 \text{ in.})(0.025 \text{ in.}) = 5.8 \times 10^6$ in.-lb/s = $(5.8 \times 10^6/12 \text{ ft.lb/s})[1.8 \times 10^{-3} \text{ hp}/(\text{ft.lb/s})] = 876$ hp. This is the horsepower that must be delivered by the rolls to the metal. (This solution neglects energy losses between the motor and the rolls.)

Deformation Zone Geometry

One of the chief concerns with mechanical working is the homogeneity of the deformation. Inhomogeneity affects the hardness distribution, residual stress patterns, and internal porosity in the final product and the tendency to crack during forming. The inhomogeneity is, in turn, dependent on the geometry of the deformation zone, which can be characterized in various processes by a parameter, Δ, defined as

$$\Delta = h/L. \tag{22.10}$$

Here, h is the height, thickness, or diameter at the middle of the deformation zone, and L is the length of contact between tools and work piece. Several examples are shown in Figure 22.6. For rolling, $R^2 = L^2 + (R - \Delta h/2)^2 = L^2 + R^2 = -2R\Delta h + \Delta h^2/4$ and $L \approx \sqrt{(R^2\Delta h)}$, so

$$\Delta \approx h/\sqrt{(R^2\Delta h)}, \tag{22.11}$$

and for extrusion and drawing,

$$\Delta = h/L = h\tan\alpha/(\Delta h/2). \tag{22.12}$$

EXAMPLE PROBLEM 22.3: To roll a sheet of 2.0 mm in thickness with 15-cm-diameter rolls, how large a reduction (% reduction of thickness) would be necessary to ensure $\Delta \geq 1$?

Solution: Take $\Delta = h/(R\Delta h)^{1/2} = h/(Rrh_o)^{1/2}$ and assume for the purpose of calculation that $h = h_o$. Then $r = (h_o/R)/\Delta^2 = (2.0/75)/\Delta^2$, so for $\Delta = 1$, $r = 0.027$ or 2.7%. (The reduction is small enough that the assumption that $h = h_o$ is justified.)

One way of describing the inhomogeneity is by a *redundant work factor*, Φ, defined as

$$\Phi = 1 - w_r/w_i. \tag{22.13}$$

Figure 22.7 shows the increase of Φ with Δ for both strip drawing (plane–strain flow) and wire drawing (axially symmetric flow). Friction has little effect on redundant strain.

Another way of characterizing inhomogeneity is by an *inhomogeneity factor*, defined as

$$IF = H_s/H_c. \tag{22.14}$$

The inhomogeneity factor increases with increased die angle, α, as shown in Figure 22.8. With high die angles, shearing at the surface causes more work hardening, thereby increasing IF.

As Δ increases, forming processes leave the surface under increasing residual tension, as shown in Figure 22.9. Under high Δ conditions, a state of hydrostatic tension develops near the centerline, and this may cause pores to open around

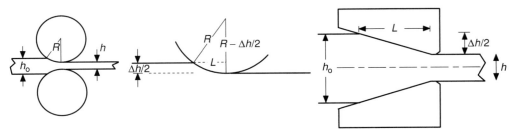

Figure 22.6. Deformation zones in rolling (*left*) and extrusion and drawing (*right*). The parameter, Δ, is defined as the ratio of the mean height or thickness to the contact length.

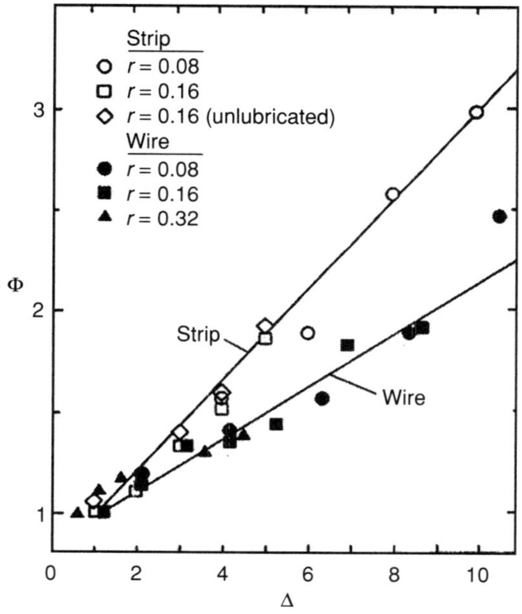

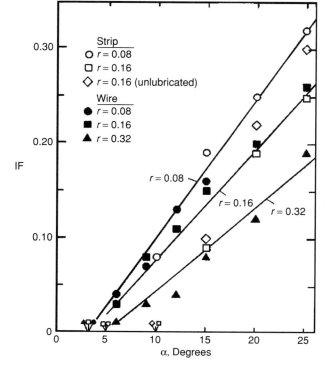

Figure 22.7. Increase of the redundancy factor, Φ, with Δ for both strip and wire drawing. Note that friction has little effect on Φ. From W. F. Hosford and R. M. Caddell, *Ibid.*

Figure 22.8. Increase of the inhomogeneity factor, $IF = 1 + H_{surf}/H_{center}$ with die angle. From W. F. Hosford and R. M. Caddell, *Ibid.* Data from J. J. Burke, ScD thesis, MIT (1968).

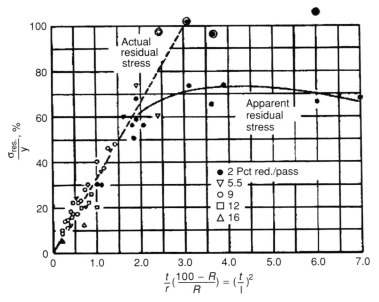

Figure 22.9. Residual stresses at the surface of cold-rolled brass strips increase with $\Delta = t/l$. From W. M. Baldwin, *Proc. ASTM*, v. 49 (1949).

inclusions (Figure 22.10). If the conditions are extreme, these pores may grow to form macroscopic cavities at the centerline, as shown in Figure 22.11.

Friction in Bulk Forming

In forging, friction can play an important role when the length of contact, L, between tools and work piece is large compared with the work piece thickness, h, becomes very large. During compression of a slab between two parallel platens, the friction tends to suppress lateral flow of the work material. Higher compressive stresses are necessary to overcome this restraint. The greater the L/h is, the greater the average compressive stress needs to be.

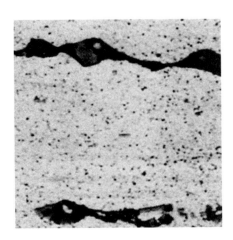

Figure 22.10. Voids opening at inclusions near the centerline during drawing under high-Δ conditions. With high-Δ, the center is under hydrostatic tension. From H. C. Rogers, R. C. Leach, and L. F. Coffin, Jr., *Final Report Contract No. 65-0097-6*, Bur. Naval Weapons (1965).

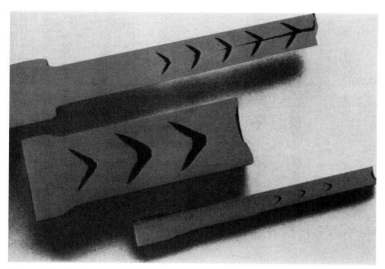

Figure 22.11. Centerline cracks in extruded steel bars. Note that the reductions were very small so Δ as very high. From D. J. Blickwede, *Metals Progress*, v. 97 (May 1970).

For compression in plane strain with a constant coefficient of friction, the ratio of the average pressure to the plane–strain flow strength, P_{av}/σ_o, is

$$P_{av}/\sigma_o = (h/\mu L)[\exp(\mu L/h) - 1], \qquad (22.15)$$

which can be approximated by $P_{av}/\sigma_o \approx 1 + \mu L/h$ for small values of $\mu L/h$.

For compression of circular discs of diameter, D, the ratio of the average pressure to the flow strength, P_{av}/Y, is

$$P_{av}/Y = 2(h/\mu D)^2[\exp(\mu D/h) - \mu D/h - 1]. \qquad (22.16)$$

For small values of $\mu L/D$ this can be approximated by

$$P_{av}/Y = 1 + (1/3)\mu D/h + (1/12)(\mu D/h)^2. \qquad (22.17)$$

In rolling of thin sheets, the pressure between the rolls and work piece can become so large that the rolls bend. Such bending would produce a sheet that is thicker in the middle than at the edges. To prevent this, the rolls may be backed up with larger rolls or ground to a barrel shape to compensate for bending.

Formability

In bulk-forming processes, other than wire drawing, formability is limited by fracture. Whether fracture occurs depends on both the material and the process. The formability of a material is related to its reduction of area in a tension test. A material with high fracture strain is likely to have a high formability. Figure 22.12 shows the correlation of edge cracking during rolling with tensile ductility for a number of materials. The factors governing a material's fracture strain were discussed in Chapter 13. High inclusion content and high strength tend to decrease ductility and formability.

Formability also depends on the level of hydrostatic stress in the process. The difference in Figure 22.12 between the fracture strains in "square-edge" and "round-edge" strips reflects this. If the edge of a rolled strip is allowed to become round

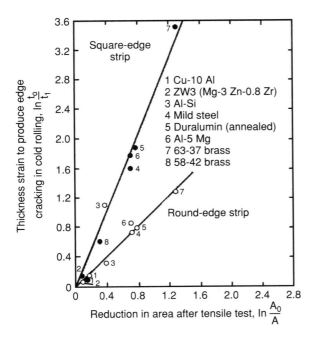

Figure 22.12. Correlation of the strain at which edge cracking occurs in flat rolling with the reduction of area in a tension test. The higher fracture strains of the "square-edge" strips reflect a higher state of compression at the edge during rolling. From M. G. Cockcroft and D. J. Latham, *J. Inst. Metals*, v. 96 (1968).

during rolling, the through-thickness compressive stress is less at the edge. Therefore, a greater tensile stress in the rolling direction is needed to cause the same elongation as in the middle of the strip. With the higher level of hydrostatic tension, the strain to fracture is lower.

Deep Drawing

A major concern in sheet forming is tensile failure by necking. Sheet-forming operations may be divided into *drawing*, where one of the principal strains in the plane of the sheet is compressive, and *stretching*, where both of the principal strains in the plane of the sheet are tensile. Compressive stresses normal to the sheet are usually negligible in both cases.

A typical drawing process is the making of cylindrical, flat-bottom cups. It starts with a circular disc blanked from a sheet. The blank is placed over a die with a circular hole, and a punch forces the blank to flow into the die cavity, as sketched in Figure 22.13. A hold-down force is necessary to keep the flange from wrinkling. As the punch descends, the blank is deformed into a hat shape and finally into a cup. Deforming the flange consumes most of the energy. The energy expended in friction and some in bending and unbending as material flows over the die lip is much less. The stresses in the flange are compressive in the hoop direction and tensile in the radial direction. The tension is a maximum at the inner lip and the compression a maximum at the outer periphery.

As with wire drawing, there is a limit to the amount of reduction that can be achieved. If the ratio of the initial blank diameter, d_o, to the punch diameter, d_1, is too large, the tensile stress required to draw the material into the die will exceed the

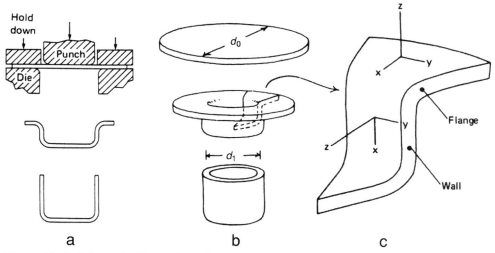

Figure 22.13. Schematic illustration of cup drawing showing the coordinate axes. As the punch descends, the outer circumference must undergo compression so that it will be small enough to flow over the die lip. From W. F. Hosford and R. M. Caddell, *Ibid.*

tensile strength of the wall, and the wall will fail by necking. It can be shown that for an isotropic material, the highest ratio of d_o/d_1 that can be drawn (*limiting drawing ratio* [LDR]) is

$$(d_o/d_1)_{max} = \exp(\eta), \qquad (22.18)$$

where η is the deformation efficiency. For an efficiency of 70%, $(d_o/d_1)_{max} = 2.01$. There is little thickening or thinning of the sheet during drawing, so this corresponds to cup with a height-to-diameter ratio of 3 : 4. Materials with R values greater than unity have somewhat higher limiting drawing ratios. This is because with a high R value, the increased thinning resistance permits higher wall stresses before necking, as well as easier flow in the plane of the sheet, which decreases the forces required. Forming cylindrical cups with a greater height-to-diameter ratio requires *redrawing*, as shown in Figure 22.14. In can making, an additional operation called *ironing* thins and elongates the walls, as illustrated in Figure 22.15.

Figure 22.14. Direct redrawing. A sleeve around the punch acts as a hold-down while the punch descends.

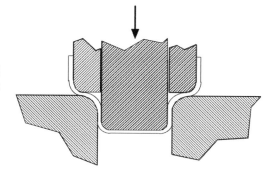

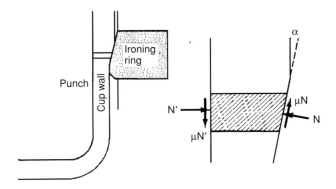

Figure 22.15. Section of a cup wall and ironing ring during ironing. The die ring causes wall thinning. Friction on opposite sides of the wall acts in opposing directions. From W. F. Hosford and R. M. Caddell, *Ibid.*

Stamping

Operations variously called *stamping*, *pressing*, or even *drawing* involve clamping the edges of the sheet and forcing it into a die cavity by a punch, as shown in Figure 22.16. The metal is not squeezed between tools. Rather, it is made to conform to the shape of the tools by stretching. Failures occur by either *wrinkling* or by *localized necking*. Wrinkling will occur if the restraint at the edges is not great enough to prevent excessive material being drawn into the die cavity. Blank holder pressure and draw beads are often used to control the flow of material into the die. If there is too much restraint, more stretching may be required to form the part than the material can withstand. The result is that there is a window of permissible restraint for any part, as illustrated in Figure 22.17. With too little blank holder force, the depth of draw is limited by wrinkling. On the other hand, if the blank holder force is too high, too little material is drawn into the die cavity to form the part, and the part fails by *localized necking*. The size of this window depends on properties of the sheet. Materials with higher strain-hardening exponents, n, can stretch more before necking failure so the right-hand limit is raised and shifted to the left. There will be more lateral contraction in materials having a high R value, thus decreasing the wrinkling tendency. Therefore, high R values increase the wrinkling limit and shift it to the left.

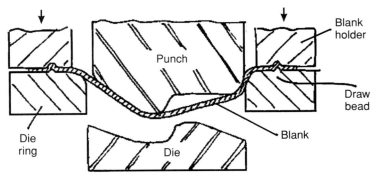

Figure 22.16. Sketch of a sheet stamping operation by J. L. Duncan. The sheet is stretched to conform to the tools rather than being squeezed between them. In this case, the lower die contacts the sheet and causes a reverse bending only after it has been stretched by the upper die. For many parts, there is not a bottom die.

Figure 22.17. The effect of blank holder force on the possible depth of draw. If the blank holder force is too low, wrinkling will result from too much material being drawn into the die cavity. Too high a blank holder force will require too much stretching of the sheet and result in a necking failure. For deep draws, there may be only a narrow window of permissible blank holder forces. High R values widen this window.

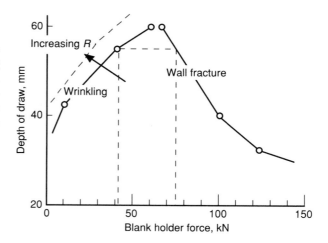

In a tension test of a ductile material, the maximum load and *diffuse necking* occur when $\sigma = d\sigma/d\varepsilon$. For a material that follows a power law ($\sigma = K\varepsilon^n$) stress–strain curve, this is when $\varepsilon = n$. This diffuse necking occurs by local contraction in both the width and thickness directions and is generally not a limitation in practical sheet forming. If the specimen is wide (as a sheet is), such localization must be very gradual. Eventually, a point is reached where lateral contraction in the plane of the sheet ceases. At this point, a *localized neck* forms in which there is only thinning. In uniaxial tension, the conditions for localized necking are $\sigma = 2d\sigma/d\varepsilon$, or $\varepsilon = 2n$. Figure 22.18 illustrates general and localized necking in a tensile specimen. The characteristic angle at which the neck forms must be such that the incremental strain in that direction, $d\varepsilon_\theta$, becomes zero.

Because $d\varepsilon_\theta = d\varepsilon_1 \cos^2 \theta + d\varepsilon_2 \sin^2 \theta$ and $d\varepsilon_\theta = 0$,

$$\tan \theta = (-d\varepsilon_1/d\varepsilon_2)^{1/2}. \qquad (22.19)$$

For uniaxial tension of an isotropic material, $d\varepsilon_2 = -d\varepsilon_1/2$ so $\tan \theta = \sqrt{2}$ and $\theta = 54.7°$.

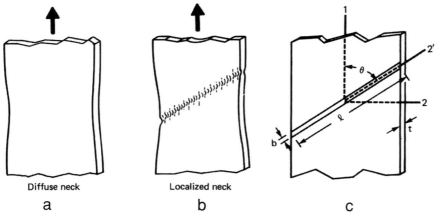

Figure 22.18. Development of a diffuse neck (a) and a localized neck (b). The coordinate axes used in the analysis are shown in (c). The characteristic angle, θ, of the local neck must be such that $\varepsilon_2' = 0$. From W. F. Hosford and R. M. Caddell, *Ibid.*

EXAMPLE PROBLEM 22.4: Find the angle between the tensile axis and the local neck form in a tension test on material with a strain ratio $R = 2$.

Solution: $R = \varepsilon_3/\varepsilon_2 = -(\varepsilon_1 + \varepsilon_2)/\varepsilon_2 = -\varepsilon_1/\varepsilon_2 - 1$ so, $-\varepsilon_1/\varepsilon_2 = 1 + R$. Substituting into equation (22.19), $\tan\theta = (1 + R)^{1/2} = \sqrt{3}$. $\theta = 60°$.

In sheet forming, ε_2 will be less negative if there is a tensile stress, σ_2. This in turn will increase the characteristic angle. For plane–strain conditions, $d\varepsilon_2 = 0$, $\theta = 90°$.

If $d\varepsilon_2$ is positive, there is no angle at which a local neck can form. Under conditions of biaxial stretching, however, a small preexisting groove perpendicular to the largest principal stress can grow gradually into a localized neck. The strain ε_2 must be the same inside and outside the groove. However, the stress σ_1 within the groove will be greater than outside the groove, so the strain ε_1 will be also be larger in the groove. As the strain rate $\dot{\varepsilon}_1$ inside the groove accelerates, the ratio of $\dot{\varepsilon}_2/\dot{\varepsilon}_1$ within the groove approaches zero, which is the condition necessary for local necking. Figure 22.19 shows how the strain path inside and outside a groove can diverge. Straining outside the groove will virtually cease once $\dot{\varepsilon}_1/\dot{\varepsilon}_2$ becomes very large. The terminal strain outside the groove is the *limit strain*. Very shallow grooves are sufficient to cause such localization. How rapidly this happens depends largely on the strain-hardening exponent, n, and to a lesser extent the strain rate exponent, m.

A plot of the combinations of strains that lead to necking failure is called a *forming limit diagram* (FLD). Figure 22.20 is such a plot for low-carbon steels. Combinations of strains below the forming limits are safe, whereas those above the limits will cause local necking. Note that the lowest failure strains correspond to plane strain, $\varepsilon_2 = 0$.

Some materials, when stretched in biaxial tension, may fail by shear fracture instead of local necking. Shear failures are also possible before necking under high

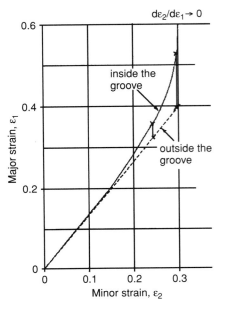

Figure 22.19. Calculated strain paths inside and outside a preexisting groove for a material with n = 0.22, m = 0.012, and $R = 1.5$. The initial thickness in the groove was assumed to be 1/2% less than outside. A strain path of $\varepsilon_2 = 0.75\varepsilon_1$ was imposed outside the groove. As $d\varepsilon_2/d\varepsilon_1 \to 0$ inside the groove, a local neck develops and deformation outside the groove virtually ceases, fixing a limit strain of $\varepsilon_1 = 0.4$ and $\varepsilon_2 = 0.3$.

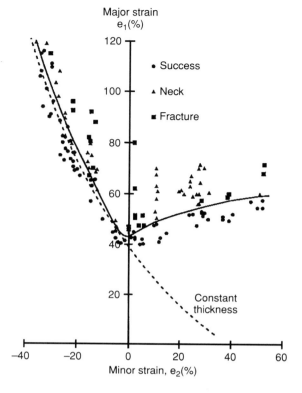

Figure 22.20. Forming limit diagram for low carbon steel. The strain combinations below the curve are acceptable, whereas those above it will cause local necking. The limiting strains here are expressed as engineering strains, although true strains could have been plotted. Data from S. S. Hecker, *Sheet Metal Ind.*, v. 52 (1975).

strains in the left-hand side of the diagram. It should be noted that if the minor strain, ε_2, is less than $-\varepsilon_1/2$, the minor stress, σ_2, must be compressive. Under these conditions, wrinkling or buckling of the sheet may occur. Because wrinkled parts are usually rejected, this, too, should be regarded as a failure mode. The possibilities of both shear fracture and wrinkling are shown in Figure 22.21.

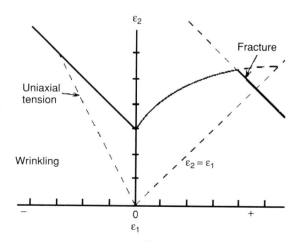

Figure 22.21. Schematic forming limit diagram showing regions where wrinkling may occur and a possible fracture limit in biaxial tension.

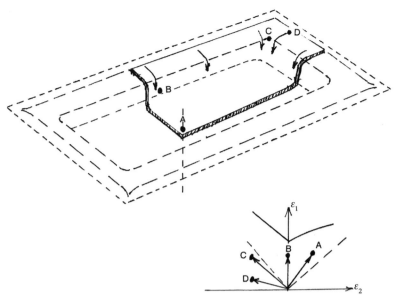

Figure 22.22. Sketch showing the strain paths in several locations during the drawing of a pan. At A, the strain state is nearly balanced biaxial tension. The deformation at point B is in plane strain. At C, there is drawing with contraction in the minor strain direction. At D, there may be enough compression in the 2-direction to cause wrinkling. Courtesy of J. L. Duncan.

The strains vary from one place to another in a given part. A pan being formed is sketched in Figure 22.22. The strain paths at several different locations are indicated schematically on a forming limit diagram.

Spinning (Figure 22.23) is a sheet-forming process that is suitable for forming axially symmetric parts. A tool forces a disc that is spinning parallel to its axis of rotation to conform to a mandrel. For pure shear, the deformation is restricted to just under the tool so that deformation does not occur elsewhere. This eliminates danger wrinkling. Because tooling costs are low and the process is relatively slow, spinning is suited to producing low production parts.

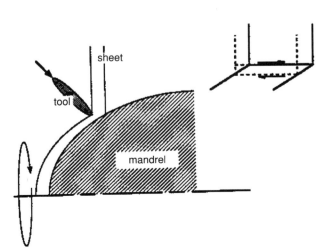

Figure 22.23. Sketch of a spinning operation. If the tool causes only shearing parallel to the axis of rotation, deformation does not occur in the flange.

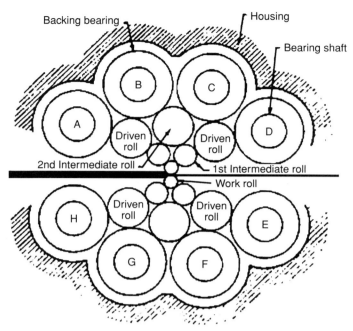

Figure 22.24. Sendzimer mill. Small diameter work rolls are used keep the ratio of L/h low. A cluster of larger backup rolls prevents bending of the work rolls. Courtesy T. Sendizimer.

REFERENCES

W. F. Hosford and R. M. Caddell, *Metal Forming: Mechanics and Metallurgy*, 3rd ed. Cambridge U. Press (2007).

W. A. Backofen, *Deformation Processing*, Addison-Wesley (1972).

E. M. Mielnick, *Metalworking Science and Engineering*, McGraw-Hill (1991).

Z. Marciniak and J. L. Duncan, *Mechanics of Sheet Forming*, Edward Arnold (1992).

G. E. Dieter, *Mechanical Metallurgy*, 2nd ed. McGraw-Hill (1976).

Notes

The first aluminum two-piece beverage cans were produced in 1963. They replaced the earlier steel cans made from three separate pieces: a bottom; the wall, which was bent into a cylinder and welded; and a top. The typical beverage can is made from a circular blank, $5\frac{1}{2}$ in. in diameter, by drawing it into a $3\frac{1}{2}$-in.-diameter cup, redrawing to 25/8 in. diameter, and then ironing the walls to achieve the desired height. There are about 200 billion made in the United States each year. Beverage cans account for about one-fifth of the total usage of aluminum.

In flat rolling of thin sheets or foils, the L/h ratio can become very large. This raises the average roll pressure to the extent that rolls elastically flatten where they are in contact with the work material. This flattening prevents further thinning. One way of overcoming this difficulty is to roll two foils at the same time, effectively doubling h. One side of commercial aluminum foil has a matte finish, whereas the other side is shiny. The sides with the matte finish were in contact during rolling. Another method of circumventing the roll-flattening problem is to use a cluster mill (Figure 22.24) developed by Sendzimer. The work roll has a very small diameter to keep L/h low. To prevent bending of the work rolls, both are backed up by two rolls of somewhat larger diameter. These are, in turn, backed up by three still larger diameter rolls, and so forth.

Problems

1. A small special alloy shop received an order for slabs 4 in. wide and $1/2$ in. thick of an experimental superalloy. The shop cast ingots 4 in. $\times$ 4 in. $\times$ 12 in. and hot rolled them in a 12-in.-diameter mill, making reductions of about 5% per pass. On the fifth pass, the first slab split longitudinally parallel to the rolling plane. The project engineer, the shop foreman, and a consultant met to discuss the problem. The consultant proposed applying forward and back tension during rolling, the project engineer suggested reducing the reduction per pass, and the shop foreman favors higher reductions per pass. With whom would you agree? Explain your reasoning.

2. A high-strength steel bar must be cold reduced from a diameter of 1.00 in. to 0.65 in. A number of schedules have been proposed. Which of the following schedules would you choose to avoid drawing failure and minimize the likelihood of centerline bursts? Explain. Assume $\eta = 0.50$.

    A. A single reduction in a die having a die angle of 8 degrees.
    B. Two passes (1.00–0.81 in. and 0.81–0.65 in. using dies with angles of $\alpha = 8°$).
    C. Three passes (1.00–0.87 in. and 0.87–0.75 in., and 0.75–0.65 in. using dies with angles of 8°).
  D–F. Same schedules as A, B, and C, except using dies with $\alpha = 15°$.

3. Your company is planning to produce niobium wire, and you have been asked to decide how many passes would be required to reduce the wire from 0.125 to 0.010 in. in diameter. In laboratory experiments with dies having the same angle as will be used in the operation, the efficiency increased with reduction, $\eta = 0.65 + \Delta\varepsilon/3$, where $\Delta\varepsilon$ is the strain in the pass. Assume that in practice the efficiency will be only 75% of that found in the laboratory experiments. To ensure no failures, stress on the drawn section of wire must never exceed 80% of its strength. Neglect work hardening.

4. One stand of a hot-rolling mill is being designed. It will reduce 60-in.-wide sheet from 0.150 to 0.120 in. thickness at an exit speed of 20 feet per second. Assume that the flow stress of the steel at the temperature and strain rate in the rolling mill is 1,500 psi. If the deformation efficiency is 82% and the efficiency of transferring energy from the motor to the mill is 85%, what horsepower motor should be used?

5. A typical aluminum beverage can is 2.6 in. in diameter and 4.8 in. high. The thickness of the bottom is 0.010 in., and the wall thickness is 0.004. The cans are produced from circular blanks 0.010 in. thick by drawing, redrawing, and ironing to a height of 5.25 in. before trimming.

    A. Calculate the diameter of the initial circular blank.
    B. Calculate the total effective strain at the top of the cup from rolling, drawing, redrawing, and ironing.

6. Figure 22.12 shows that many reductions in rolling can be achieved before edge cracking occurs if the edges are maintained square instead of being allowed to become rounded. Figure 22.25 shows the edge elements. Explain in terms of the stress state at the edge why the higher strains are possible with square edges.

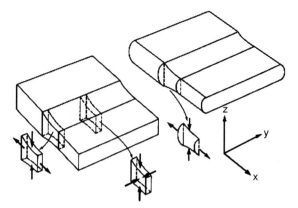

Figure 22.25. Difference between the stress states at the edges of square edge and round edge strips during rolling. From W. F. Hosford and R. M. Caddell, *Ibid.*

7. When aluminum alloy 6061-T6 is cold drawn through a series of dies with a 25% reduction per pass, a loss of density is noted, as shown in Figure 22.26. Explain why the density loss increases with higher angle dies.

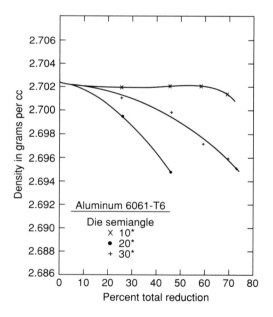

Figure 22.26. Density changes in aluminum alloy 6161-T6 during drawing. From H. C. Rogers, *General Electric Co. Report No. 69-C-260, 1969.*

8. Assuming that in drawing of cups, the thickness of the cup bottom and wall is the same as that of the original sheet, find an expression for the ratio of the cup height to diameter, h/d_1, in terms of the ratio of blank diameter to cup diameter, d_o/d_1. Evaluate h/d_1 for $d_o/d_1 = 1.5, 1.75, 2.0$, and 2.25, and plot h/d_1 versus d_o/d_1.

9. In drawing of cups with conical wall, the elements between the punch and the die must deform in such a way that their circumference shrinks. Otherwise, they will buckle or wrinkle. The tendency to wrinkle can be decreased by applying a greater blank holder force, as shown in Figure 22.27. This increases the radial tension between the punch and die. How would the R value of the material affect how much blank holder force is necessary to prevent wrinkling?

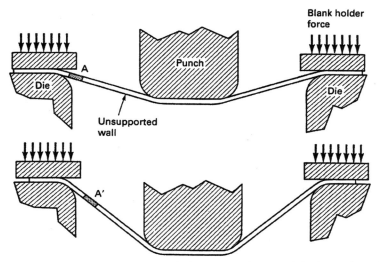

Figure 22.27. Drawing of a conical cup. As element A is drawn into the die cavity, its circumference must shrink. This requires enough tensile stretching in the radial direction. From W. F. Hosford and R. M. Caddell, *Ibid.*

10. Figure 22.28 is a forming limit diagram for a low carbon steel. This curve represents the combinations of strains that would lead to failure under plane–stress ($\sigma_3 = 0$) loading.

    A. Show the straining path inside a Marciniak defect under biaxial tension that would lead to necking at point N.

    B. Plot carefully on the diagram the strain path that corresponds to uniaxial tension ($\sigma_3 = 0$).

    C. Describe how this path would be changed for a material with a value of $R > 1$.

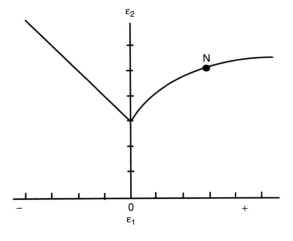

Figure 22.28. Forming limit diagram for a low-carbon steel.

11. Consider drawing a copper wire from 0.125 to 0.100 in. in diameter. Assume that $\sigma = (55 \text{ MPa}) \, \varepsilon^{0.36}$ in a die for which $\alpha = 6°$.

    A. Calculate the drawing strain.

    B. Calculate the reduction of area.

    C. Use Figure 22.23 to determine η and calculate the drawing stress. Calculate the yield strength of the drawn wire. Can this reduction be achieved?

Miller Indices

Planes

Planes in a crystal are identified by their Miller indices. The system involves;

1. Writing the intercepts on the three axes;
2. Taking the reciprocals;
3. Reducing these to the lowest set of integers in the same ratio;
4. Enclosing in parentheses.

EXAMPLE PROBLEM AI.1: Write the Miller indices of the two planes shown in Figure AI.1.

Solution: For plane a:

1. The intercepts on the three axes are 1, ∞, and 1. Note that the y intercept is taken as ∞ because the plane does not intercept the y axis.
2. Reciprocals are 1, 0, and 1.
3. These are already reduced to lowest set of integers in same ratio.
4. The plane is (101).

For plane b:

1. The intercepts on the three axes are 1, -1, and 1/2. Note that the plane has to be extended out of the cubic element to intercept the y axis at -1.
2. Reciprocals are 1, -1, and 2.
3. These are already reduced to lowest set of integers in same ratio.
4. The plane is $(1\bar{1}2)$. Note that the minus sign is indicated by an over bar and that no commas are used.

Directions

Directions are indicated by their components parallel to the three axes reduced to the lowest set of integers and enclosed in brackets, [ ].

EXAMPLE PROBLEM AI.2: Write the direction indices of the two directions shown in Figure AI.2.

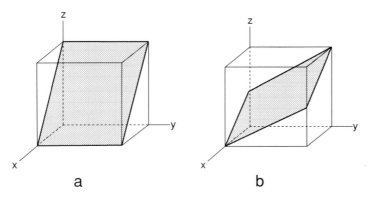

Figure AI.1. Two plane in a cubic crystal.

Solution: For A:

1. The components in the three directions are −1, 1, and 0.
2. These are already reduced to lowest set of integers in same ratio.
3. The direction is $[\bar{1}10]$.

For B:

1. The components in the three directions are $1/2$, −1, and −1
2. Reduced to lowest set of integers in same ratio they are $1,-2,-2$
3. The direction is $[1\bar{2}\bar{2}]$.

The notation {hkl} is used to describe a set of planes that are crystallographically equivalent. For example in a cubic crystal, {110} means the set (110), $(1\bar{1}0)$, (101), $(10\bar{1})$, (011) and $(01\bar{1})$ and their negatives. Likewise the notation <uvw> is used to describe a set of directions that are crystallographically equivalent. For example in a cubic crystal, <110> means the set [110], $[1\bar{1}0]$, [101], $[10\bar{1}]$, [011], and $[01\bar{1}]$.

The Miller-Bravais System for Hexagonal Crystals

The Miller-Bravais system for hexagonal crystals uses four axes rather than three. The reason is that with four axes the symmetry is more apparent, as will be illustrated below. Three of the axes, a_1, a_2, and a_3 lie in the hexagonal (basal) plane at $120°$ to one another and the fourth or c axis is perpendicular to it as shown in Figure AI.3.

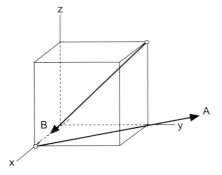

Figure AI.2. Two directions in a cubic crystal.

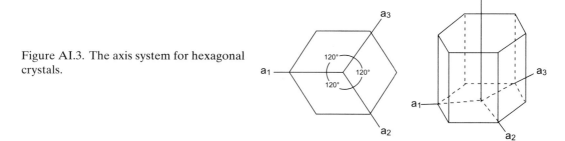

Figure AI.3. The axis system for hexagonal crystals.

Planar Indices: The rules are similar to those for Miller indices with three axes. To find the indices of a plane:

1. Write its intercepts on the four axes in order (a_1, a_2, a_3, and c),
2. Take the reciprocals of these,
3. Reduce to the lowest set of integers with the same ratios
4. Enclose in parentheses (hkiℓ).

Commas are not used except in the rare case that one of the integers is larger than a one-digit number. (This is rare because we are normally interested only in directions with low indices.) If a plane is parallel to an axis, regard its intercept as ∞ and its reciprocal as 0. If the plane contains one of the axes or the origin, either draw a parallel plane or translate the axes before finding indices. This is permissible since all parallel planes have the same indices. Figure AI.4 shows several examples.

   In the four digit system, the third digit, i, can always be deduced from the first two, $i = -h -k$, and is therefore redundant. With the three-digit systems, it may either be replaced by a dot, (hk$\cdot\ell$), or omitted entirely, (hkℓ). However the disadvantage of omitting the third index is that the hexagonal symmetry is not apparent.

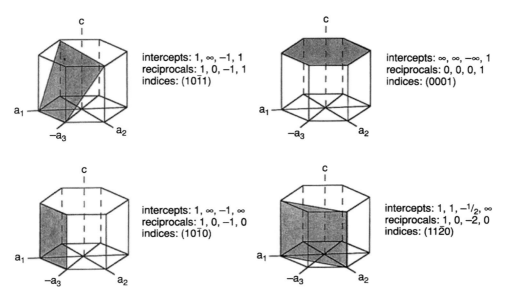

intercepts: 1, ∞, -1, 1
reciprocals: 1, 0, -1, 1
indices: ($10\bar{1}1$)

intercepts: ∞, ∞, $-\infty$, 1
reciprocals: 0, 0, 0, 1
indices: (0001)

intercepts: 1, ∞, -1, ∞
reciprocals: 1, 0, -1, 0
indices: ($10\bar{1}0$)

intercepts: 1, 1, $-1/_2$, ∞
reciprocals: 1, 0, -2, 0
indices: ($11\bar{2}0$)

Figure AI.4. Examples of planar indices for hexagonal crystals. Note that the sum of the first three indices is always zero, $h + k + i = 0$.

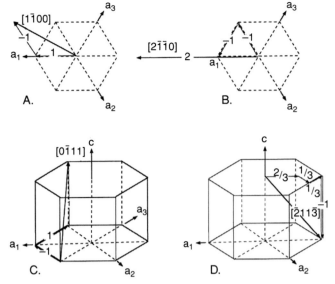

Figure AI.5. Examples of direction Indices with the Miller-Bravais system.

In the four-digit (Miller-Bravais) system a family of planes is apparent from the indices. For example;

$\{01\bar{1}0\} = (01\bar{1}0)$, $(\bar{1}010)$ and $(1\bar{1}00)$. The equivalence of the same family is not so apparent in the three-digit system, $\{010\} = (010)$, $(\bar{1}00)$ and $(1\bar{1}0)$.

Also compare $\{\bar{2}110\} = (\bar{2}110)$, $(1\bar{2}10)$ and $(11\bar{2}0)$ with $\{\bar{2}10\} = (\bar{2}10)$, $(1\bar{2}0)$, (110).

Direction Indices: The direction indices are the translations parallel to the four axes that produce the direction under consideration. The first three indices must chosen so that they sum to zero and are the smallest set of integers that will express the direction. They are enclosed without commas in brackets [hki ℓ]. Examples are shown in Figure AI.5.

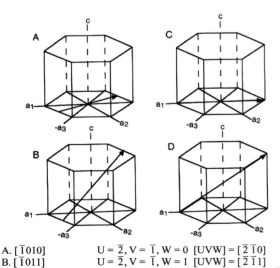

Figure AI.6. Comparison of the four- and three-digit systems.

A. $[\bar{1}010]$ $U = \bar{2}, V = \bar{1}, W = 0$ $[UVW] = [\bar{2}\,\bar{1}0]$
B. $[\bar{1}011]$ $U = \bar{2}, V = \bar{1}, W = 1$ $[UVW] = [\bar{2}\,\bar{1}1]$
C. $[\bar{2}110]$ $U = \bar{3}, V = 0, w = 0$ $[UVW\} = [\bar{3}00] = [\bar{1}00]$
D. $[\bar{2}111]$ $U = \bar{3}, V = 0, w = 1$ $[UVW\} = [\bar{3}01] = [\bar{1}01]$

There is also three-digit system for directions in hexagonal crystals. It uses the translations along the a_1, a_2, and a_3 axes (U, V, W, respectively). The four-digit [uvtw] and three-digit [UVW] systems are related by

$$U = u - t \quad u = (2U - V)/3,$$

$$V = v - t \quad v = -(2V - U)/3,$$

$$W = w \quad \text{and}$$

$$t = -(u + v) = -(U + V)/3. \tag{AI.1}$$

The four- and three-digit systems are compared in Figure AI.6.

The four-digit Miller-Bravais system is used in this text.

Problem

1. Write the correct direction indices, [ ], and planar indices, ( ), for the directions and planes Figure AI.7 sketched below.

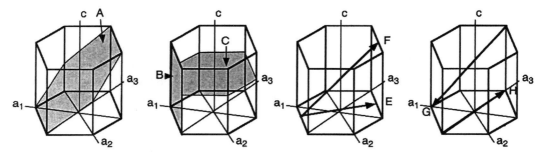

Figure AI.7. Several planes and directions for problem 1.

APPENDIX II

Stereographic Representation of Orientations

The *stereographic projection* is often used to represent the angular relations between directions and planes in a crystal. This projection system can be visualized by imagining a tiny (infinitesimal) crystal at the center of a sphere. All of the planes and directions of interest are extended until they intersect the surface of the sphere. Directions intersect the sphere as points and planes intersect it as great circles as shown in Figure AII.1. These points and great circles are the projected onto a flat surface. See Figure AII.2. The problem of plotting these on a flat surface is exactly the same as the mapmaker's problem of plotting the spherical surface of the earth. For crystals, it is necessary to plot only half of the spherical surface because the opposite hemisphere is identical. Barrett and Cullity describe the details of stereographic projection.

The standard projection of a cubic crystal can be thought of as a map with the [001] direction represented by the North Pole. The other end of [001] is [00$\bar{1}$], which is represented by the South Pole. The [100] direction is in the center of the map on the equator as shown in Figure AII.3a. The (100) plane is the reference circle at the periphery of the hemisphere that is plotted, the (001) plane is the equator, and the (010) plane is plotted as a vertical line through [100] and [001]. The [101] and [0$\bar{1}$0] directions lie in the (001) at its intersection with the equator. Note that squares are used to represent <100> directions because their 4-fold symmetry.

All of the <011> directions lie in {100} planes at ± 45° from <100> directions and are represented as ovals because their 2-fold symmetry. The [011] and [01 ≫ 1] directions lie on the reference circle ± 45° from [010]. The [101] and [10$\bar{1}$] directions

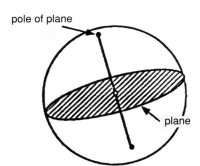

pole of plane

plane

Figure AII.1. Spherical projection of a plane and its pole. The infinitesimal crystal is at the center of the sphere. The plane intersects the sphere as a great circle and its pole (a direction) intersects the sphere at a point.

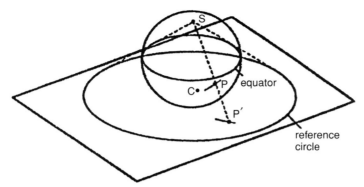

Figure AII.2. Projection of a point, P, and a great circles from a hemisphere onto a plane. From W. F. Hosford, *The Mechanics of Crystals and Textured Polycrystals*, Oxford University Press, Inc. (1993).

lie on the (010) plane $\pm$ 45° from [100] and the [110] and [1$\bar{1}$0] directions lie in the (001) plane at $\pm$45° from [100]. The (011) and (01$\bar{1}$) planes are represented by straight lines at $\pm$ 45° to [010] and [001] and the (101), (10$\bar{1}$), (110) and (1$\bar{1}$0) planes are represented by arcs of great circles (Figure AII.3b.)

There are four points at which three great circles representing {110} planes intersect. These are the <111> directions that lie in the three {110} planes. Triangles are used as symbols for the <111> directions because of their three-fold symmetry (Figure AII.3bc).

This construction divides the hemisphere into 24 spherical triangles, the corners of which are <100>, <110> and <111>directions. These triangles are crystallographically equivalent so the variation of any property can be represented in a single triangle. Conventionally the triangle having [100], [110] and [111] corners is taken as the standard triangle. (See, for example, Figure 2.9.)

EXAMPLE PROBLEM AII.1: Locate [2$\bar{1}$1] on the standard cubic projection.

Solution: See Figure AII.4. The dot product [2$\bar{1}$1]·[100] is positive, so [2$\bar{1}$1] must lie within 90° of [100] and therefore in the hemisphere represented (Figure AII.4a.) Since [2$\bar{1}$1]·[010] is negative, [2$\bar{1}$1] must lie more than 90° from [010] so it is in the left-hand side of the plot (Figure AII.4b). Since [2$\bar{1}$1]·[001] is positive, [2$\bar{1}$1] must lie less than 90° from [001] and therefore in the top half of the plot. Thus [2$\bar{1}$1] is in the upper left quadrant (Figure AII.4c.)

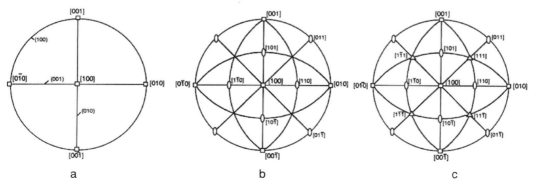

Figure AII.3. Construction of a standard cubic projection.

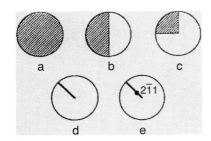

Figure AII.4. Location of $[2\bar{1}1]$ on the standard cubic projection.

The dot products $[2\bar{1}1]\cdot[0\bar{1}0]$ and $[2\bar{1}1]\cdot[001]$ are equal so $[2\bar{1}1]$ is equidistant from $[0\bar{1}0]$ and $[001]$ on the line connecting $[100]$ and $[0\bar{1}1]$ (Figure AII.4d.) Finally, the dot product of $[2\bar{1}1]$ with $[100]$ is larger than the dot products of $[2\bar{1}1]$ with either $[0\bar{1}0]$ or $[001]$. This indicates that the angle between $[2\bar{1}1]$ and $[100]$ is less than the angles between $[2\bar{1}1]$ and either $[0\bar{1}0]$ or $[001]$ (Figure AII.4e.)

REFERENCES

C. S. Barrett, *Structure of Metals*, McGraw-Hill, New York & London (1943).
B. D. Cullity, *Elements of X-ray Diffraction*, Addison-Wesley, Reading, MA (1956).

Problems

1. Sketch a standard cubic projection with $[100]$ at the center and $[001]$ at the North Pole. Locate the $[1\bar{2}1]$ direction on this projection.

2. If an fcc crystal were stressed in tension with tensile axis parallel to the $[1\bar{2}1]$ direction, on which of the $<110>\{11\bar{1}\}$ slip system (or systems) would the shear stress be the highest?

Index